PHILOSOPHICAL PROBLEMS OF SPACE AND TIME

Borzoi Books in the Philosophy of Science

GENERAL EDITOR

Sidney Morgenbesser
Columbia University

Adolf Grünbaum
PHILOSOPHICAL PROBLEMS OF SPACE AND TIME

Israel Scheffler
THE ANATOMY OF INQUIRY
Philosophical Studies in the Theory of Science

in preparation

Sidney Morgenbesser
DETERMINISM AND HUMAN BEHAVIOR
A Study of Social Scientific Theory

Michael Scriven
THE CAUSES AND REASONS FOR BEHAVIOR

Patrick Suppes
AXIOMATIC METHOD IN THE EMPIRICAL SCIENCES

Hilary Putnam
THE PHILOSOPHY OF PHYSICS AND MATHEMATICS

PHILOSOPHICAL PROBLEMS OF SPACE AND TIME

Adolf Grünbaum

Andrew Mellon Professor of Philosophy

UNIVERSITY OF PITTSBURGH

NEW YORK

Alfred · A · Knopf

1963

L. C. catalog card number: 63–18683

THIS IS A BORZOI BOOK,

PUBLISHED BY ALFRED A. KNOPF, INC.

 Manufactured in the United States of America, and distributed by Random House, Inc. Published simultaneously in Toronto, Canada, by Random House of Canada, Limited.

FIRST EDITION

To Thelma

Preface

My principal intellectual debt is to Hans Reichenbach's outstanding work *Philosophie der Raum-Zeit-Lehre* (Berlin, 1928) and to A. d'Abro's remarkable *The Evolution of Scientific Thought from Newton to Einstein* (New York, 1927).

In addition to those mentioned within the text, a number of colleagues and friends offered suggestions or criticisms which were of significant benefit to me in developing the ideas of this book. Among these, I should name especially the scientists Peter Havas, Allen Janis, Samuel Gulden, E. L. Hill, and Albert Wilansky as well as the following philosophers: Henry Mehlberg, Wilfrid Sellars, Abner Shimony, Grover Maxwell, Herbert Feigl, Hilary Putnam, Paul K. Feyerabend, Ernest Nagel, Nicholas Rescher, Sidney Morgenbesser, and Robert S. Cohen. Fruitful exchanges with some of these colleagues were made possible by the stimulating sessions of the Minnesota Center for Philosophy of Science to whose Director, Professor Herbert Feigl, I am very grateful for much encouragement.

I wish to thank Mrs. Helen Farrell of Bethlehem, Pennsylvania for typing an early draft of a portion of the manuscript, and Mrs. Elizabeth McMunn whose intelligence and conscientiousness were invaluable in the preparation of the final text for the printer. I am also indebted to Mr. Richard K. Martin for assistance in the preparation of the index and in the drawing of diagrams.

I have drawn on material which appeared in earlier versions in the following prior essays of mine: "Geometry, Chronometry and Empiricism," *Minnesota Studies in the Philosophy of Science* (ed. H. Feigl and G. Maxwell), Vol. III, Minneapolis, 1962, pp.

405–526, and "Carnap's Views on the Foundations of Geometry," in: P. A. Schilpp (ed.), *The Philosophy of Rudolf Carnap*, Open Court Publishing Company, LaSalle, 1963, pp. 599–684. I wish to thank the editors and publishers of these essays for their kind permissions.

Contents

PART I. *Philosophical Problems of the Metric of Space and Time*

PART II. *Philosophical Problems of the Topology of Time and Space*

PART III. *Philosophical Issues in the Theory of Relativity*

PART I

Philosophical Problems of the Metric of Space and Time

Chapter 1

SPATIAL AND TEMPORAL CONGRUENCE IN PHYSICS: A CRITICAL COMPARISON OF THE CONCEPTIONS OF NEWTON, RIEMANN, POINCARÉ, EDDINGTON, BRIDGMAN, RUSSELL, AND WHITEHEAD

The metrical comparisons of separate spatial and temporal intervals required for geo-chronometry involve rigid rods and isochronous clocks. Is this involvement of a transported congruence standard to which separate intervals can be referred a matter of the mere ascertainment of an otherwise intrinsic equality or inequality obtaining among these intervals? Or is reference to the congruence standard essential logically to the very existence of these relations? More specifically, we must ask the following questions:

1. What is the warrant for the claim that a solid rod remains rigid or self-congruent under transport in a spatial region free from inhomogeneous thermal, elastic, electromagnetic and other "deforming" or "perturbational" influences? The geometrically pejorative characterization of thermal and other inhomogeneities in space as "deforming" or "perturbational" is due to the fact that they issue in a dependence of the coincidence behavior of transported solid rods on the latter's chemical composition, and *mutatis mutandis* in a like dependence of the rates of clocks.

2. What are the grounds for asserting that a clock which is not

perturbed in the sense just specified is isochronous, i.e., yields equal durations for congruent time intervals?

This pair of questions and some of their far-reaching philosophical ramifications will occupy us in this chapter. I shall endeavor to evolve answers to them in the course of a critical discussion of the relevant rival conceptions of a number of major thinkers. It will first be in Chapter Four that we shall deal with the further issues posed by the logic of making corrections to compensate for deformations and rate-variations exhibited by rods and clocks respectively when employed geo-chronometrically under perturbing conditions.

(A) NEWTON

In the *Principia*[1] Newton states his thesis of *the intrinsicality of the metric* in "container" space and the corresponding contention for absolute time as follows:

> the common people conceive those quantities [i.e., time, space, place and motion] under no other notions but from the relation they bear to sensible objects. And thence arise certain prejudices, for the removing of which it will be convenient to distinguish them into absolute and relative, true and apparent, mathematical and common.[2] . . . because the parts of space cannot be seen, or distinguished from one another by our senses, therefore in their stead we use sensible measures of them. For from the positions and distances of things from any body considered as immovable, we define all places; and then with respect to such places, we estimate all motions, considering bodies as transferred from some of those places into others. And so, instead of absolute places and motions, we use relative ones; and that without any inconvenience in common affairs; but in philosophical disquisitions, we ought to abstract from our senses, and consider things themselves, distinct from what are only sensible measures of them. For it may be that there is no body really at rest, to which the places and motions of others may be referred[3] . . . those . . . defile the purity of mathematical and

[1] I. Newton: *Principia,* edited by F. Cajori (Berkeley: University of California Press; 1947).
[2] *Ibid.,* p. 6.
[3] *Ibid.,* p. 8.

philosophical truths, who confound real quantities with their relations and sensible measures.[4] . . .

I. Absolute, true, and mathematical time, of itself, and from its own nature, flows equably[5] without relation to anything external, and by another name is called duration: relative, apparent, and common time, is some sensible and external (whether accurate or unequable) measure of duration by the means of motion, which is commonly used instead of true time; such as an hour, a day, a month, a year.

II. Absolute space, in its own nature, without relation to anything external, remains always similar and immovable. Relative space is some movable dimension or measure of the absolute spaces; which our senses determine by its position to bodies; and which is commonly taken for immovable space; such is the dimension of a subterraneous, an aerial, or celestial space, determined by its position in respect of the earth. Absolute and relative space are the same in figure and magnitude; but they do not remain always numerically the same. For if the earth, for instance, moves, a space of our air, which relatively and in respect of the earth remains always the same, will at one time be one part of the absolute space into which the air passes; at another time it will be another part of the same, and so, absolutely understood, it will be continually changed.[6] . . . Absolute time, in astronomy, is distinguished from relative, by the equation or correction of the apparent time. For the natural days are truly unequal, though they are commonly considered as equal, and used for a measure of time; astronomers correct this inequality that they may measure the celestial motions by a more accurate time. It may be, that there is no such thing as an equable motion, whereby time may be accurately measured. All motions may be accelerated and retarded, but the flowing of absolute time is not liable to any change. The duration or perseverance of the existence of things remain the same, whether the motions are swift or slow, or none at all: and therefore this duration ought to be distinguished from what are only sensible

[4] *Ibid.*, p. 11.

[5] It is Newton's conception of the attributes of "equable," (i.e., congruent) time-intervals which will be subjected to critical examination and found untenable in this chapter. But in Chapter Ten below, we shall give reasons for likewise rejecting Newton's view that the concept of "flow" has relevance to the time of physics, as distinct from the time of psychology.

[6] I. Newton: *Principia, op. cit.*, p. 6.

> measures thereof; and from which we deduce it, by means of the astronomical equation.[7]

Newton's fundamental contentions here are that (a) the identity of points in the physical container space in which bodies are located and of the instants of receptacle time at which physical events occur is autonomous and not derivative: physical things and events do not first define, by their own identity, the points and instants which constitute their loci or the loci of other things and events, and (b) receptacle space and time each have their own *intrinsic metric*, which exists quite independently of the existence of material rods and clocks in the universe, devices whose function is at best the purely epistemic one of enabling us to ascertain the intrinsic metrical relations in the receptacle space and time contingently containing them. Thus, for example, even when clocks, unlike the rotating earth, run "equably" or uniformly, these periodic devices merely record but do not first define the temporal metric. And what Newton is therefore rejecting here is a *relational* theory of space and time which asserts that (a) bodies and events first *define* (individuate) points and instants by conferring their identity upon them, thus enabling them to serve as the loci of other bodies and events, and (b) instead of having an intrinsic metric, physical space and time are metrically amorphous pending explicit or tacit appeal to the bodies which are first to define their respective metrics.

To be sure, Newton would also reject quite emphatically any identification or isomorphism of absolute space and time, on the one hand, with the psychological space and time of conscious awareness whose respective metrics are given by unaided ocular congruence and by psychological estimates of duration, on the other. But one overlooks the essential point here, if one is led to suppose with F. S. C. Northrop[8] that the relative, apparent and common space and time which Newton contrasts with absolute, true and mathematical space and time are the private visual space and subjective psychological time of immediate sensory experience. For Newton makes it unambiguously clear, as shown

[7] *Ibid.*, pp. 7–8.

[8] Cf. e.g., F. S. C. Northrop: *The Meeting of East and West* (New York: The Macmillan Company; 1946), pp. 76–77.

by the quoted passages, that his relative space and time are indeed that public space and time which is defined by the system of relations between material bodies and events, and not the egocentrically private space and time of phenomenal experience. The "sensible" measures discussed by Newton as constitutive of "relative" space and time are those furnished by the public bodies of the physicist, not by the unaided ocular congruence of one's eyes or by one's mood-dependent psychological estimates of duration. This interpretation of Newton is fully attested by the following specific assertions of his:

(1) "Absolute and relative space are the same in figure and magnitude," a declaration which is incompatible with Northrop's interpretation of *relative* space as "the immediately sensed spatial extension of, and relation between, sensed data (which is a purely private space, varying with the degree of one's astigmatism or the clearness of one's vision)."[9]

(2) As examples of merely "relative" times, Newton cites any "sensible and external (whether accurate or unequable [non-uniform]) measure of duration" such as "an hour, a day, a month, a year."[1] And he adds that the apparent time commonly used as a measure of time is based on natural days which are "truly unequal," true equality being allegedly achievable by astronomical corrections compensating for the non-uniformity of the earth's rotational motion caused by tidal friction, etc.[2] But Northrop erroneously takes Newton's relative time to be the "immediately sensed time" which "varies from person to person, and even for a single person passes very quickly under certain circumstances and drags under others" and asserts incorrectly that Newton identified with absolute time the public time "upon which the ordinary time of social usage is based."

(3) Newton illustrates *relative* motion by reference to the kinematic relation between a body on a moving ship, the ship, and the earth, these relations being defined in the customary manner of physics without phenomenal space or time.

---

[9] *Ibid.*, p. 76.

[1] I. Newton: *Principia, op. cit.*, p. 6.

[2] The logical status of the criterion of uniformity implicitly invoked here will be discussed in some detail in Chapter Two.

Northrop is entirely right in going on to say that Einstein's conceptual innovations in the theory of relativity cannot be construed, as they have been in certain untutored quarters, as the abandonment of the distinction between physically public and privately or egocentrically sensed space and time. But Northrop's misinterpretation of the Newtonian conception of "relative" space and time prevents him from pointing out that Einstein's philosophical thesis can indeed be characterized epigrammatically as the enthronement of the very relational conception of the space-time framework which Newton sought to interdict by his use of the terms "relative," "apparent," and "common" as philosophically disparaging epithets!

(B) RIEMANN

Long before the theory of relativity was propounded, a relational conception of the metric of space and time diametrically opposite to Newton's was enunciated by Riemann in the following words as part of his famous "Inaugural Lecture":

> Determinate parts of a manifold, distinguished by a mark or by a boundary, are called quanta. Their comparison as to quantity comes in discrete magnitudes by counting, in continuous magnitude by measurement.[3] Measuring consists in superposition of the magnitudes to be compared; for measurement there is requisite some means of carrying forward one magnitude as a measure for the other. In default of this, one can compare two magnitudes only when the one is a part of the other, and even then one can only decide upon the question of more and less, not upon the question of how many. . . . in the question concerning the ultimate basis of relations of size in space. . . . the above remark is applicable, namely that while in a discrete manifold the principle of metric relations is implicit in the notion of this manifold, it must come from somewhere else in the case of a continuous manifold. Either then the actual things forming the groundwork of a space must constitute a discrete

[3] Riemann apparently does not consider denumerable dense sets, which are neither discrete nor continuous, but we shall find below that his pre-Cantorean treatment of discrete and continuous types of order as *jointly exhaustive* does not vitiate the essential core of his analysis for our purposes.

manifold, or else the basis of metric relations must be sought for outside that actuality, in colligating forces that operate upon it. . .[4]

Although we shall see below that Riemann was mistaken in supposing that the first part of this statement will bear critical scrutiny as a characterization of continuous manifolds in general, he does render here a fundamental feature of the continua of *physical space* and *time,* which are manifolds whose elements, taken singly, all have zero magnitude. And this basic feature of the spatio-temporal continua will presently be seen to invalidate decisively the Newtonian claim of the intrinsicality of the metric in empty space and time. When now proceeding to state the upshot of Riemann's declaration for the spatio-temporal congruence issue before us, we shall not need to be concerned with the inadequacies arising from Riemann's pre-Cantorean treatment of discrete and continuous types of order as *jointly exhaustive.* And we shall defer until Chapters Fourteen and Fifteen consideration of the significance of Riemann's suggestion that "the basis of metric relations must be sought for outside [the groundwork of space] . . . in colligating forces that operate upon it" for Einstein's original quest to implement Mach's Principle in the general theory of relativity.[5]

Construing Riemann's statement as applying not only to lengths but also, *mutatis mutandis,* to areas and to volumes of higher dimensions, he gives the following sufficient condition for the intrinsic definability and non-definability of a metric without claiming it to be necessary as well: in the case of a discretely ordered set, the "distance" between two elements can be defined intrinsically in a rather natural way by the cardinality of the (least) number of intervening elements.[6] By contrast, upon

[4] B. Riemann: "On the Hypotheses Which Lie at the Foundations of Geometry," in: David E. Smith (ed.) *A Source Book in Mathematics* (New York: Dover Publications, Inc.; 1959), Vol. II, pp. 413 and 424–25.

[5] A. Einstein: "Prinzipielles zur allgemeinen Relativitätstheorie," *Annalen der Physik,* Vol. LV (1918), p. 241.

[6] The *basis* for the discrete ordering is not here at issue: it can be conventional, as in the case of the letters of the alphabet, or it may arise from special properties and relations characterizing the objects possessing the specified order.

confronting the extended continuous manifolds of physical space and time—their continuity being postulated in modern physical theories apart from programs of space and time quantization—we see that neither the cardinality of intervals nor any of their other topological properties provide a basis for an *intrinsically* defined metric. The first part of this conclusion was tellingly emphasized by Cantor's proof of the equi-cardinality of all positive intervals independently of their length. Thus, there is no *intrinsic* attribute of the space between the endpoints of a line-segment AB, or any relation between these two points themselves, in virtue of which the interval AB could be said to contain the same amount of space as the space between the termini of another interval CD not coinciding with AB. Corresponding remarks apply to the time continuum. Accordingly, the continuity we postulate for physical space and time furnishes a sufficient condition for their intrinsic metrical amorphousness. And in this sense then metric geometry is concerned not with space itself but with the relations between bodies.

Clearly, this does not preclude the existence of sufficient conditions *other than continuity* for the intrinsic metrical amorphousness of sets. But one cannot invoke densely ordered, *denumerable* sets of points (instants) in an endeavor to show that discontinuous sets of such elements may likewise lack an intrinsic metric: as we shall see in Chapter Six, even without measure theory ordinary analytic geometry allows the deduction that the length of a *denumerably* infinite point set is intrinsically zero. More generally, the measure of a denumerable point set is always zero,[7] unless one succeeds in developing a very restrictive intuitionistic measure theory of some sort.

These considerations show incidentally that space-intervals cannot be held to be merely *denumerable* aggregates within the context of the usual mathematical theory. Hence in the context of our post-Cantorean meaning of "continuous," it is actually not as damaging to Riemann's statement as it might seem prima facie that he neglected the *denumerable* dense sets by incorrectly treating the discrete and continuous types of order as *jointly exhaustive*. Moreover, since the distinction between denumer-

[7] E. W. Hobson: *The Theory of Functions of a Real Variable* (New York: Dover Publications, Inc.; 1957), Vol. I, p. 166.

able and super-denumerable dense sets was almost certainly unknown to Riemann, it is likely that by "continuous" he merely intended the property which we now call "dense." Evidence of such an earlier usage of "continuous" is found as late as 1914.[8]

The intrinsic metric amorphousness of the spatial continuum is made further evident by reference to the axioms for spatial congruence *after* it has been specified that they are to be given a *spatial* interpretation in terms of intervals of physical space.[9] These axioms pre-empt "congruent" (for intervals) to be a *spatial equality predicate* by assuring the reflexivity, symmetry, and transitivity of the congruence relation in the class of spatial intervals. But although having thus pre-empted the use of "congruent" and no longer being an uninterpreted axiom system, the congruence axioms still allow an *infinitude* of *mutually exclusive* congruence classes of spatial intervals, where it is to be understood that any *particular* congruence class is a *class of classes* of congruent intervals whose lengths are specified by a *particular* distance function $ds^2 = g_{ik}dx^i dx^k$. And we just saw that there are no intrinsic metric attributes of intervals which could be invoked to single out one of these congruence classes as unique.

How then can we speak of the assumedly continuous physical space as having *a* metric or *mutatis mutandis* suppose that the physical time continuum has a unique metric? The answer can be none other than the following:[1] Only the choice of a particular congruence standard which is *extrinsic* to the continuum itself can determine a unique congruence class, the *rigidity* or self-congruence of that standard under transport being *decreed by convention,* and similarly for the periodic devices which are

[8] B. Russell: *Our Knowledge of the External World* (London: George Allen and Unwin, Ltd.; 1926), p. 138.

[9] For a statement of these axioms, see A. N. Whitehead: *The Principle of Relativity* (Cambridge: Cambridge University Press; 1922), Chap. iii, pp. 42–50.

[1] The conclusion which is about to be stated will appear unfounded to those who follow A. N. Whitehead in rejecting the "bifurcation of nature," which is assumed in its premises. But later in this chapter, the reader will find a detailed rebuttal of the Whiteheadian contention that *perceptual* space and time do have an intrinsic metric such that once the allegedly illegitimate distinction between physical and perceptual space (or time) has been jettisoned, an intrinsic metric can hence be meaningfully imputed to physical space and time.

held to be *isochronous* (uniform) clocks. Thus the role of the spatial or temporal congruence standard cannot be construed with Newton or Russell[2] to be the mere ascertainment of an otherwise intrinsic equality obtaining between the intervals belonging to the congruence class specified by it. Unless one of two segments is a subset of the other, the obtaining of the congruence relation between two segments is a matter of convention, stipulation, or definition and not a factual matter concerning which empirical findings could show one to have been mistaken. And hence there can be no question at all of an empirically or factually determinate metric geometry or chronometry until after a physical stipulation of congruence.[3]

In the case of geometry, the specification of the intervals which are stipulated to be congruent is given by the distance function $ds = \sqrt{g_{ik}dx^i dx^k}$, congruent intervals being those which are assigned equal lengths ds by this function. Whether the intervals defined by the coincidence behavior of a transported rod not subject to "deforming influences" are those to which the distance function assigns equal lengths ds or not will depend on our selection of the functions g_{ik}. Thus, if the components of the metric tensor g_{ik} are suitably chosen in any given coordinate system, then the transported rod will have been stipulated to

---

[2] Cf. B. Russell: "Sur les Axiomes de la Géométrie," *Revue de Métaphysique et de Morale,* Vol. VII (1899), pp. 684–707.

[3] A. d'Abro (A. d'Abro: *The Evolution of Scientific Thought from Newton to Einstein* [New York: Dover Publications, Inc.; 1950], p. 27) has offered an unsound illustration of the thesis that the metric in a continuum is conventional: he considers a stream of sounds of varying pitch and points out that a congruence criterion based on the successive auditory octaves of a given musical note would be at variance with the congruence defined by equal differences between the associated frequencies of vibration, since the frequency differences between successive octaves are not equal. But instead of constituting an example of the alternative metrizability of the same mathematically continuous manifold of elements, d'Abro's illustration involves the metrizations of two different manifolds only one of which is continuous in the mathematical sense. For the auditory contents sustaining the relation of being octaves of one another are elements of a merely *sensory* "continuum." Moreover, we shall see later in this chapter that while holding for the mathematical continua of physical space and time, whose elements (points and instants) are respectively alike both qualitatively and in magnitude, the thesis of the conventionality of the metric cannot be upheld for all kinds of mathematical continua, Riemann and d'Abro to the contrary notwithstanding.

be congruent to itself everywhere independently of its position and orientation. Contrariwise, by an appropriately different choice of the functions g_{ik}, the length ds of the transported rod will be made to vary with position or orientation instead of being constant. Once congruence has been specified via the distance function ds, the geodesics (straight lines)[4] associated with that choice of congruence are determined, since the family of geodesics is defined by the variational requirement $\delta \int ds = 0$, which takes the form of a differential equation whose solution is the equation of the family of geodesics.[5] The geometry char-

---

[4] The geodesics are called "straight lines" when discussing their relations in the context of synthetic geometry. But this identification must not be taken to entail that on surfaces other than the Euclidean plane, every geodesic connection between any two points is a line of shortest distance between them. For once we abandon the restriction to Euclidean geometry, being a geodesic connection is only a necessary and not also a sufficient condition for being the shortest distance: on the surface of a sphere, for example, "it is true that the shortest distance between two points *P* and *Q* on a sphere is along a geodesic, which on the sphere is a great circle. But there are two arcs of a great circle between two of their points, and only one of them is the curve of shortest distance, except when *P* and *Q* are at the end points of a diameter, when both arcs have the same length. This example of the sphere also shows that it is not always true that through two points only one geodesic passes: when *P* and *Q* are the endpoints of a diameter any great circle through *P* and *Q* is a geodesic and a solution of the problem of finding the shortest distance between two points." Cf. D. J. Struik: *Classical Differential Geometry* (Cambridge: Addison-Wesley Publishing Company, Inc.; 1950), p. 140. It is the case, however, that "If two points in a surface are such that only one geodesic passes through them, the length of the segment of the geodesic is the shortest distance in the surface between the two points." (L. P. Eisenhart: *An Introduction to Differential Geometry* [Princeton: Princeton University Press; 1947], Sec. 32, p. 175.)

For an account of sufficient conditions for a geodesic connection being a minimum or shortest distance, see O. Bolza: *Lectures on the Calculus of Variations* (New York: G. E. Stechert; 1946), Chap. III, §§ 17–23 inclusive, and N. I. Akhiezer: *The Calculus of Variations* (New York: Blaisdell Publishing Company, 1962), Sections 3, 4, and 15.

[5] In the *differential* calculus, there is the problem of finding maxima and minima (extrema) of a function $y = f(x)$. And a necessary condition for the occurrence of an extremum at $x = a$ is that $dy/dx = 0$ or $dy = 0$ at $x = a$. Now, in our context the *calculus of variations* deals with the following similar but more complicated problem: Find the function $y = f(x)$—this equation representing a family of geodesic paths here—such that the definite integral taken over the function $\int ds$ of the function $y = f(x)$ shall be a minimum or (relative) maximum, i.e., an extremum for small variations which vanish at the limits of integration. We regard $\int ds$ as a function of the function $y = f(x)$, since the former will depend on the particular kind

acterizing the relations of the geodesics in question is likewise determined by the distance function ds, because the Gaussian curvature K of every surface element at any point in space is fixed by the functions g_{ik} ingredient in the distance function ds.

There are therefore alternative metrizations of the same factual coincidence relations sustained by a transported rod, and some of these alternative definitions of congruence will give rise to different metric geometries than others. Accordingly, via an appropriate definition of congruence we are free to choose as the description of a given body of spatial facts any metric geometry compatible with the existing topology. Moreover, we shall find in Chapter Three, Section B that there are infinitely many incompatible definitions of congruence which will implement the choice of any one metric geometry, be it the Euclidean one or one of the non-Euclidean geometries.

We have been speaking of alternative "definitions" of congruence. In particular, we shall have occasion to refer to one of these "definitions" as being given by unperturbed solid rods and hence as the "customary definition" of congruence. But it has been objected that such concepts as the customary concept of spatial congruence are "multiple-criterion" concepts as opposed

---

of path $y = f(x)$ over which the integral is taken. In analogy to the condition $dy = 0$ for an extremum in the differential calculus, the defining condition for the family of geodesics in the calculus of variations is $\delta \int ds = 0$. Expressing ds as I dx, it is shown in the calculus of variations (cf. H. Margenau and G. M. Murphy: *The Mathematics of Physics and Chemistry* [New York: D. Van Nostrand Company; 1943], pp. 193–95) that this defining condition requires the following differential equation, which is known as Euler's equation, to be satisfied:

$$\frac{\partial I}{\partial y} - \frac{d}{dx}\frac{\partial I}{\partial\left(\frac{dy}{dx}\right)} = 0,$$

where the symbol "∂" refers to partial differentiation as distinct from the variational symbol "δ."

As a simple illustration, consider the problem of finding the geodesics of the Euclidean plane that are associated with the standard metric $ds^2 = dx^2 + dy^2$, which can be written as $ds = \sqrt{1 + \left(\frac{dy}{dx}\right)^2}\, dx \equiv I\, dx$. If $\int_a^1 ds$ is to be a minimum, Euler's equation must be satisfied for the case of $I \equiv \sqrt{1 + \left(\frac{dy}{dx}\right)^2}$. Hence we have $dy/dx = m$, where m is a constant, or $y = mx + b$. As expected, this is the equation of a family of straight lines.

to "single-criterion" concepts:[6] congruent space intervals in inertial systems could be "defined," for example, as those for which light rays require equal (one-way or round-trip) transit times no less than they can be "defined" as the intervals given by the coincidence behavior of unperturbed transported solid rods. Thus, it has been objected that it is unsound logically to speak of giving a "definition" of congruence in the spirit of Reichenbach's "coordinative definition," since no one physical criterion, such as the one based on the solid rod, can exhaustively render the actual and potential physical meaning of the concept of spatial congruence in physics. But this objection overlooks that our reference to one or another "definition" of congruence within a set of *mutually-exclusive* "definitions" of congruence does not commit us at all to the crude operationist claim that any particular "definition" chosen by the physicist exhaustively renders "*the* meaning" of spatial congruence in physical theory. For our concern is with the alternative metrizability of the spatial continuum noted by Riemann and with the resulting conventionality of congruence. Hence when we speak of "defining" congruence in this context, all that we understand by a "definition" is a specification which employs one or another criterion to single out one particular congruence class from among an infinitude of mutually-exclusive congruence classes. Thus, we shall speak of a "definition" of congruence throughout this book entirely without prejudice to the fact that spatial congruence is an open multiple-criterion concept in physics in the following sense: there is a potentially growing multiplicity of compatible physical criteria rather than only one criterion by which any one spatial congruence class (e.g., the one familiar from elementary physics) can be specified to the exclusion of every other congruence class. And it is now evident that much as attention to the multiple-criterion character of concepts in physics may be philosophically salutary in other contexts, it constitutes an intrusion of a pedantic irrelevancy in the consideration of the consequences of alternative metrizability.

In pointing out earlier in this section that the status of spatial

---

[6] Cf. H. Putnam: "The Analytic and the Synthetic," in: H. Feigl and G. Maxwell (eds.) *Minnesota Studies in the Philosophy of Science* (Minneapolis: University of Minnesota Press; 1962), Vol. III, esp. pp. 376–81.

and temporal congruence is decisively illuminated by Riemann's theory of continuous manifolds, I stated that this theory will not bear critical scrutiny as a characterization of continuous manifolds in general. To justify and clarify this indictment, we shall now see that continuity cannot be held with Riemann to furnish a sufficient condition for the intrinsic metric amorphousness of any manifold *independently of the character of its elements.* For, as Russell saw correctly,[7] there are continuous manifolds, such as that of colors (in the physicist's sense of spectral frequencies), in which the individual elements differ qualitatively from one another and have inherent magnitude, thus allowing for metrical comparison of the elements themselves. By contrast, in the continuous manifolds of *space* and of *time,* neither points nor instants have any inherent magnitude allowing an individual metrical comparison between them, since all points are alike, and similarly for instants. Hence in these manifolds metrical comparisons can be effected only among the *intervals* between the elements, not among the homogeneous elements themselves. And the continuity of these manifolds then assures the nonintrinsicality of the metric for their intervals.

To exhibit further the bearing of the character of the elements of a continuous manifold on the feasibility of an intrinsic metric in it, I shall contrast the status of the metric in space and time, on the one hand, with its status in both (a) the continuum of real numbers, arranged according to magnitude and (b) the quasi-continuum of masses, mass being assumed to be a property of bodies in the Newtonian sense clarified by Mach's definition.[8]

The assignment of real numbers to points of physical space in the manner of the introduction of generalized curvilinear coordinates effects only a *coordinatization* but *not* a *metrization* of the manifold of physical space. No informative metrical comparison among individual points could be made by comparing the magnitudes of their real number coordinate-*names.* However, within the continuous manifold formed by the real numbers

---

[7] B. Russell: *The Foundations of Geometry* (New York: Dover Publications, Inc.; 1956), Secs. 63 and 64.

[8] For a concise account of that definition, cf. L. Page: *Introduction to Theoretical Physics* (New York: D. Van Nostrand Company, Inc.; 1935), pp. 56–58.

themselves, when arranged according to magnitude, every real number is singly distinguished from and metrically comparable to every other by its inherent magnitude. And the measurement of mass can be seen to constitute a counter-example to Riemann's metrical philosophy from the following considerations.

In the Machian definition of Newtonian (gravitational and inertial) mass, the *mass ratio* of a particle *B* to a standard particle *A* is given by the magnitude ratio of the acceleration of *A* due to *B* to the acceleration of *B* due to *A*. Once the space-time metric and thereby the accelerations are fixed in the customary way, this ratio for any particular body *B* is *independent,* among other things, of how far apart *B* and *A* may be during their interaction. Accordingly, any affirmations of the mass equality (mass-"congruence") or inequality of two bodies will hold independently of the extent of their spatial separation. Now, the set of medium-sized bodies form a *quasi*-continuum with respect to the dyadic relations "being more massive than," and "having the same mass," i.e., they form an array which is a continuum except for the fact that several bodies can occupy the same place in the array by sustaining the mass-"congruence" relation to each other. Without having such a relation of mass-equality *ab initio,* the set of bodies do not even form a quasi-continuum. We complete the metrization of this quasi-continuum by choosing a unit of mass (e.g., one gram) and by availing ourselves of the numerical mass-ratios obtained by experiment. There is no question here of the lack of an intrinsic metric in the sense of a choice of making the mass *difference* between a given pair of bodies equal to that of another pair or not. In the resulting continuum of real mass numbers, the elements themselves have inherent magnitude and can hence be compared individually, thus defining an intrinsic metric. Unlike the point-elements of space, the elements of the set of bodies are not all alike masswise, and hence the metrization of the quasi-continuum which they form with respect to the relations of being more massive and having the same mass can take the form of directly comparing the individual elements of that quasi-continuum rather than only intervals between them.

If one did wish for a spatial (or temporal) analogue of the metrization of masses, one should take as the set to be metrized *not* the continuum of points (or instants) but the quasi-continuum of all spatial (or temporal) *intervals*. To have used such intervals as the elements of the set to be metrized, we must have had a prior criterion of spatial congruence and of "being longer than" in order to arrange the intervals in a quasi-continuum which can then be metrized by the assignment of length numbers. This metrization would be the space or time analogue of the metrization of masses.

(C) POINCARÉ

An illustration will serve to give concrete meaning to the general formulation of the conventionality of spatial congruence which we presented in Section B as the direct outgrowth of Riemann's analysis of the status of the metric in the spatial continuum. Consider a physical surface such as part or all of an infinite blackboard and suppose it to be equipped with a network of Cartesian coordinates. The customary metrization of such a surface is based on the congruence defined by the coincidence behavior of transported rods: line segments whose termini have coordinate differences dx and dy are assigned a length ds given by $ds = \sqrt{dx^2 + dy^2}$, and the geometry associated with this metrization of the surface is, of course, Euclidean. But we are also at liberty to employ a different metrization in part or all of this space. Thus, for example, we could equally legitimately metrize the portion *above* the x-axis on our blackboard by means of the new metric $ds = \sqrt{\frac{dx^2 + dy^2}{y^2}}$. This alternative metrization is incompatible with the customary one: for example, it makes the lengths $ds = dx/y$ of horizontal segments whose termini have the same coordinate differences dx depend on *where* they are along the y-axis. Consequently, the new metric would commit us to regard a segment for which $dx = 2$ at $y = 2$ as *congruent* to a segment for which $dx = 1$ at $y = 1$, although the customary metrization would regard the length ratio of these segments to be 2:1. But, of course, the new metric does not say that a transported solid rod will coincide successively with the intervals belonging to the congruence class defined by that metric; instead it allows for this *non*-coincidence by making the length of the rod a suitably non-constant function of its position: lying parallel to the x-axis at $y = 2$, the rod would now be assigned only one half of the length which is ascribed to it in a like orientation at $y = 1$.

Since our new metric, which was introduced by Poincaré, yields a congruence class of *line segments* different from the customary one, one wonders whether it also issues in a non-customary congruence class of *angles.* To deal with this question we note the following requisite mathematical results. The angle θ determined by the directions A and B in a Riemann space, which are defined by the displacements A^i and B^i respectively, is defined by the equation

$$\cos \theta = \frac{g_{ij}A^iB^j}{\sqrt{g_{ab}A^aA^b} \cdot \sqrt{g_{ab}B^aB^b}},$$

where the g_{ij} represent the metric coefficients[9] of the metric $ds^2 = g_{ij}dx^idx^j$ for line segments. Now introduce a *new* metric which has the property that its metric coefficients g^*_{ij} are related to the original coefficients g_{ij} by the following socalled "conformal" transformation: $g^*_{ij} = f(x^i)\, g_{ij}$, where $f(x^i)$ is an analytic multiplying function of the coordinates x^i. It is then evident from the above expression for $\cos \theta$ that the angles θ and hence the *congruence* relations among angles will be left *unchanged* by any new metrization whose metric coefficients g^*_{ij} are related by a conformal transformation to the coefficients of the original metric.[1]

This result enables us to see that Poincaré's metric

$$ds^2 = \frac{1}{y^2}\,(dx^2 + dy^2)$$

yields the *same* congruence class of *angles* as the original standard metric $ds^2 = dx^2 + dy^2$: the coefficients of Poincaré's metric are related to those of the standard metric by the multiplying function $1/y^2$.

To determine what particular paths in the semi-blackboard are the *geodesics* of our non-standard Poincaré metric

$$ds = \frac{\sqrt{dx^2 + dy^2}}{y}$$

[9] Cf. T. Y. Thomas: *The Differential Invariants of Generalized Spaces* (Cambridge: Cambridge University Press; 1934), p. 12.

[1] *Ibid.*, p. 20.

one must substitute this ds and carry out the variation in the defining condition $\delta\int ds = 0$ for the family of geodesics. As will be recalled from footnote 5, page 14 in Section B above, the desired geodesics of our new metric must therefore be given by Euler's equation with

$$I \equiv \frac{\sqrt{1 + \left(\frac{dy}{dx}\right)^2}}{y}.$$

Upon substituting this value of I in Euler's equation, we obtain the differential equation of the family of geodesics:

$$\frac{d^2y}{dx^2} + \frac{1}{y}\left[1 + \left(\frac{dy}{dx}\right)^2\right] = 0.$$

The solution of this equation is of the form

$$(x - k)^2 + y^2 = R^2,$$

where k and R are constants of integration. This solution represents the family of *straights* (geodesics) associated with Poincaré's metric, but—in the Euclidean language appropriate to the standard Cartesian metric—it represents a family of "circles" centered on and perpendicular to the x-axis, the upper "semicircles" being the *geodesics* of Poincaré's remetrized half-plane above the x-axis.

The reader can convince himself that Poincaré's metrization issues in a *hyperbolic non*-Euclidean geometry on the semi-blackboard by using the new metric coefficients $g_{11} = 1/y^2$, and $g_{22} = 1/y^2$ to obtain a *negative* value of the Gaussian curvature K via Gauss's formula.[2] That Poincaré's metric confers a hyperbolic geometry on the very same semi-blackboard $y > 0$ which is a *Euclidean* plane relatively to the standard metrization becomes palpable geometrically upon noting that the new geodesics of Poincaré's half plane have the following properties: First, their infinitude is assured by the behavior of Poincaré's metric as $y \rightarrow 0$, and second, they satisfy the hyperbolic parallel postulate that there exists more than one coplanar parallel

[2] For a statement of this formula, see, for example, F. Klein: *Vorlesungen über Nicht-Euklidische Geometrie* (Berlin: Springer-Verlag; 1928), p. 281.

straight line, since they *also* qualify as Euclidean semicircles and hence exemplify the Euclidean attribute that through a point outside a semicircle more than one semicircle can be drawn not intersecting the given semicircle.

It is apparent that the supplanting of the standard Cartesian metric by that of Poincaré has had the effect of *renaming* various paths on the semi-blackboard such that the language of hyperbolic geometry describes the very same facts of coincidence of a rod under transport on the semi-blackboard which are customarily rendered in the language of Euclidean geometry. And in the light of Riemann's searching account of spatial (and temporal) metrics, we must conclude that the hyperbolic metrization of the semi-blackboard possesses not only mathematical but also philosophical credentials as good as those of the Euclidean one.

Yet it might be objected that although non-standard metrics are legitimate philosophically, there is a pedantic artificiality and even perverse complexity in all congruence definitions which do not assign equal lengths ds to the intervals defined by the coincidence behavior of an unperturbed solid rod. The grounds of this objection would be that (a) there are no convenient and familiar natural objects whose coincidence behavior under transport furnishes a physical realization of the bizarre, non-customary congruences, and (b) after correcting for the chemically dependent distortional idiosyncrasies of various kinds of solids in inhomogeneous thermal, electric, and other fields, all transported solid bodies furnish the same physical intervals, thereby realizing one of the infinitude of incompatible mathematical congruences. *Mutatis mutandis,* the same objection might be raised to any definition of temporal congruence which is non-standard in virtue of not according with the cycles of standard unperturbed material clocks. The reply to this criticism is twofold:

(1) The prima facie plausibility of the demand for *simplicity* in the choice of the congruence definition gives way to second thoughts the moment it is realized that the desideratum of simplicity requires consideration not only of the congruence definition but also of the latter's bearing on the form of the associated system of geometry and physics. And our discussions in Chapters

Two and Four will show that a bizarre definition of congruence may well have to be countenanced as the price for the attainment of the over-all simplicity of the total theory. Specifically, we anticipate Chapter Two by just mentioning here that although Einstein merely alludes to the possibility of a non-customary definition of *spatial* congruence in the general theory of relativity without actually availing himself of it,[3] he does indeed utilize in that theory what our putative objector deems a highly artificial definition of *temporal* congruence, since it is not given by the cycles of standard material clocks;

(2) It is particularly instructive to note that the cosmology of E. A. Milne[4] postulates the actual existence in nature of two metrically different kinds of clocks whose respective periods constitute physical realizations of *incompatible* mathematical congruences. Specifically, Milne's assumptions lead to the result that there is a *non-linear* relation

$$\tau = t_0 \log \left(\frac{t}{t_0}\right) + t_0$$

between the time τ defined by periodic astronomical processes and the time t defined by atomic ones, t_0 being an appropriately chosen arbitrary constant. The non-linearity of the relation between these two kinds of time is of paramount importance here, because it assures that two intervals which are congruent in one of these two time-scales will be incongruent in the other, as is evident from the fact that the derivative $\frac{d\tau}{dt}$ is not a constant. We can visualize the relation of the two time-scales geometrically as follows: let a half open line have the t-scale point $t = 0$ as an extremity, and use *equal* spatial intervals on the line to represent equal time intervals of the t-scale. Then the τ-scale would be represented on this same line by metrizing it so as to chop it up into spatial intervals that get progressively *shorter* in the direction of the point $t = 0$, a point which does not, how-

---

[3] A. Einstein: "The Foundations of the General Theory of Relativity," *The Principle of Relativity, A Collection of Original Memoirs* (New York: Dover Publications, Inc.; 1952), p. 161.

[4] E. A. Milne: *Kinematic Relativity* (Oxford: Oxford University Press; 1948), p. 22.

ever, belong to the τ-scale since $\tau \rightarrow -\infty$ as $t \rightarrow 0$. Thus, in the direction of $t = 0$ (toward the past), equal time intervals on the τ-scale correspond to ever smaller intervals on the t-scale.

Clearly, it would be utterly gratuitous to regard one of Milne's two congruences as bizarre, since each of them is presumed to have a physical realization. And the choice between these scales is incontestably conventional, for it is made quite clear in Milne's theory that their associated different metric descriptions of the world are factually equivalent and hence equally true.

What would be the verdict of the Newtonian proponent of the intrinsicality of the metric on the examples of alternative metrizability which we gave both for space (Poincaré's hyperbolic metrization of the half-plane) and also for time (general theory of relativity and Milne's cosmology)? He would first note correctly that once it is understood that the term "congruent," as applied to intervals, is to denote a reflexive, symmetrical and transitive relation in this class of geometrical configurations, then the use of this term is restricted to designating a spatial equality relation. But then the Newtonian would proceed to claim unjustifiably that the spatial equality obtaining between congruent line segments of physical space (or between regions of surfaces and of 3-space respectively) consists in their each containing *the same intrinsic amount of space*. And having introduced this false premise, he would feel entitled to contend that first, it is *never* legitimate to choose arbitrarily what specific intervals are to be regarded as congruent, and second, as a corollary of this lack of choice, there is no room for selecting the lines which are to be regarded as straight and hence no choice among alternative geometric descriptions of actual physical space for the following reason: the geodesic requirement $\delta \int ds = 0$ must be satisfied by the straight lines subject to the restriction that only the members of the unique class of *intrinsically equal* line segments may ever be assigned the same length ds. By the same token, the Newtonian asserts that only "truly" (intrinsically) equal time intervals may be regarded as congruent, and he therefore holds that there exists only one admissible congruence class in the time continuum, a conclusion which he then attempts to buttress further by adducing certain causal considerations from Newtonian dynamics which will be refuted in Chapter Two below.

(D) EDDINGTON

Poincaré's view that the epistemological status of *congruence* is pivotal for the philosophical assessment of the issue of Euclideanism *vs.* non-Euclideanism has been rejected by Eddington. According to Eddington, the thesis that congruence (for line segments or time intervals) is conventional is true only in the trivial sense that "the meaning of every word in the language is conventional."[5] Commenting on Poincaré's statement that we can always avail ourselves of alternative metrizability to give a Euclidean interpretation of any results of stellar parallax measurements[6] Eddington writes:

> Poincaré's brilliant exposition is a great help in understanding the problem now confronting us. He brings out the interdependence between geometrical laws and physical laws, which we have to bear in mind continually.[7] We can add on to one set of laws that which we subtract from the other set. I admit that space is conventional—for that matter, the meaning of every word in the language is conventional. Moreover, we have actually arrived at the parting of the ways imagined by Poincaré, though the crucial experiment is not precisely the one he mentions. But I deliberately adopt the alternative, which, he takes for granted, everyone would consider less advantageous. I call the space thus chosen *physical space,* and its geometry *natural geometry,* thus admitting that other conventional meanings of space and geometry are possible. If it were only a question of the meaning of space—a rather vague term—these other possibilities might have some advantages. But the meaning assigned to length and distance has to go along with the meaning assigned to space. Now these are quantities which the physicist has been accustomed to measure with great accuracy; and they enter fundamentally into the whole of our experimental knowledge of the world. . . . Are we to be robbed of the terms in which we are accustomed to describe that knowledge?[8]

---

[5] A. S. Eddington: *Space, Time and Gravitation* (Cambridge: Cambridge University Press; 1953), p. 9.

[6] Cf. H. Poincaré: *The Foundations of Science* (Lancaster: The Science Press; 1946), p. 81, and the discussion in Chapter Four below.

[7] This interdependence will be analyzed in Chapter Four below.

[8] A. S. Eddington: *Space, Time and Gravitation, op. cit.,* pp. 9–10.

Eddington maintains that instead of being an insight into the status of spatial or temporal equality, the conventionality of congruence is a semantical platitude expressing our freedom to decree the referents of the *word* "congruent," a freedom which we can exercise in regard to any linguistic symbols whatever which have not already been pre-empted semantically. Thus, we are being told that though the conventionality of congruence is merely an unenlightening triviality holding for the language of any field of inquiry whatever, it has been misleadingly inflated into a philosophical doctrine about the relation of spatio-temporal equality purporting to codify fundamental features endemic to the materials of geo-chronometry. In particular, Eddington objects to Poincaré's willingness to guarantee the retention of Euclidean geometry by resorting to an alternative metrization: in the context of general relativity, the retention of Euclideanism would indeed require a congruence definition different from the customary one, as we shall see in Chapter Three. Regarding the possibility of a remetrizational retention of Euclidean geometry as merely illustrative of being able to avail oneself of the conventionality of all language, Eddington would rule out such a procedure on the grounds that the customary definition of spatial congruence which would be supplanted by the remetrization retains its usefulness.

Eddington's conclusion that only the use of the *word* "congruent" but not the ascription of the congruence *relation* can be held to be a matter of convention has also been defended by a cognate argument which invokes the theory of models of uninterpreted formal calculi as follows: (1) physical geometry is a spatially interpreted abstract calculus, and this interpretation of a formal system was effected by semantical rules which are all equally conventional and among which the definition of the relation term "congruent" (for line segments) does *not* occupy an epistemologically distinguished position, since we are just as free to give a *non*-customary interpretation of the abstract sign "point" as of the sign "congruent"; (2) this model-theoretic conception makes it apparent that there can be no basis at all for an epistemological distinction *within* the system of *physical* geo-chronometry between factual statements, on the one hand, and supposedly conventional assertions of rigidity and isochronism on the other; (3) the factual credentials of physical geometry or chronometry can no more be impugned by adducing the alleged conventionality of rigidity and isochronism than one could gainsay the factuality of genetics by incorrectly affirming the conventionality of the relation of uniting which obtains between two gametes when forming a zygote.

When defending the alternative metrizability of space and time and the resulting possibility of giving either a Euclidean or a non-Euclidean description of the same spatial facts, Poincaré had construed the conventionality of congruence as an epistemological discovery about the status of the relation of spatial or temporal equality. The proponent of the foregoing model-theoretic argument therefore indicts Poincaré's defense of the feasibility of *choosing* the metric geometry as amiss, misleading, and unnecessary, deeming this choice to be automatically assured by the theory of models. And, by the same token, this critic maintains that there is no more reason for Poincaré's inquiry into the *empirical* credentials of metric geometry as such than there would be for instituting a philosophical inquiry as to the sense in which genetics as such can be held to have an empirical warrant.

In order to discern the basic error in Eddington's critique, it is of the utmost importance to realize clearly that the thesis of the conventionality of congruence is, in the first instance, a claim concerning *structural properties of physical space and time;* only the semantical corollary of that thesis concerns the language of the geo-chronometric description of the physical world. Having failed to appreciate this fact, Eddington and those who invoke the theory of models were misled into giving a shallow caricature of the debate between the Newtonian, who affirms the factuality of congruence on the strength of the alleged intrinsicality of the metric, and his Riemannian conventionalistic critic such as Poincaré. According to the burlesqued version of this controversy, Poincaré is offering no more than a semantical truism. More specifically, the detractors suppose that their trivialization of the congruence issue can be vindicated by pointing out that we are, of course, free to decree the referents of the *unpre-empted word* "congruent," because such freedom can be exercised with respect to any as yet semantically uncommitted term or string of symbols whatever. And, in this way, they misconstrue the conventionality of congruence as merely an inflated special case of a semantical banality holding for any and all linguistic signs or symbols, a banality which we shall call "trivial semantical conventionalism" or, in abbreviated form, "TSC."

No one, of course, will wish to deny that *qua uncommitted signs,* the terms "spatially congruent" and "temporally congruent" are fully on a par in regard to the *trivial* conventionality of the semantical rules governing their use with all linguistic symbols whatever. And thus a sensible person would hardly wish to contest that the unenlightening affirmation of the conventionality of the use of the *unpre-empted word* "congruent" is indeed a subthesis of TSC. But it is a serious obfuscation to identify the Riemann-Poincaré doctrine that the *ascription* of the congruence or equality *relation* to space or time intervals is conventional with the platitude that the use of the *unpre-empted word* "congruent" is conventional. And it is therefore totally incorrect to conclude that the Riemann-Poincaré tenet is merely a gratuitously emphasized special case of TSC. For what these mathematicians are advocating is not a doctrine about the semantical freedom we have in the use of the uncommitted sign "congruent." Instead, they are putting forward the initially non-semantical claim that the continua of physical space and time each lack an intrinsic metric. And the metric amorphousness of these continua then serves to *explain* that even *after* the word "congruent" has been pre-empted semantically as a spatial or temporal *equality* predicate by the axioms of congruence, congruence remains ambiguous in the sense that these axioms still allow an infinitude of mutually exclusive congruence classes of intervals. Accordingly, fundamentally *non*-semantical considerations are used to show that only a conventional choice of one of these congruence classes can provide a unique standard of length equality. In short, the conventionality of congruence is a claim not about the noise "congruent" but about the character of the conditions relevant to the *obtaining* of the kind of equality relation denoted by the term "congruent." For alternative metrizability is not a matter of the freedom to use the semantically uncommitted noise "congruent" as we please; instead, it is a matter of the *non-*

uniqueness of a relation term already *pre-empted* as the *physico-spatial (or temporal)* equality predicate. And *this non-uniqueness arises from the lack of an intrinsic metric in the continuous manifolds of physical space and time.*

The philosophical status of the Riemann-Poincaré conventionality of congruence is fully analogous to that of Einstein's conventionality of simultaneity. And if the reasoning used by Eddington in an endeavor to establish the banality of the former were actually sound, then, as we shall now show, it would follow by a precisely similar argument that Einstein's enunciation of the conventionality of simultaneity[9] was no more than a turgid statement of the platitude that the uncommitted word "simultaneous" (or "gleichzeitig") may be used as we please. In fact, due to the complete philosophical affinity of the conventionality of congruence with the conventionality of simultaneity, which we are about to exhibit, it will be useful subsequently to combine these two theses under the name "geo-chronometric conventionalism" or, in abbreviated form, "GC."

We saw in the case of spatial and temporal congruence that congruence is conventional in a sense other than that prior to being pre-empted semantically, the *sign* "congruent" can be used to denote anything we please. *Mutatis mutandis,* we now wish to show that precisely the same holds for the conventionality of metrical simultaneity. Once we have furnished this demonstration as well, we shall have established that neither of the component claims of conventionality in our compound GC thesis is a subthesis of TSC.

We proceed in Einstein's manner in the special theory of relativity and first decree that the *noise* "topologically simultaneous" denotes the relation of *not* being connectible by a physical causal (signal) chain, a relation which may obtain between two physical events. We now ask: is this definition unique in the sense of assuring that one and only one event at a point Q will be topologically simultaneous with a given event occurring at a point P elsewhere in space? The answer to this question depends on facts of nature, namely on the range of the causal chains existing in

---

[9] A. Einstein: "On the Electrodynamics of Moving Bodies," *The Principle of Relativity, A Collection of Original Memoirs,* pp. 37–65. (New York: Dover Publications, Inc.; 1952), §1. For details, see Chapter Twelve below.

the physical world. Thus, once the above definition is given, its uniqueness is not a matter subject to regulation by semantical convention. If now we assume with Einstein as a fact of nature that light *in vacuo* is the fastest causal chain, then this postulate entails the non-uniqueness of the definition of "topological simultaneity" given above and thereby also prevents *topological* simultaneity from being a transitive relation. By contrast, if the facts of the physical world had the structure assumed by Newton, this *non*-uniqueness would not arise. Accordingly, the structure of the facts of the world postulated by relativity prevents the above definition of "topological simultaneity" from also serving, as it stands, as a metrical synchronization rule for clocks at the spatially separated points P and Q. Upon coupling this result with the relativistic assumption that transported clocks do not define an absolute metrical simultaneity, we see that the facts of the world leave the equality relation of metrical simultaneity *indeterminate,* for they do not confer upon topological simultaneity the uniqueness which it would require to serve as the basis of metrical simultaneity as well. Therefore, the assertion of that indeterminateness and of *the corollary that metrical simultaneity is made determinate by convention* is in no way tantamount to the purely semantical assertion that the mere uncommitted *noise* "metrically simultaneous" must be given a physical interpretation before it can denote, and that this interpretation is trivially a matter of convention.

Far from being a claim that a mere linguistic noise is still awaiting an assignment of semantical meaning, the assertion of the factual indeterminateness of metrical simultaneity concerns *facts of nature* which find expression in the residual non-uniqueness of the definition of "topological simultaneity" once the latter has already been given. And it is thus impossible to construe this residual non-uniqueness as being attributable to taciturnity or tight-lippedness on Einstein's part in telling us what he means by the noise "simultaneous." Here, then, we are confronted with a kind of logical gap needing to be filled by definition which is precisely analogous to the case of congruence, where the continuity of space and time issued in the residual non-uniqueness of the congruence axioms *after* they are given a spatial or temporal interpretation. When I say that metrical simultaneity is not wholly factual but contains a conventional ingredient, what am I asserting? I am claiming none other than that the residual non-uniqueness or logical gap cannot be removed by an appeal to facts but only by a conventional choice of a unique pair of events at P and at Q as *metrically* simultaneous from within the class of pairs that are topologically simultaneous. And when I assert that it was a great philosophical (as well as physical) achievement for Einstein to have discovered the conventionality of metrical simultaneity, I am crediting Einstein *not* with the triviality of having decreed semantically the meaning of the *noise* "metrically simultaneous" (or "gleichzeitig") but with the recognition that, contrary to earlier belief, the facts of nature are such as to deny the required kind of semantical univocity to the already pre-empted term "simultaneous" ("gleichzeitig"). In short, Einstein's insight that metrical simultaneity is conventional is a contribution to the theory of time rather than to semantics, because *it concerns the character of the conditions relevant to the obtaining of the relation denoted by the term "metrically simultaneous."*

The conventionality of metrical simultaneity has just been formulated without any reference whatever to the relative motion of different Galilean frames, and does not depend upon there being a relativity or non-concordance of simultaneity as between different Galilean frames. On the contrary, as Chapter Twelve will show in detail, it is the conventionality of simultaneity which provides the logical framework within which the relativity of simultaneity can first be understood: if each Galilean observer adopts the particular metrical synchronization rule adopted by Einstein in Section 1 of his fundamental paper[1]—a rule which corresponds to the value $\varepsilon = \frac{1}{2}$ in the Reichenbach notation[2]—then the relative motion of Galilean frames issues in their choosing as metrically simultaneous *different pairs of events* from within the class of topologically simultaneous events at P and Q, a result embodied in the familiar Minkowski diagram.

In discussing the definition of simultaneity,[3] Einstein italicized

---

[1] *Ibid.*

[2] H. Reichenbach: *The Philosophy of Space and Time* (New York: Dover Publications, Inc.; 1957), p. 127.

[3] A. Einstein: "On the Electrodynamics of Moving Bodies," *The Principle of Relativity, A Collection of Original Memoirs, op. cit.*, §1.

the words "by definition" in saying that the *equality* of the to and fro velocities of light between two points *A* and *B* is a matter of definition. Thus, he is asserting that metrical simultaneity is a matter of definition or convention. Do the detractors really expect anyone to believe that Einstein put these words in italics to convey to the public that the *noise* "simultaneous" can be used as we please? Presumably they would recoil from this conclusion. But how else could they solve the problem of making Einstein's avowedly conventionalist conception of metrical simultaneity compatible with their semantical trivialization of GC? H. Putnam, one of the advocates of the view that the conventionality of congruence is a subthesis of TSC, has sought to meet this difficulty along the following lines: in the case of the congruence of intervals, one would never run into trouble upon using the *customary* definition;[4] but in the case of simultaneity, actual contradictions would be encountered upon using the customary classical definition of metrical simultaneity, which is based on the transport of clocks and is vitiated by the dependence of the clock rates (readings) on the transport velocity.

But Putnam's retort will not do. For the appeal to Einstein's recognition of the inconsistency of the classical definition of metrical simultaneity accounts only for his abandonment of the latter but does not illuminate—as does the thesis of the conventionality of simultaneity—the logical status of the *particular set* of definitions which Einstein put in its place. Thus, the Putnamian retort does not recognize that the logical status of Einstein's synchronization rules is not at all adequately rendered by saying that whereas the classical definition of metrical simultaneity was inconsistent, Einstein's rules have the virtue of consistency. For what needs to be elucidated is *the nature of the logical step* leading to Einstein's particular synchronization scheme within the wider framework of the *alternative consistent sets of rules* for metrical simultaneity any one of which is allowed by the non-uniqueness of topological simultaneity. Precisely this elucidation is given, as we have seen, by the thesis of the conventionality of metrical simultaneity.

We see therefore that the philosophically illuminating con-

[4] We shall see in our discussion of time measurement on the rotating disk of the general theory of relativity in Chapter Two that there is one sense of "trouble" for which Putnam's statement would not hold.

ventionality of an affirmation of the congruence of two intervals or of the metrical simultaneity of two physical events does not inhere in the arbitrariness of what linguistic sentence is used to express the proposition that a relation of equality obtains among intervals, or that a relation of metrical simultaneity obtains between two physical events. Instead, the important conventionality lies in the fact that even *after* we have specified what respective linguistic sentences will express these propositions, a convention is ingredient in each of the propositions expressed, i.e., in the very obtaining of a congruence relation among intervals or of a metrical simultaneity relation among events.

These considerations enable us to articulate the misunderstanding of the conventionality of congruence to which Eddington and contemporary proponents of its *model-theoretic* trivialization fell prey. It will be recalled that the latter critics argued somewhat as follows: "The theory of models of uninterpreted formal calculi shows that there can be no basis at all for an epistemological distinction *within* the system of physical geometry (or chronometry) between factual statements, on the one hand, and supposedly conventional statements of rigidity (or isochronism), on the other. For we are just as free to give a *non*-customary spatial interpretation of, say, the abstract sign 'point' in the formal geometrical calculus as of the sign 'congruent,' and hence the physical interpretation of the relation term 'congruent' (for line segments) cannot occupy an epistemologically distinguished position among the semantical rules effecting the interpretation of the formal system, all of which are on a par in regard to conventionality."

But this objection overlooks that (a) the obtaining of the spatial congruence relation provides scope for the role of convention because, independently of the particular formal geometrical calculus which is being interpreted, the term "congruent" functions as a spatial *equality predicate* in its *non*-customary spatial interpretations no less than in its customary ones; (b) consequently, suitable alternative spatial interpretations of the term "congruent" and correlatively of "straight line" ("geodesic") show that it is always a live option (subject to the restrictions imposed by the existing topology) to give either a Euclidean or a *non*-Euclidean description of the same body of physico-geometrical facts; and (c) by contrast, the possibility of alternative spatial interpretations of such other primitives of rival geometrical calculi as "point" does not generally issue in this option. Our concern is to note that, even disregarding inductive imprecision, the empirical facts themselves do not uniquely dictate the truth of either Euclidean geometry or of one of its *non*-Euclidean rivals in virtue of the lack of an intrinsic metric. Hence in this context the different spatial interpretations of the term "congruent" (and hence of "straight line") in the respective geometrical calculi play a philosophically different role than do the interpretations of such other primitives of these calculi as "point," since the latter generally have the same spatial meaning in both the Euclidean and non-Euclidean descriptions. The pre-eminent status occupied by the interpretation of "congruent" in this context becomes apparent once we cease to look at physical geometry as a spatially interpreted system of abstract *synthetic* geometry and regard it instead as an interpreted system of abstract *differential* geometry of the Gauss-Riemann type: by choosing a particular distance function $ds = \sqrt{g_{ik}dx^i dx^k}$ for the line element, we specify not only what segments are congruent and what lines are straights (geodesics) but the entire geometry, since the metric tensor g_{ik} fully determines the Gaussian curvature K. To be sure, if one were discussing not the alternative between a Euclidean and a non-Euclidean description of the same spatial facts but rather the set of all models (including non-spatial ones) of a given calculus, say the Euclidean one, then indeed the physical interpretation of "congruent" and of "straight line" would not merit any more attention than that of other primitives like "point."[5]

H. Putnam and P. K. Feyerabend have elaborated the following corollary of Eddington's charge of triviality: GC must be a

[5] We have been speaking of certain *uninterpreted* formal calculi as systems of synthetic or differential *geometry*. It must be understood, however, that *prior* to being given a *spatial* interpretation, these abstract deductive systems no more qualify as *geometries*, strictly speaking, than as systems of genetics or of anything else; they are called "geometries," it would seem, only "because the name seems good, on emotional and traditional grounds, to a sufficient number of competent people." (O. Veblen and J. H. C. Whitehead: *The Foundations of Differential Geometry*, No. 29 of "Cambridge Tracts in Mathematics and Mathematical Physics" [Cambridge: Cambridge University Press; 1932], p. 17.)

subthesis of TSC because GC has bona fide analogues in every branch of human inquiry, such that GC cannot be construed as an insight into the structure of space or time. As Eddington puts it:

> The law of Boyle states that the pressure of a gas is proportional to its density. It is found by experiment that this law is only approximately true. A certain mathematical simplicity would be gained by conventionally redefining *pressure* in such a way that Boyle's law would be rigorously obeyed. But it would be high-handed to appropriate the word pressure in this way, unless it had been ascertained that the physicist had no further use for it in its original meaning.[6]

P. K. Feyerabend has noted that what Eddington seems to have in mind here is the following: instead of revising Boyle's law

$$pv = RT$$

in favor of van der Waals's law

$$(p + \frac{a}{v^2})(v - b) = RT,$$

we could preserve the statement of Boyle's law by merely redefining "pressure"—now to be symbolized by "P" in its new usage —putting

$$P =_{\text{Def}} (p + \frac{a}{v^2})\ (1 - \frac{b}{v}).$$

In the same vein, H. Putnam maintains that instead of using phenomenalist (naive realist) color words as we do customarily in English, we could adopt a new usage for such words—to be called the "Spenglish" usage—as follows: we take a white piece of chalk, for example, which is moved about in a room, and we lay down the rule that depending (in some specified way) upon the part of the visual field which its appearance occupies, its color will be called "green," "blue," "yellow," etc. rather than "white" under constant conditions of illumination.

It is a fact, of course, that whereas actual scientific practice in the general theory of relativity, for example, countenances and uses remetrizational procedures based on non-customary

[6] A. S. Eddington, *Space, Time, and Gravitation, op. cit.,* p. 10.

definitions of temporal congruence,[7] scientific practice does not contain any examples of Putnam's "Spenglish" *space-dependent* (or *time-dependent*) use of phenomenalist (naive realist) *color predicates* to denote the color of a given object in various places (or at various times) under like conditions of illumination. According to Eddington and Putnam, the existence of non-customary usages of "congruent" in the face of there being no such usages of color predicates is no more than a fact about the linguistic behavior of the members of our linguistic community. We saw that the use of linguistic alternatives in the specifically geo-chronometric contexts reflects fundamental *structural properties* of the facts to which these alternative descriptions pertain. And we must now demonstrate that Eddington's and Putnam's alleged analogues of GC are *pseudo*-analogues in the sense of failing to show that every empirical domain possesses analogues to GC. I should emphasize, however, that my rejection of the aforementioned purported analogues is not intended to deny the existence of one or another genuine analogue but to deny only that GC may be deemed to be trivial on the strength of such relatively few bona fide analogues as may obtain.

The essential point in assessing the cogency of the purported analogues is the following: do the domains from which they are drawn (e.g., phenomenalist or naive realist color properties, pressure phenomena, etc.) exhibit *structural* counterparts to (a) those factual properties of the world postulated by relativity which entail the non-uniqueness of topological simultaneity, and (b) the postulated topological properties of physical space and time which make for the non-uniqueness of the respective spatial and temporal interpretations allowed by the congruence axioms? Or are the examples cited by Eddington and Putnam analogues of the conventionality of metrical simultaneity or of congruence

[7] The proponents of ordinary language usage in science, to whom the "ordinary man" seems to be the measure of all things, may wish to rule out *non*-customary congruence definitions as linguistically illegitimate. But they would do well to remember that it is no more incumbent upon the scientist (or philosopher of science) to use the customary scientific definition of congruence in *every* geo-chronometric description than it is obligatory for, say, the student of mechanics to be bound by the familiar common sense meaning of "work," which contradicts the mechanical meaning as given by the space integral of the force.

only in the impoverished, trivial sense that they feature linguistically alternative equivalent descriptions while lacking the following decisive property of the geo-chronometric cases: the alternative metrizations are the linguistic renditions or reverberations, as it were, of the *structural properties* assuring the aforementioned two kinds of non-uniqueness enunciated by GC? If the examples given are analogues only in the superficial, impoverished sense—as indeed I shall show them to be—then what have Eddington and Putnam accomplished by their examples? In that case they have merely provided unnecessary illustrations of the correctness of TSC without proving their examples to be on a par with the geo-chronometric ones. In short, their examples will then have served in no way to make good their claim that GC is a subthesis of TSC.

We shall find presently that their examples fail because (a) the domains to which they pertain do not exhibit structural counterparts to those features of the world which make the definitions of topological simultaneity and the axioms for spatial or temporal congruence *non-unique*, and (b) Putnam's example in "Spenglish" is indeed an illustration only of the trivial conventionality of all language: no structural property of the domain of phenomenal color (e.g., in the appearances of chalk) is rendered by the feasibility of the Spenglish description.

To state my objections to the Eddington-Putnam thesis, I call attention to the following two sentences:

(A) Person *X* does not have a gall bladder.

(B) The platinum-iridium bar in the custody of the Bureau of Weights and Measures in Paris (Sèvres) is one meter long everywhere rather than some other number of meters (after allowance for "differential forces").

I maintain that there is a *fundamental difference* between the senses in which each of these statements can possibly be held to be conventional, and I shall refer to these respective senses as "*A*-conventional" and "*B*-conventional": in the case of statement (A), what is conventional is only the use of the given sentence to render the proposition of *X*'s not having a gall bladder, not the factual proposition expressed by the sentence. This *A*-conventionality is of the trivial weak kind affirmed by TSC. On the other hand, (B) is conventional not merely in the trivial sense that the English sentence used could have been replaced by one in French or some other language but in the much stronger and deeper sense that it is not a factual proposition that the Paris bar has everywhere a length unity in the meter scale even after we have specified what sentence or string of noises will express this proposition. In brief, in (A), semantic conventions are *used,* whereas, in (B), a semantic convention is *mentioned.* Now I claim that the alleged analogues of Eddington and Putnam illustrate conventionality only in the sense of *A*-conventionality and therefore cannot score against my contention that geo-chronometric conventionality is *non*-trivial by having the character of *B*-conventionality.

Specifically, I assert that statements about phenomenalist colors are empirical statements pure and simple in the sense of being only *A*-conventional and not *B*-conventional, while an important class of statements of geo-chronometry possess a different, deeper conventionality of their own by being *B*-conventional. What is it that is conventional in the case of the color of a given piece of chalk, which appears white in various parts of the visual field? I answer: only our customary decision to use the same word to refer to the various qualitatively same white chalk appearances in different parts of the visual field. But it is not conventional whether the various chalk appearances do have the same phenomenal color property (to within the precision allowed by vagueness) and thus are "color-congruent" to one another or not! Only the color-words are conventional, not the obtaining of specified color-properties and of color-congruence. And the obtaining of color-congruence is non-conventional quite independently of whether the various occurrences of a particular shade of color are denoted by the same color-word or not. In other words, there is no convention in whether two objects or two appearances of the same object under like optical conditions have the same phenomenal color property of whiteness (apart from vagueness) but only in whether the noise "white" is applied to both of these objects or appearances, to one of them and not to the other (as in Putnam's chalk example) or to neither. And the alternative color descriptions *do not render any structural facts of the color domain* and are therefore purely trivial.

Though failing in this decisive way, Putnam's chalk color case is falsely given the *semblance* of being a bona fide analogue to the spatial congruence case by the device of laying down a rule which makes the use of color names *space-dependent:* the rule is that different noises (color names) will be used to refer to the same de facto color property occurring in different portions of visual space. But this stratagem cannot overcome the fact that while the assertion of the possibility of assigning a space-dependent length to a transported rod reflects linguistically the objective non-existence of an intrinsic metric, the space-dependent use of color names does not reflect a corresponding property of the domain of phenomenal colors in visual space. In short, the phenomenalist color of an appearance is an intrinsic, objective property of that appearance, and phenomenal color-congruence is an objective relation (to within the precision allowed by vagueness). But the length of a body and the congruence of non-coinciding intervals are not similarly non-conventional. And we shall see in our critique of Whitehead that this conclusion cannot be invalidated by the fact that two non-coinciding intervals can *look* spatially-congruent no less than two color patches can appear color-congruent.

Next consider Eddington's example of the preservation of the language of Boyle's law to render the *new facts* affirmed by van der Waals's law by the device of giving a new meaning to the word "pressure" as explained earlier. The customary concept of pressure has geo-chronometric ingredients (force, area), and any alterations made in the geo-chronometric congruence definitions will, of course, issue in changes as to what pressures will be held to be equal. But the conventionality of the geo-chronometric ingredients is not of course at issue, and we ask: of what *structural feature* of the domain of pressure phenomena does the possibility of Eddington's above linguistic transcription render testimony? The answer clearly is: *of none.* Unlike GC, the thesis of the "conventionality of pressure," if put forward on the basis of Eddington's example, concerns only *A*-conventionality and is thus merely a special case of TSC. We observe incidentally, that two pressures which are equal on the customary definition will also be equal (congruent) on the suggested redefinition of that term: apart from the distinctly geo-chronometric ingredients not here at issue, the domain of pressure phenomena does not present us with any structural property as the counterpart of the lack of an intrinsic metric of space which would be reflected by the alternative definitions of "pressure."

The absurdity of likening the conventionality of spatial or temporal congruence to the conventionality of the choice between the two above meanings of "pressure" or between English and Spenglish color discourse becomes patent upon considering the expression given to the conventionality of congruence by the Klein-Lie group-theoretical treatment of congruences and metric geometries. For their investigations likewise serve to show, as we shall now indicate, how far removed from being a semantically uncommitted noise the term "congruent" is while still failing to single out a unique congruence class of intervals, and how badly amiss it is for Eddington and Putnam to maintain that this non-uniqueness is merely a special example of the semantical non-uniqueness of all uncommitted noises.

Felix Klein's Erlangen Program (1872) of treating geometries from the point of view of groups of spatial transformations was rooted in the following two observations: first, the properties in virtue of which spatial congruence has the logical status of an equality relation depend upon the fact that displacements are given by a group of transformations, and second, the congruence of two figures consists in their being intertransformable into one another by means of a certain transformation of points. Continuing Klein's reasoning, Sophus Lie then showed that, in the context of this group-theoretical characterization of metric geometry, the conventionality of congruence issues in the following results: first, the set of all the continuous groups in space having the property of displacements in a bounded region fall into three types which respectively characterize the geometries of Euclid, Lobachevski-Bolyai, and Riemann,[8] and second, for each of these metrical geometries, there is not one but an infinitude of different congruence classes,[9] a result which we shall demonstrate in Chapter Three without group-theoretical devices. On the Eddington-Putnam thesis, Lie's profound and justly celebrated results

[8] R. Bonola: *Non-Euclidean Geometry* (New York: Dover Publications, Inc.; 1955), p. 153.

[9] Cf. A. N. Whitehead: *The Principle of Relativity, op. cit.*, Chapter iii, p. 49.

no less than the relativity of simultaneity and the conventionality of temporal congruence must be consigned absurdly to the limbo of trivial semantical conventionality along with Spenglish color discourse!

As previously noted, these objections against the Eddington-Putnam claim that GC has bona fide analogues in every empirical domain are not intended to deny the existence of one or another genuine analogue but to deny only that GC may be deemed to be trivial on the strength of such relatively few bona fide analogues as may obtain. Putnam has given one example which does seem to qualify as a bona fide analogue. This example differs from his color case in that not merely the name given to a property but the sameness of the property named is dependent on spatial position as follows: when two bodies are at essentially the same place, their sameness with respect to a certain property is a matter of fact, but when they are (sufficiently) apart spatially, no objective relation of sameness or difference with respect to the given property obtains between them. And, in the latter case, therefore, it becomes a matter of convention whether sameness or difference is ascribed to them in this respect.

Specifically, suppose that we do not aim at a definition of mass adequate to classical mechanics and thus ignore Mach's definition of mass.[1] Then we can consider Putnam's hypothetical definition of "mass-equality," according to which two bodies balancing one another on a suitable scale at what is essentially the same place in space have equal masses. Whereas on the Machian definition mass equality obtains between two bodies *as a matter of fact* independently of the extent of their spatial separation, on Putnam's definition such separation leaves the relation of mass equality at a distance *indeterminate*. Hence, on Putnam's definition, it would be a matter of convention whether (a) we would say that two masses which balance at a given place remain equal to one another in respect to mass after being spatially separated, or (b) we would make the masses of two bodies space-dependent such that two masses that balance at one place would have different masses when separated, as specified by a certain function of the coordinates. The conventionality arising

[1] For a concise account of that definition, see L. Page: *Introduction to Theoretical Physics, op. cit.*, pp. 56–58.

in Putnam's mass example is not a consequence of GC but is logically independent of it. For it is not spatial congruence of non-coinciding intervals but spatial position that is the source of conventionality here.

The bona fide character of Putnam's mass analogue cannot, however, invalidate our earlier conclusion that we must attach a very different significance to alternative metric geometries or chronometries as equivalent descriptions of the same facts than to alternate types of visual color discourse as equivalent descriptions of the same phenomenal data. By the same token, we must attach much greater significance to being able to render factually different geo-chronometric states of affairs by the same geometry or chronometry, coupled with appropriately different congruence definitions, than to formulating both Boyle's law and van der Waals's law, which differ in factual content, by the same law-statement coupled with appropriately different semantical rules. In short, there is an important respect in which physical geo-chronometry is less empirical than all or almost all of the non-geo-chronometric portions (ingredients) of other sciences.

(E) BRIDGMAN

We have grounded the conventionality of spatial and temporal congruence on the continuity of the manifolds of space and time. And, in thus arguing that "true," absolute, or intrinsic rigidity and isochronism are non-existing properties in these respective continua, we did not adduce any homocentric-operationist criterion of factual meaning. For we did not say that the actual and potential failure of human testing operations to disclose "true rigidity" constitutes either its non-existence or its meaninglessness. It will be well, therefore, to compare our Riemannian espousal of the conventionality of rigidity and isochronism with the reasoning of those who arrive at this conception of congruence by arguing from non-testability or from an operationist view of scientific concepts. Thus W. H. Clifford writes: "we have defined length or distance by means of a measure which can be carried about *without changing its length.* But how then is this property of the measure to be tested? . . . Is it possible . . . that lengths do really change by mere moving about, without our knowing it? Whoever likes to meditate seriously upon this question will find that it is wholly devoid of meaning."[2]

We saw that within our Riemannian framework of ideas, length is *relational* rather than absolute in a twofold sense: First, length obviously depends numerically on the units used and is thus arbitrary to within a constant factor, and second, in virtue of the lack of an intrinsic metric, sameness or change of the length possessed by a body in different places and at different times consist in the sameness or change respectively of the ratio (relation) of that body to the conventional standard of congruence. Whether or not this ratio changes is quite independent of any human discovery of it: the number of times which a given body *B* will contain a certain unit rod is a property of *B* that is not first conferred on *B* by human operations. As Reichenbach has noted: "*The objective character of the physical statement* [concerning the geometry of physical space] *is thus shifted to a statement about relations . . . it is a statement about a relation between the universe and rigid rods.*"[3] And thus the relational character of length derives, in the first instance, not from how we human beings measure lengths but from the failure of the continuum of physical space to possess an intrinsic metric, a failure obtaining quite independently of our measuring activities. In fact, it is this relational character of length which prescribes and regulates the kinds of human operations appropriate to its discovery. Since, to begin with, there exists no property of true rigidity to be disclosed by any human test, no test could possibly reveal its presence. Accordingly, the unascertainability of true rigidity by us humans is a consequence of its non-existence in physical space and evidence for that non-existence but not constitutive of it.

On the basis of this non-homocentric relational conception of length, the utter vacuousness of the following assertion is evident at once: overnight *everything* has *expanded* (i.e., increased its length) but such that all length *ratios* remained unaltered. That

---

[2] W. K. Clifford: *The Common Sense of the Exact Sciences* (New York: Dover Publications, Inc.; 1955), pp. 49–50.

[3] H. Reichenbach: *The Philosophy of Space and Time, op. cit.*, p. 37; italics are in the original.

such an alleged "expansion" will elude any and all human tests is then obviously explained by its not having obtained: the increase in the ratios between all bodies and the congruence standard which would have constituted the expansion avowedly did not materialize. And in the absence of an intrinsic metric, it is sheer nonsense to say that overnight the Paris meter ceased to be truly one meter and is now only apparently so, since it has actually expanded.

We see that the relational theory of length and hence the particular assertion of the vacuousness of a universal nocturnal expansion do not depend on a grounding of the meaning of the metrical concepts of length and duration on human testability or manipulations of rods and clocks in the manner of Bridgman's homocentric operationism. Moreover, there is a further sense in which the Riemannian recognition of the need for a specification of the congruence criterion does not entail an operationist conception of congruence and length. As we noted preliminarily at the beginning of this chapter and will see in detail in Chapter Four, the definition of "congruence" on the basis of the coincidence behavior common to all kinds of transported solid rods provides a rule of correspondence (coordinative definition) *through the mediation of hypotheses and laws* that are *collateral* to the abstract geometry receiving a physical interpretation. For the physical laws used to compute the corrections for thermal and other substance-specific deformations of solid rods made of different kinds of materials enter integrally into the definition of "congruent." Thus, in the case of "length" no less than in most other cases, operational definitions (in any distinctive sense of the term "operational") are a quite idealized and limiting species of correspondence rule even though the definition of "length" is often adduced as the prototype of all "operational" definitions in Bridgman's sense.

Further illustrations of this fact are given by Reichenbach, who cites the definitions of the unit of length on the basis of the wave length of cadmium light and also in terms of a certain fraction of the circumference of the earth and writes: "Which distance serves as a unit for actual measurements can ultimately be given only by reference to some actual distance. . . . We say with regard to the measuring rod . . . that only 'ultimately' the reference is to be conceived in this form, because we know that by means of the interposition of conceptual relations the reference may be rather remote."[4] An even stronger repudiation of the operationist account of the definition of "length" because of its failure to allow for the role of auxiliary theory is presented by K. R. Popper, who says: "As to the doctrine of operationalism —which demands that scientific terms, such as length . . . should be defined in terms of the appropriate experimental procedure— it can be shown quite easily that all so-called operational definitions will be circular . . . the circularity of the operational definition of length . . . may be seen from the following facts: (a) the 'operational' definition of *length* involves *temperature* corrections, and (b) the (usual) operational definition of *temperature* involves measurements of length."[5]

(F) RUSSELL

During the years 1897–1900, B. Russell and H. Poincaré had a controversy which was initiated by Poincaré's review[6] of Russell's *Foundations of Geometry* of 1897, and pursued in Russell's reply[7] and Poincaré's rejoinder.[8] Russell criticized Poincaré's conventionalist conception of congruence and invoked the existence of an intrinsic metric as follows:

> It seems to be believed that since measurement [i.e., comparison by means of the congruence standard] is necessary to *discover* equality or inequality, these cannot exist without measurement. Now the proper conclusion is exactly the opposite. Whatever one can discover by means of an operation must exist independently of that operation: America existed before Chris-

[4] *Ibid.*, p. 128.

[5] K. R. Popper: *The Logic of Scientific Discovery* (London: Hutchinson and Co., Ltd.; 1959), pp. 440 and 440n. In Chapter Four, we shall see how the circularity besetting the operationist conception of length is handled within our framework when making allowance for thermal and other deformations in the statement of the definition of congruence.

[6] H. Poincaré: "Des Fondements de la Géométrie, à propos d'un Livre de M. Russell," *Revue de Métaphysique et de Morale,* Vol. VII (1899), pp. 251–79.

[7] B. Russell: "Sur les Axiomes de la Géométrie," *Revue de Métaphysique et de Morale,* Vol. VII (1899), pp. 684–707.

[8] H. Poincaré: "Sur les Principes de la Géométrie, Réponse à M. Russell," *Revue de Métaphysique et de Morale,* Vol. VIII (1900), pp. 73–86.

> topher Columbus, and two quantities of the same kind must *be* equal *or* unequal before being measured. Any method of measurement [i.e., any congruence definition] is good or bad according as it yields a result which is true or false. Mr. Poincaré, on the other hand, holds that measurement creates equality and inequality. It follows [then] . . . that there is nothing left to measure and that equality and inequality are terms devoid of meaning.[9]

The congruence issue between Russell and Poincaré here is, of course, distinct from the controversy between the neo-Kantian and the empiricist conceptions of geometry, which plays a major role in Russell's *Foundations of Geometry* of 1897.

As we recall from our citation of Riemann's "Inaugural Lecture" in Section B, if space were discrete in some specified sense, then the "distance" between two elements could be defined intrinsically in a rather natural way by the cardinality of the least number of intervening elements. In that eventuality, both the "length" of any given interval and the obtaining of spatial congruence among separated intervals would be wholly independent of the behavior of any transported standard. And if space is thus granular, the logic of the discovery of length would be analogous to that of Columbus's discovery of America in Russell's example, the role of the measuring rod then being no more than a purely epistemic one. Moreover, in the case of discreteness, the measuring rod is even dispensable for epistemic purposes since a separate determination of the number of chunks or space-atoms contained in each of two bodies would yield a verdict on their spatial congruence before any comparison of them would need to be effected via a transported congruence standard. Russell overlooked that once we assume the continuity of our physical space, the congruence of two line segments cannot derive from their respective possession of an intrinsic metric attribute and that their congruence depends for its very obtaining and not merely for its human ascertainment on a relation to an *extrinsic* standard whose "rigidity" under transport is decreed conventionally. It is the bodies or segments themselves, but not their relations of spatial equality or inequality which exist independently of the coincidence behavior of a

[9] B. Russell: "Sur les Axiomes de la Géométrie," *op. cit.*, pp. 687–88.

transported congruence standard. And *to recognize the dependence of the very obtaining of spatial congruence among separated intervals on the mediation of the transported congruence standard is not to confuse measurement in its epistemic function of discovery with the facts ascertained by it.* Hence we see that Poincaré was not guilty of the following error: what makes the property of length in our supposedly continuous physical space different from those discovered by Columbus in Russell's example is first generated by the difference in the respective operational procedures used by us humans in their discovery. Instead, Poincaré rested his case on the pre-existing difference in the properties to be discovered, a difference that determines and lends significance to the operational procedures appropriate to their discovery.

Although we therefore reject Russell's argument against Poincaré, our critique of the model-theoretic trivialization of the conventionality of congruence shows that we must likewise reject as inadequate the following kind of criticism of Russell's position, which he would have rightly regarded as a *petitio principii:* "Russell's claim is an absurdity, because it is the *denial* of the truism that we are at liberty to give whatever physical interpretations we like to such abstract signs as 'congruent line segments' and 'straight line' and then to inquire whether the system of objects and relations thus arbitrarily named is a model of one or another abstract geometric axiom system. Hence, purely linguistic considerations suffice to show that there can be no question, as Russell would have it, whether two non-coinciding segments are truly equal or not and whether measurement is being carried out with a standard yielding results that are true in that sense. Accordingly, awareness of the model-theoretic conception of geometry would have shown Russell that (1) alternative metrizability of spatial and temporal continua should never have been either startling or a matter for dispute, (2) the stake in his controversy with Poincaré was no more than the pathetic one that Russell was advocating the customary linguistic usage of the term "congruent" (for line segments) while Poincaré was maintaining that we need not be bound by the customary usage but are at liberty to introduce bizarre ones as well."

Since this model-theoretic argument altogether fails to come to grips with Russell's root-assumption of an intrinsic metric, he would have been entitled to dismiss it as a shallow *petitio* by raising exactly the same objections against the alternative metrizability of space and time, which we attributed to the Newtonian at the end of Section C above. And Russell might have gone on to point out that the model-theoretician cannot evade the spatial equality issue by (i) noting that there are axiomatizations of each of the various geometries dispensing with the abstract relation term "congruent" (for line segments), and (ii) claiming that there can then be no problem as to what physical interpretations of that relation term are permissible. For a metric geometry makes metrical comparisons of equality and inequality, however covertly or circuitously these may be rendered by its language. It is quite immaterial, therefore, whether the relation of spatial equality between line segments is designated by the term "congruent" or by some other term or terms. Thus, for example, Tarski's axioms for elementary Euclidean geometry[1] do not employ the term "congruent" for this purpose, using instead a quaternary predicate denoting the equidistance relation between four points. Also, in Sophus Lie's group-theoretical treatment of metric geometries, the congruences are specified by groups of point transformations.[2] But just as Russell invoked his conception of an intrinsic metric to restrict the permissible spatial interpretations of "congruent line segments," so also he would have maintained that it is never arbitrary what quartets of physical points may be regarded as the denotata of Tarski's quaternary equidistance predicate. And he would have imposed corresponding restrictions on Lie's transformations, since the displacements defined by these groups of transformations have the logical character of spatial congruences. These considerations show that it will not suffice in this context simply to take the model-theoretic conception of geometry for granted and thereby to dismiss Russell's claim peremptorily in favor of alternative metrizability. Rather what is needed is a *refutation of Russell's root-assumption of an intrinsic metric:* to exhibit the untenability

---

[1] A. Tarski: "What is Elementary Geometry?" *The Axiomatic Method,* ed. by L. Henkin, P. Suppes, and A. Tarski (Amsterdam: North Holland Publishing Company; 1959), pp. 16–29.

[2] R. Bonola: *Non-Euclidean Geometry, op. cit.,* pp. 153–54.

of that assumption as we have endeavored to do earlier is to provide the physical justification of the model-theoretic affirmation that a given set of physico-spatial facts of coincidence of a transported rod may be held to be as much a realization of a Euclidean calculus as of a non-Euclidean one yielding the same topology.

(G) WHITEHEAD

A. N. Whitehead has given a *perceptualistic* version of Russell's argument by attempting to ground an intrinsic metric of physical space and time on the deliverances of sense. We therefore now turn to an examination of Whitehead's conception of congruence.

Commenting on the Russell-Poincaré controversy, Whitehead[3] maintains the following: First, Poincaré's argument on behalf of alternative metrizability is unanswerable only if the philosophy of physical geometry and chronometry is part of an epistemological framework resting on an illegitimate bifurcation of nature; second, consonant with the rejection of bifurcation, we must ground our metric account of the space and time of nature not on the relations between material bodies and events as fundamental entities but on the more ultimate metric deliverances of sense perception; and third, perceptual time and space exhibit an intrinsic metric. Specifically, by countenancing the bifurcation of nature to begin with, Riemann was driven to the conclusion that the very meaning of spatial (or temporal) congruence depends on a standard which cannot make any claim to uniqueness. On his bifurcationist view that the congruence standard must "come from somewhere else," the congruence thus defined can enjoy only a conventional pre-eminence over alternative congruences that might have been selected with equal mathematical justification. Contrariwise, an antibifurcationist theory of nature relies on the immediate deliverances of sense awareness which furnish and justify a unique set of physically relevant congruence relations for both space and time (to within the precision allowed by sensory vagueness). And by thus taking cognizance of the disclosures of sense perception, we can confer

[3] A. N. Whitehead: *The Concept of Nature* (Cambridge: Cambridge University Press; 1926), pp. 121–24. Hereafter this work will be cited as *CN*.

intelligibility on the unique role of the familiar congruence criteria for space and time in the face of the mathematically competing claims of "the indefinite herd" of other, mutually exclusive congruences.[4]

Clarity may be served by first stating in turn the gist of the several arguments given by Whitehead in support of his contentions. After thus obtaining a synoptic view of his polemic, we shall cite and examine each of his arguments in some detail.

According to Whitehead, the thesis of alternative metrizability can be seen to be absurd for two reasons: First, the conventionalist conception defines temporal congruence by the requirement that Newton's laws (as modified by the small corrections for the relativistic motion of the perihelia) are true. But "uniformity in change is directly perceived"[5] and "the measurement of time was known to all civilized nations long before the laws [of Newton] were thought of. It is this time as thus measured that the laws are concerned with. . . . It is for science to give an intellectual account of what is so evident in sense awareness."[6] Second, just as it is an objective datum of experience that two phenomenal color patches have the same color, i.e., are "color-congruent" to within the precision allowed by vagueness, so also we *see* that a given rod has the same length in different positions, thereby showing that spatial congruence or *matching* is no less objective a relation than phenomenal color-"congruence."[7] Hence, "it is a fact of nature that a distance of thirty miles is a long walk for anyone. There is no convention about that."[8] And it is "no slight recommendation"[9] of the antibifurcationist theory of nature that it removes the following difficulty besetting the bifurcationist version of classical nineteenth-century geometry and physics: There is a "breakdown of the uniqueness of congruence for space [because of alternative metrizability] and of its very existence for time," because "time, in itself, according

[4] *Ibid.*, p. 124.

[5] *Ibid.*, p. 137.

[6] *Ibid.*, p. 140.

[7] Cf. A. N. Whitehead: *The Principles of Natural Knowledge* (Cambridge: Cambridge University Press; 1955), p. 56.

[8] Cited from Whitehead in R. M. Palter: *Whitehead's Philosophy of Science* (Chicago: University of Chicago Press; 1960), p. 93.

[9] A. N. Whitehead: *CN*, *op. cit.*, p. 124.

to the classical theory, presents us with no qualifying [i.e., congruence] class at all.[1] Furthermore, contrary to the "modern doctrine that 'congruence' *means* the possibility of coincidence," the correct account of the matter is that "although 'coincidence' is used as a *test* of congruence, it is not the *meaning* of congruence."[2] Also "immediate judgments of congruence are presupposed in measurement, and the process of measurement is merely a procedure to extend the recognition of congruence to cases where these immediate judgments are not available. Thus we cannot define congruence by measurement."[3]

Turning now to the detailed examination of Whitehead's argumentation, we note that he proposes to point out "the factor in nature which issues in the pre-eminence of one [spatial] congruence relation over the indefinite herd of other such relations"[4] as follows:

> The reason for this result is that nature is no longer confined within space at an instant. Space and time are now interconnected; and this peculiar factor of time which is so immediately distinguished among the deliverances of our sense-awareness, relates itself to one particular congruence relation in space.[5] . . . Congruence depends on motion, and thereby is generated the connexion between spatial congruence and temporal congruence.[6]

Whitehead's argument is thus seen to turn on his ability to show that *temporal* congruence cannot be regarded as conventional in physics as understood by Riemann and Poincaré. He believes to have justified this crucial claim by the following reasoning in which he refers to the conventionalist conception as "the prevalent view" and to his opposing thesis as "the new theory":

> The new theory provides a definition of the congruence of periods of time. The prevalent view provides no such definition.

---

[1] A. N. Whitehead: *The Principle of Relativity* (Cambridge: Cambridge University Press; 1922), p. 49. Hereafter this work will be cited as *R*.

[2] A. N. Whitehead: *Process and Reality* (New York: The Macmillan Co.; 1929), p. 501. Hereafter this work will be cited as *PR*.

[3] A. N. Whitehead: *CN*, *op. cit.*, p. 121.

[4] *Ibid.*, p. 124.

[5] *Ibid.*

[6] *Ibid.*, p. 126.

Its position is that if we take such time-measurements so that certain familiar velocities which seem to us to be uniform are uniform, then the laws of motion are true. Now in the first place no change could appear either as uniform or non-uniform without involving a definite determination of the congruence for time-periods. So in appealing to familiar phenomena it allows that there is some factor in nature which we can intellectually construct as a congruence theory. It does not however say anything about it except that the laws of motion are then true. Suppose that with some expositors we cut out the reference to familiar velocities such as the rate of rotation of the earth. We are then driven to admit that there is no meaning in temporal congruence except that certain assumptions make the laws of motion true. Such a statement is historically false. King Alfred the Great was ignorant of the laws of motion, but knew very well what he meant by the measurement of time, and achieved his purpose by means of burning candles. Also no one in past ages justified the use of sand in hour glasses by saying that some centuries later interesting laws of motion would be discovered which would give a meaning to the statement that the sand was emptied from the bulbs in equal times. Uniformity in change is directly perceived, and it follows that mankind perceives in nature factors from which a theory of temporal congruence can be formed. The prevalent theory entirely fails to produce such factors.[7] . . . On the orthodox theory the position of the equations of motion is most ambiguous. The space to which they refer is completely undetermined and so is the measurement of the lapse of time. Science is simply setting out on a fishing expedition to see whether it cannot find some procedure which it can call the measurement of space and some procedure which it can call the measurement of time, and something which it can call a system of forces, and something which it can call masses, so that these formulae may be satisfied. The only reason—on this theory—why anyone should want to satisfy these formulae is a sentimental regard for Galileo, Newton, Euler, and Lagrange. The theory, so far from founding science on a sound observational basis, forces everything to conform to a mere mathematical preference for certain simple formulae.

I do not for a moment believe that this is a true account of the real status of the Laws of Motion. These equations want some slight adjustment for the new formulae of relativity. But

[7] *Ibid.*, p. 137.

with these adjustments, imperceptible in ordinary use, the laws deal with fundamental physical quantities which we know very well and wish to correlate.

The measurement of time was known to all civilised nations long before the laws were thought of. It is this time as thus measured that the laws are concerned with. Also they deal with the space of our daily life. When we approach to an accuracy of measurement beyond that of observation, adjustment is allowable. But within the limits of observation we know what we mean when we speak of measurements of space and measurements of time and uniformity of change. It is for science to give an intellectual account of what is so evident in sense-awareness. It is to me thoroughly incredible that the ultimate fact beyond which there is no deeper explanation is that mankind has really been swayed by an unconscious desire to satisfy the mathematical formulae which we call the Laws of Motion, formulae completely unknown till the seventeenth century of our epoch.[8]

After commenting that purely mathematically, an infinitude of incompatible spatial congruence classes of intervals satisfy the congruence axioms, Whitehead says:

This breakdown of the uniqueness of congruence for space and of its very existence for time is to be contrasted with the fact that mankind does in truth agree on a congruence system for space and a congruence system for time which are founded on the direct evidence of its senses. We ask, why this pathetic trust in the yard measure and the clock? The truth is that we have observed something which the classical theory does not explain.

It is important to understand exactly where the difficulty lies. It is often wrongly conceived as depending on the inexactness of all measurements in regard to very small quantities. According to our methods of observation we may be correct to a hundredth, or a thousandth, or a millionth of an inch. But there is always a margin left over within which we cannot measure. However, this character of inexactness is *not* the difficulty in question.

Let us suppose that our measurements can be ideally exact; it will be still the case that if one man uses one qualifying [i.e., congruence] class γ and the other man uses another qualifying [i.e., congruence] class δ, and if they both admit the

[8] *Ibid.*, pp. 139–40.

standard yard kept in the exchequer chambers to be their unit of measurement, they will disagree as to what other distances [at other] places should be judged to be equal to that standard distance in the exchequer chambers. Nor need their disagreement be of a negligible character.[9] . . .

When we say that two stretches match in respect to length, what do we mean? Furthermore we have got to include time. When two lapses of time match in respect to duration, what do we mean? We have seen that measurement presupposes matching, so it is of no use to hope to explain matching by measurement.[1] . . .

Our physical space therefore must already have a structure and the matching must refer to some qualifying class of qualities inherent in this structure.[2]

. . . there will be a class of qualities γ one and only one of which attaches to any stretch on a straight line or on a point, such that matching in respect to this quality is what we mean by congruence.

The thesis that I have been maintaining is that measurement presupposes a perception of matching in quality. Accordingly in examining the meaning of any particular kind of measurement we have to ask, What is the quality that matches?[3]

. . . a yard measure is merely a device for making evident the spatial congruence of the [extended] events in which it is implicated.[4]

Let us begin by inquiring whether Whitehead's historical observation that the human race possessed a time-metric prior to the enunciation of Newton's laws during the seventeenth century can serve to invalidate Poincaré's contentions[5] that (1) time-congruence in physics is conventional, (2) the definition of temporal congruence used in refined physical theory is given by Newton's laws, and (3) we have no direct intuition of the temporal congruence of non-adjacent time-intervals, the belief in the existence of such an intuition resting on an illusion.

---

[9] A. N. Whitehead: *R, op. cit.*, pp. 49–50.
[1] *Ibid.*, pp. 50–51.
[2] *Ibid.*, p. 51.
[3] *Ibid.*, p. 57.
[4] *Ibid.*, p. 58.
[5] H. Poincaré: "La Mesure du Temps," *Revue de Métaphysique et de Morale*, Vol. VI (1898), pp. 1–13.

To see how unavailing Whitehead's historical argument is, consider first a hypothetical spatial analogue of his reasoning. We shall see in Chapter Three, Section B that although the demand that Newton's laws be true does uniquely define temporal congruence in the one-dimensional time-continuum, the following *non*-analogous spatial result obtains: the requirement of the Euclideanism of a tabletop does *not* similarly yield a *unique* definition of spatial congruence for that two-dimensional space, there being an infinitude of incompatible congruences issuing in a Euclidean geometry on the tabletop. For the sake of constructing a *hypothetical spatial analogue* to Whitehead's historical argument, however, let us assume that, contrary to fact, it were the case that the requirement of the Euclideanism of the tabletop did uniquely determine the customary definition of rigidity. And now suppose that a philosopher were to say that the latter definition of spatial congruence, like all others, is conventional. What then would be the force of the following kind of putative Whiteheadian assertion: "Well before Hilbert rigorized Euclidean geometry and even much before Euclid less perfectly codified the geometrical relations between the bodies in our environment, men used not only their own limbs but also diverse kinds of solid bodies to certify spatial equality"? Ignoring now refinements required to allow for substance-specific distortions, it is clear that, under the assumed hypothetical conditions, we would be confronted with logically-independent definitions of spatial equality issuing in the same congruence-class of intervals.[6] The hypothetical concordance of these definitions would indeed be an impressive empirical fact, but, were it to obtain, it could not possibly refute the claim that the one congruence defined alike by all of them is conventional.

Precisely analogous considerations serve to invalidate Whitehead's historical argument regarding time-congruence. Let us grant his implicit assumption of agreement, after allowance for substance-specific idiosyncrasies, between the time congruences defined by various kinds of "clock" devices, so that we can discount hypotheses like that of Milne, which assert the incom-

[6] In Chapter Four, we shall see in what sense the criterion of rigidity based on the solid body can be regarded as logically-independent of Euclidean geometry when cognizance is taken of substance-specific distortions.

patibility of the congruences defined by "atomic" and "astronomical" clocks. What are some of the clocks adduced by Whitehead as furnishing the "true" time-metric independently of Newton's laws? A candle always made of the same material, of the same size, and having a wick of the same material and size burns very nearly the same number of inches each hour. Hence as early as during the reign of King Alfred (872–900 A.D.), burning candles were used as rough timekeepers by placing notches or marks at such a distance apart that a certain number of spaces would burn each hour.[7] Ignoring the relatively small variations of the rate of flow of water with the height of the water column in a vessel, the water clock or clepsydra served the ancient Chinese, Byzantines, Greeks, and Romans,[8] as did the sand-clock, keeping very roughly the same time as burning candles. Again, an essentially frictionless pendulum oscillating with constant amplitude at a point of given latitude on the earth defines the same time-metric as do "natural clocks," i.e., quasi-closed periodic systems.[9] And, ignoring various refinements, similarly for the rotation of the earth, the oscillations of crystals, the successive round-trips of light over a fixed distance in an inertial system, and the time based on the natural periods of vibrating atoms or "atomic clocks."[1]

Thus, unless a hypothesis like that of Milne is right, we find a striking concordance between the time-congruence defined by Newton's amended laws and the temporal equality furnished by several kinds of definitions logically independent of that Newtonian one. This presumed agreement obtains as a matter of presumed *empirical fact* for which the general theory of relativity (hereafter called GTR) has sought to provide an explana-

---

[7] W. I. Milham: *Time and Timekeepers* (New York: The Macmillan Co.; 1929), pp. 53–54. A more recent edition appeared in 1941.

[8] D. J. Price: "The Prehistory of the Clock," *Discovery,* Vol. XVII (1956), pp. 153–57.

[9] C. Brouwer: "The Accurate Measurement of Time," *Physics Today,* Vol. IV (1951), pp. 7–15.

[1] F. A. B. Ward: *Time Measurement* (4th edition, London: Royal Stationery Office; 1958), Part I, Historical Review; P. Hood: *How Time is Measured* (London: Oxford University Press; 1955); J. J. Baruch: "Horological Accuracy: Its Limits and Implications," *American Scientist,* Vol. XLVI (1958), pp. 188A–196A, and H. Lyons: "Atomic Clocks," *Scientific American,* Vol. CXCVI (1957), pp. 71–82.

tion through its conception of the metrical field, just as it has endeavored to account for the corresponding concordance in the coincidence behavior of various kinds of solid rods.[2] No one, of course, would wish to deny that of all the definitions of temporal congruence which yield the same time-metric as the (amended) Newtonian laws, some were used by man well before these laws could be invoked to provide such a definition. Moreover, it can readily be agreed that it was only because it was possible to measure time in one or another of these pre-Newtonian ways that the discovery and statement of Newton's laws became possible. But what is the bearing of these genetic considerations and of the (presumed) fact that the *same* congruence class of time intervals is furnished alike by each of the aforementioned logically independent definitions on the issue before us? It seems quite clear that they cannot serve as grounds for impugning the thesis that the equality obtaining among the time-intervals belonging to this *one* congruence-class is conventional in the Riemann-Poincaré sense which we have articulated: this particular equality is no less conventional in virtue of being defined by a plethora of physical processes in addition to Newton's laws than if it were defined merely by one of these processes alone or by Newton's laws alone.

Can this conclusion be invalidated by adducing such agreement as does obtain under appropriate conditions between the metric of psychological time and the physical time-congruence under discussion? We shall now see that the answer is decidedly in the negative.

Prior attention to the source of such concordance as does exist between the psychological and physical time-metrics will serve our endeavor to determine whether the metric deliverances of psychological time furnish any support for Whitehead's espousal of an intrinsic metric of physical time.[3]

---

[2] A. d'Abro: *The Evolution of Scientific Thought from Newton to Einstein* (New York: Dover Publications, Inc.; 1950), pp. 78–79.

[3] It will be noted that Whitehead does not rest his claim of the intrinsicality of the temporal metric on his thesis of the *atomicity of becoming*. We therefore need not deal here with the following of his contentions: First, becoming or the transiency of "now" is a feature of the time of physics, and second, there is no continuity of becoming but only becoming of continuity. (Cf. A. N. Whitehead: *PR, op. cit.*, p. 53.) But the reader is referred to my demonstration

It is well-known that in the presence of strong emotional factors such as anxiety, exhilaration, and boredom, the psychological time-metric exhibits great variability as compared to the Newtonian one of physics. But there is much evidence that when such factors are not present, physiological processes which are geared to the periodicities defining physical time-congruence impress a metric upon man's psychological time and issue in rhythmic behavior on the part of a vast variety of animals. There are two main theories at present as to the source of such concordance as obtains between the metrics of physical and psychobiological time. The older of these was put forward by W. Pfeffer and maintains that men and animals are equipped with an internal "biological clock" *not* dependent for its successful operation on the conscious or unconscious reception of sensory cues from outside the organism.[4] Instead the success of the biological clock is held to depend only on the occurrence of metabolic processes whose rate is *steady in the metric of physical clock time.*[5] As applied to humans, this hypothesis was supported by experiments of the following kind. People were asked to tap on an electric switch at a rate which they judged to be a fixed number of times per second. It was found over a relatively small range of body temperatures that the temperature coefficient

---

of the irrelevance of becoming to physical (as distinct from psychological) time in Chapter Ten below, my critique (cf. A. Grünbaum: "Relativity and the Atomicity of Becoming," *The Review of Metaphysics,* Vol. IV [1950], pp. 143–86) of Whitehead's use of the "Dichotomy" paradox of Zeno of Elea to prove that time intervals are only potential and not actual continua, and, more generally, to F.S.C. Northrop's rebuttal to Whitehead's attack on bifurcation (cf. F.S.C. Northrop: "Whitehead's Philosophy of Science," in: P. A. Schilpp (ed.) *The Philosophy of Alfred North Whitehead* (New York: Tudor Publishing Co.; 1941), pp. 165–207.

[4] W. Pfeffer: "Untersuchungen über die Entstehung der Schlafbewegungen der Blattorgane," *Abhandlungen der sächsischen Akademie der Wissenschaften, Leipzig, Mathematisch-Physikalische Klasse,* Vol. III (1907), p. 257; *ibid.,* Vol. XXXIV (1915), p. 3, and C. P. Richter: "Biological Clocks in Medicine and Psychiatry: Shock-Phase Hypothesis," *Proceedings of the National Academy of Sciences,* Vol. XLVI (1960), pp. 1506–30.

[5] C. B. Goodhard: "Biological Time," *Discovery* (December, 1957), pp. 519–21, and H. Hoagland: "The Physiological Control of Judgments of Duration: Evidence for a Chemical Clock," *The Journal of General Psychology,* Vol. IX (1933), pp. 267–87, and "Chemical Pacemakers and Physiological Rhythms," in: J. Alexander (ed.) *Colloid Chemistry* (New York: Reinhold Publishing Corp.; 1944), Vol. V, pp. 762–85.

of counting was much the same as the one characteristic of chemical reactions: a two-or-threefold increase in rate for a 10°C. rise in temperature. The defenders of the conception that the biological clock is purely internal further adduce observations of the behavior of bees: both outdoors on the surface of the earth and at the bottom of a mine, bees learned to visit at the correct time each day a table on which a dish of syrup was placed daily for a short time at a fixed hour. Since the bees had been found to be hungry for sugar all day long, some investigators hold that neither the assumption that the bees experience periodic hunger, nor the appearance of the sun nor yet the periodicities of the cosmic ray intensity can explain the bees' success in timekeeping. But dosing them with substances like thyroid extract and quinine, which affect the rate of chemical reactions in the body, was found to interfere with their ability to appear at the correct time.[6]

More recently, however, doubt has been cast on the adequacy of the hypothesis of the purely internal clock. A series of experiments with fiddler crabs and other cold-blooded animals[7] showed that these organisms hold rather precisely to a twenty-four-hour coloration cycle (lightening-darkening rhythm) regardless of whether the temperature at which they are kept is 26 degrees, 16 degrees or 6 degrees Centigrade, although at temperatures near freezing, the color clock changes. It was therefore argued that if the rhythmic timing mechanism were indeed a biochemical one wholly inside the organism, then one would expect the rhythm to speed up with increasing temperature and to slow down with decreasing temperature. And the

---

[6] C. S. Pittendrigh and V. G. Bruce: "An Oscillator Model for Biological Clocks," in D. Rudnick (ed.) *Rhythmic and Synthetic Processes in Growth* (Princeton: Princeton University Press; 1957), pp. 75–109, and C. S. Pittendrigh and V. G. Bruce: "Daily Rhythms as Coupled Oscillator Systems and their Relation to Thermoperiodism and Photoperiodism," *Photoperiodism and Related Phenomena in Plants and Animals* (Washington, D.C.: The American Association for the Advancement of Science; 1959).

[7] F. A. Brown, Jr.: "Biological Clocks and the Fiddler Crab," *Scientific American,* Vol. CXC (1954), pp. 34–37; "The Rhythmic Nature of Animals and Plants," *American Scientist,* Vol. XLVII (1959), pp. 147–68; "Living Clocks," *Science,* Vol. CXXX (1959), pp. 1535–44; and "Response to Pervasive Geophysical Factors and the Biological Clock Problem," *Cold Spring Harbor Symposia on Quantitative Biology,* Vol. XXV (1960), pp. 57–71.

exponents of this interpretation maintain that since the period of the fiddler crab's rhythm remained twenty-four hours through a wide range of temperature, the animals must possess a means of measuring time which is independent of temperature. This, they contend, is "a phenomenon quite inexplicable by any currently known mechanism of physiology, or, in view of the long period-lengths, even of chemical reaction kinetics."[8] The extraordinary further immunity of certain rhythms of animals and plants to many powerful drugs and poisons which are known to slow down living processes greatly is cited as additional evidence to show that organisms have daily, lunar, and annual rhythms impressed upon them by external physical agencies, thus having access to outside information concerning the corresponding physical periodicities.[9] The authors of this theory admit, however, that the daily and lunar-tidal rhythms of the animals studied do not depend upon any presently known kind of external cues of the associated astronomical and geophysical cycles.[1] And it is postulated[2] that these physical cues are being received because living things are able to respond to additional kinds of stimuli at energy levels so low as to have been previously held to be utterly irrelevant to animal behavior. The assumption of such sensitivity of animals is thought to hold out hope for an explanation of animal navigation.

We have dwelled on the two current rival theories regarding the source of the ability of man (and of animals) to make successful estimates of duration introspectively in order to show that, on either theory, the metric of psychological time is tied causally to those physical cycles which serve to define time congruence in physics. Hence when we make the judgment that two intervals of physical time which are equal in the metric of standard clocks also *appear* congruent in the psychometry of mere sense awareness, this justifies only the following innocuous conclusion in regard to physical time: the two intervals in ques-

---

[8] F. A. Brown, Jr.: "The Rhythmic Nature of Animals and Plants," *op. cit.*, p. 159.

[9] F. A. Brown, Jr.: "Living Clocks," *op. cit.*, and "Response to Pervasive Geophysical Factors and the Biological Clock Problem," *op. cit.*

[1] F. A. Brown, Jr.: "The Rhythmic Nature of Animals and Plants," *op. cit.*, pp. 153, 166.

[2] *Ibid.*, p. 168.

tion are congruent by the physical criterion which had furnished the psychometric standard of temporal equality both genetically and epistemologically. How then can the metric deliverances of psychological time possibly show that the time of physics possesses an intrinsic metric, if, as we saw, no such conclusion was demonstrable on the basis of the cycles of physical clocks?

As for *spatial* congruence, we must appraise Whitehead's argument from *matching* in the quotations above after dealing with the following contention by him:[3] just as it is an objective datum of experience that two phenomenal color patches have the same color, i.e., are "color-congruent," so also we *see* that a given rod has the same length in different positions, thus making the latter congruence as objective a relation as the former. Says he: "It is at once evident that all these tests [of congruence by means of steel yard-measures, etc. are] dependent on a direct intuition of permanence."[4] I take Whitehead to be claiming here that in the accompanying diagram, for example,

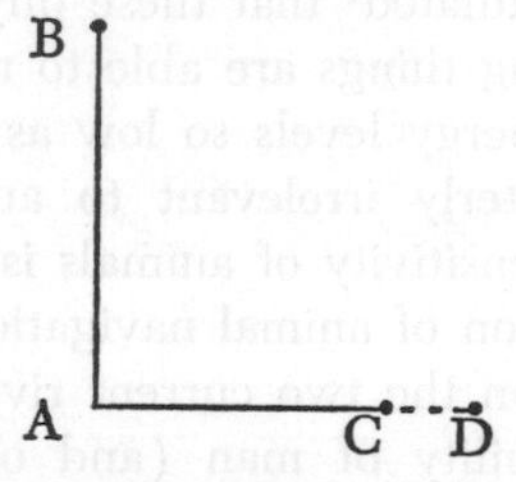

the horizontal segment *AC* could not be stipulated to be congruent to the vertical segment *AB*. His grounds for this claim are that the deliverances of our visual intuition unequivocally show *AC* to be shorter than *AB*, and *AB* to be congruent to *AD*, a fact also attested by the finding that a solid rod coinciding with *AB* to begin with and then rotated into the horizontal position would extend over *AC* and coincide with *AD*.

On this my first comment is to ask: what is the significance for the status of the metric of *physical* as distinct from *visual* space of these observational deliverances? And I answer that their significance is entirely consonant with the conventionalist view of physical congruence. The criterion for ocular congruence in

[3] A. N. Whitehead: *The Principles of Natural Knowledge, op. cit.*, p. 56.
[4] A. N. Whitehead: *PR, op. cit.*, p. 501.

our visual field was presumably furnished both genetically and epistemologically by ocular adaptation to the behavior of transported solids. For when pressed as to what it is about two congruent-*looking* physical intervals that enables them to sustain the relation of spatial equality, our answer will inevitably have to be: the fact of their capacity to coincide successively with a transported solid rod. Hence when we make the judgment that two intervals of physical space with which transported solid rods coincide in succession also *look* congruent when compared frontally purely by inspection, what this proves in regard to *physical space* is only that these intervals are congruent on the basis of the criterion of congruence which had furnished the genetic and epistemological basis for the ocular congruence to begin with, a criterion given by solid rods. But the visual deliverance of congruence does not constitute an ocular test of the "true" rigidity of solids under transport in the sense of establishing the factuality of the congruence defined by this class of bodies on the basis of an intrinsic metric. Thus, it is a *fact* that in the diagram, *AD* extends over (includes) *AC*, thus being longer. And it will be recalled that Riemann's views on the status of measurement in a spatial continuum require that every definition of "congruent" be consistent with this kind of inclusional fact. How then can visual data possibly interdict our stipulating *AC* to be congruent to *AB* and then allowing for the de facto coincidence of the rotated rod with *AB* and *AD* by assigning to the rod in the horizontal position a length which is suitably *greater* than the one assigned to it in the vertical orientation?

As for his argument from spatial *matching* and its relation to coincidence and measurement, the important issue posed by Whitehead is not whether an operationist account of congruence is adequate. Instead, it is whether spatial congruence derives from intrinsic properties of the intervals concerned rather than wholly from their relation to a transported standard of some kind. For, as we noted earlier, the conventionalist conception of congruence which he is attacking here does not require and is not adequately rendered by the operationist claim that "*the* meaning" of "congruence" is given by some operation of producing coincidence under transport. Just as Einstein's conventionalist view of simultaneity is fully justified only by the ontology of temporal relatedness postulated by him rather than operationally,[5] so the conventionalist conception of congruence obtains its philosophical credentials, as Riemann saw, from the assumed continuity of space (and of time). Thus, the question before us again concerns the intrinsicality and uniqueness of congruence, whatever the operations or test conditions by which it may be specified. Hencc wc again ignore here considerations pertaining to the fact that—on the conventionalist no less than on the Whiteheadian view—congruence is an open-cluster concept in the sense that no one criterion such as coincidence with a transported rod can exhaustively specify its entire actual and potential physical meaning. And then the reply to Whitehead is the following.

If there were a basis for the ascription of an intrinsic metric to space, then he would indeed be entitled to regard coincidence as only a test of congruence *in his sense* that coincidence merely ascertains the equality or matching of the separated intervals concerned with respect to the *intrinsic* amount of space contained by each of them. But without having established the existence of an intrinsic metric, the congruence or *matching* of spatially separated intervals is first *constituted* by the relation which they each bear to the behavior of a transported standard such as a rod or such as the round-trip times of light in an inertial system. And the conventionality of the self-congruence of the latter standard at different places is not at all invalidated by the fact, correctly noted by Whitehead, that measurement presupposes a congruence criterion in terms of which its results are formulated.

That a distance of thirty miles is indeed a long walk for anyone is due to our gait's being tied to the congruence defined by yard (or meter) sticks, thus making it an objective fact that an interval which measures thirty miles in the metric of the yardstick will contain a great many of our steps. But how can this fact possibly show in the absence of an intrinsic metric that the self-congruence of the yardstick under transport is nonconventional?

Moreover, it is plainly false and inconsistent for Whitehead to declare that, according to the classical theory, there is a

---

[5] Cf. Chap. Twelve.

"breakdown" of the "very existence" of congruence for time in contrast to the repleteness of space with mutually exclusive congruence classes. For as he himself had noted, on the classical theory a congruence class can be specified for the time continuum by the requirement that Newton's (amended) laws hold. And an infinitude of additional *alternative* time congruences can be given by metrizations based on the values of time variables that are non-linear, one-to-one functions of the Newtonian time variable t.[6] In his book on Whitehead, Palter is sympathetic to Whitehead to the point of believing that a viable interpretation can generally be given even of those of Whitehead's utterances which are either prima facie false or irritatingly obscure. Hence Palter[7] seeks to defend Whitehead's claim by construing it as referring to the following fact: time, being only a one-dimensional continuum, presents us with no analogue to the higher-dimensional distinction between Euclidean and non-Euclidean geometries, and hence time differs from higher-dimensional space by not possessing a structure corresponding to a distinctive metric geometry. But Palter's defense is unsuccessful, since it rests on the confusion of time's lack of the analogue of a distinctive metric geometry with its alleged lack of a congruence class: the lack of the former does not entail the lack of the latter, although the converse entailment does obtain.

It is significant, however, that there are passages in Whitehead where he comes close to the admission that the pre-eminent role of certain classes of physical objects as our standards of rigidity and isochronism is *not* tantamount to their making evident the *intrinsic* equality of certain spatial and temporal intervals. Thus speaking of the space-time continuum, he says:

> This extensive continuum is one relational complex in which all potential objectifications find their niche. It underlies the whole world, past, present, and future. Considered in its full generality, apart from the additional conditions proper only to the cosmic epoch of electrons, protons, molecules, and star-systems, the properties of this continuum are very few and do not include the relationships of metrical geometry.[8]

---

[6] For details, see Chap. Two.

[7] R. M. Palter: *Whitehead's Philosophy of Science, op. cit.*, pp. 90–92.

[8] A. N. Whitehead: *PR, op. cit.*, p. 103.

And he goes on to note that there are competing systems of measurement giving rise to alternative families of straight lines and correspondingly alternative systems of metrical geometry of which no one system is more fundamental than any other.[9] It is in our present cosmic epoch of electrons, protons, molecules, and star-systems that "more special defining characteristics obtain"[1] and that "the ambiguity as to the relative importance of competing definitions of congruence" is resolved in favor of "one congruence definition."[2] Thus Whitehead maintains that among competing congruence definitions, "That definition which enters importantly into the internal constitutions of the dominating . . . entities is the important definition for the cosmic epoch in question."[3] This important concession thus very much narrows the gap between Whitehead's view and the Riemann-Poincaré conception defended in this book: the question as to which of the rival metric geometries is true of physical space is one of objective physical fact only after a congruence definition has been given conventionally by means of the customary rigid body (or otherwise), assuming the usual physical interpretation of the remainder of the geometrical vocabulary.

That the gap between the two views is narrowed by Whitehead's concession here becomes clear upon reading the following statement by him in the light of that concession. Speaking of Sophus Lie's treatment of congruence classes and their associated metric geometries in terms of groups of transformations between points, Whitehead cites Poincaré and says:

> The above results, in respect to congruence and metrical geometry, considered in relation to existent space, have led to the doctrine that it is intrinsically unmeaning to ask which system of metrical geometry is true of the physical world. Any one of these systems can be applied, and in an indefinite number of ways. The only question before us is one of convenience in respect to simplicity of statement of the physical laws. This point of view seems to neglect the consideration that science is to be relevant to the definite perceiving minds of men; and that (neglecting the ambiguity introduced by the invariable

[9] *Ibid.*, p. 149.
[1] *Ibid.*
[2] *Ibid.*
[3] *Ibid.*, p. 506.

> slight inexactness of observation which is not relevant to this special doctrine) we have, in fact, presented to our senses a definite set of transformations forming a congruence-group, resulting in a set of measure relations which are in no respect arbitrary. Accordingly our scientific laws are to be stated relevantly to that particular congruence-group. Thus the investigation of the type (elliptic, hyperbolic or parabolic) of this special congruence-group is a perfectly definite problem, to be decided by experiment.[4]

[4] A. N. Whitehead: *Essays in Science and Philosophy* (New York: The Philosophical Library; 1947), p. 265.

Chapter 2

THE SIGNIFICANCE OF ALTERNATIVE TIME METRIZATIONS IN NEWTONIAN MECHANICS AND IN THE GENERAL THEORY OF RELATIVITY

(A) NEWTONIAN MECHANICS

On the conception of time congruence as conventional, the preference for the customary definition of isochronism—a preference not felt by Einstein in the general theory of relativity (GTR), as we shall see in Section B—can derive only from considerations of convenience and elegance so long as the resulting form of the theory is not prescribed. Hence, the thesis that isochronism is conventional precludes a difference in factual import (content) or in *explanatory power* between two descriptions one of which employs the customary isochronism while the other is a "translation" (transcription) of it into a language employing a time congruence incompatible with the customary one. As a test case for this thesis of *explanatory parity*, the general outline of a counter-argument has been suggested which we shall be able to state after some preliminaries.

On the Riemannian analysis, congruence must be regarded as conventional in the time continuum of Newtonian dynamics no less than in the theory of relativity. We shall therefore wish to compare in regard to explanatory capability the two forms of Newtonian dynamics corresponding to two different time congruences as follows.

The first of these congruences is defined by the requirement that Newton's laws hold, as modified by the addition of very small corrective terms expressing the so-called relativistic motion of the perihelia. This time congruence will be called "Newtonian," and the time variable whose values represent Newtonian time after a particular unit has been chosen will be denoted by "t." The second time congruence is defined by the rotational motion of the earth. It does not matter for our purpose whether we couple the latter congruence with a unit given by the mean solar second, which is the 1/86400 part of the mean interval between two consecutive meridian passages of the fictitious mean sun, or with a different unit given by the sidereal day, which is the interval between successive meridian passages of a star. What matters is that both the mean solar second and the sidereal day are based on the periodicities of the earth's rotational motion. Assume now that one or another of these units has been chosen, and let T be the time variable associated with that metrization, which we shall call "diurnal time." The important point is that the time variables t and T are non-linearly related and are associated with *incompatible* definitions of isochronism, because the speed of rotation of the earth *varies relatively to the Newtonian time-metric* in several distinct ways.[1] Of these, the best known is the relative slowing down of the earth's rotation by the tidal friction between the water in the shallow seas of the earth and the land under it. Upon calculating the positions of the moon, for example, via the usual theory of celestial mechanics, which is based on the Newtonian time-metric, the observed positions of the moon in the sky would be found to be *ahead* of the calculated ones *if* we were to identify the time defined by the earth's rotation with the Newtonian time of celestial mechanics. And the same is true of the positions of the planets of the solar system and of the moons of Jupiter in amounts all corresponding to a slowing down on the part of the earth.

Now consider the following argument for the lack of explanatory parity between the two forms of the dynamical theory respectively associated with the t- and T-scales of time: "*Dynamical*

[1] G. M. Clemence: "Time and its Measurement," *American Scientist,* Vol. XL (1952), pp. 264–67.

facts will discriminate in favor of the t-scale as opposed to the T-scale. It is granted that it is *kinematically* equivalent to say

(a) the earth's rotational motion has slowed down relatively to the "clocks" constituted by various revolving planets and satellites of the solar system,

or

(b) the revolving celestial bodies speed up their periodic motions relatively to the earth's uniform rotation.

But these two statements are not on a par explanatorily in the context of the dynamical theory of the motions in the solar system. For whereas the slowing down of the earth's rotation in formulation (a) *can* be understood as the dynamical effect of nearby masses (the tidal waters and their friction), no similar dynamical *cause* can be supplied for the accelerations in formulation (b). And the latter fact shows that a theory incorporating formulation (a) has greater explanatory power or factual import than a theory containing (b)." In precisely this vein, d'Abro, though stressing on the one hand that apart from convenience and simplicity there is nothing to choose between different metrics,[2] on the other hand adduces the provision of *causal* understanding by the t-scale as an argument in its favor and thus seems to construe such differences of simplicity as involving factually *non*-equivalent descriptions:

> if in mechanics and astronomy we had selected at random some arbitrary definition of time, if we had defined as congruent the intervals separating the rising and setting of the sun at all seasons of the year, say for the latitude of New York, our understanding of mechanical phenomena would have been beset with grave difficulties. As measured by these new temporal standards, free bodies would no longer move with constant speeds, but would be subjected to periodic accelerations *for which it would appear impossible to ascribe any definite cause,* and so on. As a result, the law of inertia would have to be abandoned, and with it the entire doctrine of classical mechanics, together with Newton's law. Thus a change in our understanding of congruence would entail far-reaching consequences.

---

[2] A. d'Abro: *The Evolution of Scientific Thought from Newton to Einstein, op. cit.*, p. 53.

> Again, in the case of the vibrating atom, had some arbitrary definition of time been accepted, we should have had to assume that the same atom presented the most capricious frequencies. *Once more it would have been difficult to ascribe satisfactory causes* to these seemingly haphazard fluctuations in frequency; and a simple understanding of the most fundamental optical phenomena would have been well-nigh impossible.[3]

To examine this argument, let us set the two formulations of dynamics corresponding to the t- and T-scales respectively before us mathematically in order to have a clearer statement of the issue.

The differences between the two kinds of temporal congruence with which we are concerned arise from the fact that the functional relationship

$$T = f(t)$$

relating the two time scales is *non-linear*, so that time-intervals which are congruent on the one scale are generally incongruent on the other. It is clear that this function is monotone-increasing, and thus we know that permanently

$$\frac{dT}{dt} \neq 0.$$

Moreover, in view of the non-linearity of $T = f(t)$, we know that dT/dt is *not* constant. Since the function f has an inverse, it will be possible to translate any set of laws formulated on the basis of either of the two time-scales into the corresponding other scale. In order to see what form the customary Newtonian force law assumes in diurnal time, we must express the acceleration ingredient in that law in terms of diurnal time. But in order to derive the transformation law for the accelerations, we first treat the velocities. By the chain rule for differentiation, we have, using "*r*" to denote the position vector,

$$(1) \qquad \frac{dr}{dt} = \frac{dr}{dT}\frac{dT}{dt}.$$

Suppose a body is *at rest* in the coordinate system in which *r* is measured, *when Newtonian time is employed;* then this body

[3] *Ibid.*, p. 78, my italics.

will also be held to be *at rest diurnally*: since we saw that the second term on the right-hand side of equation (1) cannot be zero, the left-hand side of (1) will vanish if and only if the first term on the right-hand side of (1) is zero. Though rest in a given frame in the t-scale will correspond to rest in that frame in the T-scale as well, equation (1) shows that the *constancy* of the non-vanishing Newtonian velocity dr/dt will *not* correspond to a constant diurnal velocity dr/dT, since the derivative dT/dt changes with both Newtonian and diurnal time. Now, differentiation of equation (1) with respect to the Newtonian time t yields

$$(2) \qquad \frac{d^2r}{dt^2} = \frac{dr}{dT}\frac{d^2T}{dt^2} + \frac{dT}{dt}\frac{d}{dt}\left(\frac{dr}{dT}\right).$$

But, applying the chain-rule to the second factor in the second-term on the right-hand side of (2), we obtain

$$(2a) \qquad \frac{d}{dt}\left(\frac{dr}{dT}\right) = \frac{d^2r}{dT^2}\frac{dT}{dt}.$$

Hence (2) becomes

$$(3) \qquad \frac{d^2r}{dt^2} = \frac{dr}{dT}\frac{d^2T}{dt^2} + \frac{d^2r}{dT^2}\left(\frac{dT}{dt}\right)^2.$$

Solving for the diurnal acceleration, and using equation (1) as well as the abbreviations

$$f'(t) \equiv \frac{dT}{dt} \text{ and } f''(t) \equiv \frac{d^2T}{dt^2}, \text{ we find}$$

$$(4) \qquad \underbrace{\frac{d^2r}{dT^2}}_{\substack{\text{diurnal}\\ \text{accelera-}\\ \text{tion}}} = \frac{1}{[f'(t)]^2}\underbrace{\frac{d^2r}{dt^2}}_{\substack{\text{Newtonian}\\ \text{accelera-}\\ \text{tion}}} - \overbrace{\frac{f''(t)}{[f'(t)]^3}\underbrace{\frac{dr}{dt}}_{\substack{\text{New-}\\ \text{tonian}\\ \text{velocity}}}}^{\textit{secular term}}.$$

Several ancillary points should be noted briefly in regard to equation (4) before seeing what light it throws on the form assumed by causal explanation within the framework of a diurnal description. When the Newtonian force on a body is *not* zero because the body is accelerating under the influence of masses, the diurnal acceleration will generally also not be zero, save in the unusual case when

$$\frac{d^2r}{dt^2} = \frac{f''(t)}{f'(t)} \frac{dr}{dt}. \tag{5}$$

Thus the causal influence of masses, which gives rise to the Newtonian accelerations in the usual description, is seen in (4) to make a definite contribution to the diurnal acceleration as well. But the new feature of the diurnal description of the facts lies in the possession of a *secular acceleration* by all bodies not at rest, even when no masses are inducing Newtonian accelerations, so that the first term on the right-hand side of (4) vanishes. And this secular acceleration is numerically *not* the same for all bodies but depends on their velocities dr/dt in the given reference frame and thus also on the reference frame.

The character and existence of this secular acceleration calls for several kinds of comment.

Its dependence on the velocity and on the reference frame should neither occasion surprise nor be regarded as a difficulty of any sort. As to the velocity-dependence of the secular acceleration, consider a simple numerical example which removes any surprise: if instead of calling two successive hours on Big Ben equal, we remetrized time so as to assign the measure one half hour to the second of these intervals, then all bodies having uniform speeds in the usual time metric will double their speeds on the new scale after the first interval, and the *numerical* increase or *acceleration* in the speeds of initially faster bodies will be greater than that in the speeds of the initially slower bodies. Now as for the dependence of the secular acceleration on the reference frame, in the context of the physical facts asserted by the Newtonian theory *apart from* its metrical philosophy, it is a mere prejudice to require that, to be admissible, every formulation of that theory must agree with the customary one in making the acceleration of a body at any given time be the same in all Galilean reference frames ("Galilean relativity"). For not a single bona fide physical fact of the Newtonian world is overlooked or contradicted by a kinematics not featuring this Galilean relativity. It is instructive to be aware in this connection that even in the *customary* rendition of the *kinematics of special relativity*, a constant acceleration in a frame S′ would not generally correspond to a constant acceleration in a frame S, because the component accelerations in S depend not only on the accelerations in S′ but also on the component velocities in that system which would be changing with the time.

But what are we to say, apart from the dependence on the velocity and reference system, about the very presence of this "dynamically-unexplained" or causally baffling *secular acceleration?* To deal with this question, we first observe merely for comparison that in the *customary* formulation of Newtonian mechanics, constant *speeds* (as distinct from constant *velocities*) fall into *two* classes with respect to being attributable to the dynamical action of perturbing masses: constant *rectilinear* speeds are affirmed to prevail in the absence of any mass influences, while constant *curvilinear* (e.g., circular) speeds are related to the (centripetally) accelerating actions of masses. Now in regard to the presence of a secular acceleration in the diurnal description, it is fundamental to see the following: Whereas on the version of Newtonian mechanics employing the *customary* metrizations (of time and space), *all* accelerations whatsoever in Galilean frames are of *dynamical* origin by being attributable to the action of specific masses, *this feature of Newton's theory is made possible not only by the facts but also by the particular time-metrization chosen to codify them.* As equation (4) shows upon equating d^2r/dt^2 to zero, the dynamical character of *all* accelerations is not vouchsafed by any causal facts of the world with which every theory would have to come to terms. For the diurnal description encompasses the objective behavior of bodies (point-events and coincidences) as a function of the presence or absence of other bodies no less than does the Newtonian one, thereby achieving full explanatory parity with the latter in all logical (as distinct from pragmatic!) respects.

Hence the provision of a dynamical basis for all accelerations should not be regarded as an *inflexible epistemological requirement* in the elaboration of a theory explaining mechanical phenomena. Disregarding the pragmatically decisive consideration of convenience, there can therefore be no valid explanatory objection to the diurnal description, in which accelerations fall into *two* classes by being the superpositions, in the sense of equation

(4), of a dynamically grounded *and* a kinematically grounded quantity. And, most important, since there is no *slowing down* of the earth's rotation on the diurnal metric, there can be no question in that description of specifying a *cause* for such a nonexistent deceleration; instead, a frictional *cause* is now specified for the earth's diurnally-*uniform* rotation *and* for the liberation of heat accompanying this kind of uniform motion. For in the T-scale description, it is *uniform* rotation which requires a dynamical cause constituted by masses interacting (frictionally) with the uniformly rotating body, and it is now a law of nature or entailed by such a law that all diurnally-uniform rotations issue in the dissipation of heat. Of course, the mathematical representation of the frictional interaction will not have the customary Newtonian form: to obtain the diurnal account of the frictional dynamics of the tides, one would need to apply transformations of the kind given in our equation (4) to the quantities appearing in the relevant Newtonian equations for this case.[4]

But, it will be asked, what of the Newtonian conservation principles, if the T-scale of time is adopted? It is readily demonstrable by reference to the simple case of the motion of a free particle that while the Newtonian kinetic energy will be constant in this case, its formal diurnal homologue (as opposed to its diurnal equivalent!) will *not* be constant. Let us denote the constant *Newtonian* velocity of the free particle by "v_t," the subscript "t" serving to represent the use of the t-scale, and let "v_T" denote the diurnal velocity corresponding to v_t. Since we know from equation (1) above that

$$v_t = v_T \frac{dT}{dt},$$

where v_t is constant but dT/dt is *not*, we see that the diurnal *homologue* $\frac{1}{2}mv_T^2$ of the Newtonian kinetic energy cannot be constant in this case, although the diurnal *equivalent* $\frac{1}{2}m(v_T\frac{dT}{dt})^2$ of the constant Newtonian kinetic energy $\frac{1}{2}mv_t^2$ is necessarily constant. Just as in the case of the Newtonian equations of motion themselves, so also in the case of the Newtonian conservation

[4] For these equations, cf. H. Jeffreys: *The Earth* (3rd ed.; Cambridge: Cambridge University Press; 1952), Chap. 8, and G. I. Taylor: "Tidal Friction in the Irish Sea," *Philosophical Transactions of the Royal Society*, A., Vol. CCXX (1920), pp. 1–33.

principle of mechanical energy, the diurnal equivalent or transcription explains all the facts explained by the Newtonian original. Hence our critic can derive no support at all from the fact that the formal diurnal homologues of Newtonian conservation principles generally do not hold. And we see, incidentally, that the time-invariance of a physical quantity and hence the appropriateness of singling it out from among others as a form of "energy," etc., will depend not only on the facts but also on the time-metrization used to render them. It obviously will not do, therefore, to charge the diurnal description with inconsistency via the *petitio* of grafting onto it the requirement that it incorporate the homologues of Newtonian conservation principles which are incompatible with it: a case in point is the charge that the diurnal description violates the conservation of energy because in its metric the frictional generation of heat in the tidal case is not compensated by any reduction in the speed of the earth's rotation! Whether the diurnal time-metrization permits the deduction of conservation principles of a *relatively simple type* involving diurnally based quantities is a rather involved mathematical question whose solution is not required to establish our thesis that, apart from pragmatic considerations, the diurnal description enjoys explanatory parity with the Newtonian one.

We have been disregarding pragmatic considerations in assessing the explanatory capabilities of two descriptions associated with different time-metrizations as to parity. But it would be an error to infer that in pointing to the equivalence of such descriptions in regard to factual content, we are committed to the view that there is no criterion for choosing between them and hence no reason for preferring any one of them to the others. Factual adequacy (truth) is, of course, the cardinal *necessary* condition for the acceptance of a scientific theory, but it is hardly a *sufficient* condition for accepting any one particular formulation of it which satisfies this necessary condition. As well say that a person pointing out that equivalent descriptions can be given in the decimal (metric) and English system of units cannot give telling reasons for preferring the former! Indeed, after first commenting on the *factual basis* for the existence of the Newtonian time congruence, we shall see that there are weighty *pragmatic* reasons for preferring that metrization of the time continuum. And these reasons will turn out to be entirely consonant with our twin contention that alternative metrizability allows linguistically different, equivalent descriptions *and* that geo-chronometric conventionalism is *not* a subthesis of TSC.

The *factual basis* for the existence of the Newtonian time-metrization will be appreciated by reference to the following two considerations: first, as we shall prove presently, it is a highly fortunate empirical fact, and not an a priori truth, that there exists a time-metrization *at all* in which *all* accelerations with respect to inertial systems are of dynamic origin, as claimed by the Newtonian theory, and second, it is a further empirical fact that the time-metrization having this remarkable property (i.e., "ephemeris time") is furnished physically by the earth's annual revolution around the sun (not by its diurnal rotation) albeit not in any observationally simple way, since due account must be taken computationally of the irregularities produced by the gravitational influences of the other planets.[5] That the existence of a time-metrization in which *all* accelerations with respect to inertial systems are of *dynamical* origin cannot be guaranteed a priori is demonstrable as follows.

Suppose that, *contrary to actual fact,* it *were* the case that a *free* body did accelerate with respect to an inertial system when its motion is described in the metric of ephemeris time t, it thus being assumed that there are accelerations in the customary time metric which are *not* dynamical in origin. More particularly, let us now posit that, contrary to actual fact, a *free* particle were to execute *one-dimensional* simple harmonic motion of the form

$$r = \cos \omega t,$$

where r is the distance from the origin. In that hypothetical eventuality, the acceleration of a *free* particle in the t-scale would have the time-dependent value

$$\frac{d^2r}{dt^2} = - \omega^2 \cos \omega t.$$

And our problem is to determine whether there would then exist some other time-metrization $T = f(t)$ possessing the Newtonian

[5] G. M. Clemence: "Astronomical Time," *Reviews of Modern Physics,* Vol. XXIX (1957), pp. 2–8; "Dynamics of the Solar System," ed. by E. Condon and H. Odishaw, *Handbook of Physics* (New York: McGraw-Hill Book Company, Inc.; 1958), p. 65; "Ephemeris Time," *Astronomical Journal,* Vol. LXIV (1959), pp. 113–15 and *Transactions of the International Astronomical Union,* Vol. X (1958); "Time and Its Measurement," *op. cit.*

property that our *free* particle has a *zero* acceleration. We shall now find that the answer is definitely negative: under the hypothetical empirical conditions which we have posited, there would indeed be no admissible single-valued time-metrization T at all in which all accelerations with respect to an inertial system would be of dynamical origin.

For let us now regard T in equation (4) of this chapter as the time variable associated with the sought-after metrization $T = f(t)$ in which the acceleration d^2r/dT^2 of our free particle would be *zero*. We recall that equation (5) of this chapter was obtained from equation (4) by equating the T-scale acceleration d^2r/dT^2 to zero. Hence *if* our sought-after metrization exists at all, it would have to be the solution $T = f(t)$ of the scalar form of equation (5) as applied to our one-dimensional motion. That equation is

$$\text{(6)} \qquad \frac{d^2r}{dt^2} = \frac{f''(t)}{f'(t)} \frac{dr}{dt}.$$

Putting $v \equiv dr/dt$ and noting that

$$\frac{d}{dt}\log f'(t) = \frac{f''(t)}{f'(t)}, \text{ and } \frac{d}{dt}\log v = \frac{1}{v}\frac{dv}{dt},$$

equation (6) becomes

$$\frac{d}{dt}\log v = \frac{d}{dt}\log f'(t).$$

Integrating, and using log c as the constant of integration, we obtain

$$\log v = \log cf'(t),$$

or

$$v = cf'(t),$$

which is

$$\frac{dr}{dt} = c\,\frac{dT}{dt}.$$

Integration yields

$$\text{(7)} \qquad r = cT + d,$$

where d is a constant of integration. But, by our earlier hypothesis, $r = \cos \omega t$. Hence (7) becomes

$$\text{(8)} \qquad T = \frac{1}{c}\cos \omega t - \frac{d}{c}.$$

It is evident that the solution $T = f(t)$ given by equation (8) is *not* a *one-to-one* function: the *same* time T in the sought-after metrization would correspond to all those *different* times on the t-scale at which the oscillating particle would return to the *same place* $r = \cos \omega t$ in the course of its periodic motion. And by thus violating the basic topological requirement that the function $T = f(t)$ be one-to-one, the T-scale which does have the sought-after Newtonian property under our *hypothetical* empirical conditions is physically a quite inadmissible and hence unavailable metrization.

It follows that there is no a priori assurance of the existence of at least one time-metrization possessing the Newtonian property that the acceleration of a free particle with respect to inertial systems is zero. So much for the *factual basis* of the existence of the Newtonian time-metrization.

Now, inasmuch as the employment of the time-metrization based on the earth's annual revolution issues in Newton's relatively simple laws, there are powerful reasons of mathematical tractability and convenience for greatly preferring the time-metrization in which *all* accelerations with respect to inertial systems are of dynamical origin. In fact, the various refinements which astronomers have introduced in their physical standards for temporal congruence have been dictated by the demand for a definition of temporal congruence (or of a so-called "invariable" time standard) for which Newton's laws will hold in the solar system, including the relatively simple conservation laws interconnecting diverse kinds of phenomena (mechanical, thermal, etc.). And thus, as Feigl and Maxwell have aptly put it, one of the important criteria of descriptive simplicity which greatly restrict the range of "reasonable" conventions is seen to be the *scope* which a convention will allow for mathematically tractable laws.

(B) THE GENERAL THEORY OF RELATIVITY

In the special theory of relativity, only the *customary* time-metrization is employed in the following sense: At any given point A in a Galilean frame, the length of a time-interval between two events at the point A is given by the *difference* between the time-coordinates of the two events as furnished by the readings of a standard clock at A whose periods are defined to be congruent. This is, of course, the precise analogue of the customary definition of spatial congruence which calls the rod congruent to itself everywhere, when at relative rest, after allowance for substance-specific perturbations. On the other hand, as we shall now see, there are contexts in which the *general* theory of relativity (GTR) utilizes a criterion of *temporal* congruence which is an analogue of a *non*-customary kind of spatial congruence in the following sense: the length of a time-interval separating two events at a clock depends not only on the *difference* between the time-coordinates which the clock assigns to these events but also on the *spatial* location of the clock (though not on the time itself at which the time-interval begins or ends).

A case in point from the GTR involves a *rotating disk* to which we apply those principles which the GTR takes over from the special theory of relativity. Let a set of standard material clocks be distributed at various points on such a disk. The infinitesimal application of the special relativity clock retardation then tells us the following: a clock at the *center* O of the disk will maintain the rate of a contiguous clock located in an inertial system I with respect to which the disk has angular velocity ω, *but* the same does not hold for clocks located at other points A of the disk which are at positive distances r from O. Such A-clocks have various linear velocities ωr relatively to I in virtue of their common angular velocity ω. Accordingly, all A-clocks (whatever their chemical constitution) will have readings lagging behind the corresponding readings of the respective I-system clocks adjacent to them by a *factor* of $\sqrt{1 - \frac{r^2\omega^2}{c^2}}$, where c is the velocity of light. What would be the consequence of using the *customary* time-metrization everywhere on the rotating disk and letting the duration (length) of a time-interval elapsing at a given point A be given by the *difference* between the time-coordinates of the termini of that interval as furnished by the readings of the standard clock at A? The adoption of the customary time-metric would saddle us with a *most complicated* description of the propagation of light in the rotating system having the following undesirable features: (i) time would enter the description of nature explicitly in the sense that the one-way velocity of light would depend on the time, since the lagging *rate* of the clock at A issues in a temporal change in the magnitude of the one-way transit-time of a light ray for journeys between O and A, and (ii) the number of light waves emitted at A during a unit of time on the A-clock is *greater* than the number of waves arriving at the center O in one unit of time on the O-clock.[6] To avoid the undesirably complicated laws entailed by the use of the simple customary definition of time congruence, the GTR jettisoned the latter. In its stead, it adopted the following more complicated, non-customary congruence definition for the sake of the simplicity of the resulting laws: at any point A on the disk the length (duration) of a time-interval is given *not* by the *difference* between the A-clock coordinates of its termini but by the product of this increment *and* the rate factor $\frac{1}{\sqrt{1-\frac{r^2\omega^2}{c^2}}}$, which depends on the spatial coordinate r of the point A. This rate factor serves to assign a *greater* duration to time-intervals than would be obtained from the customary procedure of letting the length of time be given by the increment in the clock readings. In view of the *dependence* of the metric on the *spatial position r,* via the rate factor entering into it, we are confronted here with a *non*-customary time-metrization fully as consonant with the temporal *order* of the events at A as is the customary metric.

A similarly non-standard time-metric is used by Einstein in his tentative GTR paper of 1911[7] in treating the effect of gravitation on the propagation of light. Analysis shows that the very same complexities in the description of light propagation which are encountered on the rotating disk arise here as well, if the standard time-metric is used. These complexities are eliminated here in quite analogous fashion by the use of a non-customary time-metric. Thus, if we are concerned with light emitted on

---

[6] C. Møller: *The Theory of Relativity* (Oxford: Oxford University Press; 1952), pp. 225–26.

[7] A. Einstein: "On the Influence of Gravitation on the Propagation of Light," in *The Principle of Relativity*, a collection of memoirs (New York: Dover Publications, Inc.; 1953), Sec. 3.

the sun and reaching the earth and if "$-\Phi$" represents the negative difference in gravitational potential between the sun and the earth, then we proceed as follows: prior to being brought from the earth to the sun, a clock is set to have a rate *faster* than that of an adjoining terrestrial clock by a factor of $\frac{1}{1-\frac{\Phi}{c^2}}$ (to a first approximation), where $\frac{\Phi}{c^2} < 1$.

Chapter 3

CRITIQUE OF REICHENBACH'S AND CARNAP'S PHILOSOPHY OF GEOMETRY

(A) THE STATUS OF "UNIVERSAL FORCES"

In *Der Raum*,[1] Carnap begins his discussion of *physical* space by inquiring whether and how a line in this space can be identified as straight. Arguing from *testability* and not, as we did in Chapter One, from the continuity of that manifold, he answers this inquiry as follows: "It is impossible in principle to ascertain this, if one restricts oneself to the unambiguous deliverances of experience and does not introduce freely chosen conventions in regard to objects of experience."[2] And he then points out that the most important convention relevant to whether certain physical lines are to be regarded as straights is the specification of the metric ("Mass-setzung"), which is conventional because it could "never be either confirmed or refuted by experience." Its statement takes the following form: "A particular body and two fixed points on it are chosen, and it is then agreed what length is to be assigned to the interval between these points under various conditions (of temperature, position, orientation, pressure, electrical charge, etc.). An example of the choice of a metric is the stipulation that the two marks on the Paris standard meter bar define an interval of 100 · f(T; ϕ, λ, h; . . .) cm; . . . a

[1] R. Carnap: *Der Raum* (Berlin: Reuther and Reichard; 1922), p. 33.
[2] *Ibid.*

unit must also be chosen, but that is not our concern here which is with the choice of the body itself and with the function f(T, . . .)."[3]

Once a particular function f has been chosen, the coincidence behavior of the selected transported body permits the determination of the metric tensor g_{ik} appropriate to that choice, thereby yielding a *congruence* class of intervals and the associated geometry. Accordingly, Carnap's thesis is that the question as to the geometry of physical space is indeed an *empirical* one but subject to an important proviso: it becomes empirical only *after* a physical definition of congruence for line segments has been given *conventionally* by stipulating (to within a constant factor depending on the choice of unit) what length is to be assigned to a transported solid rod in different positions of space.

Like Carnap, Reichenbach invokes testability[4] to defend this qualified empiricist conception of geometry and speaks of "the relativity of geometry"[5] to emphasize the dependence of the geometry on the definition of congruence. Carnap had lucidly conveyed the conventionality of congruence by reference to our freedom to choose the function f in the metric. But Reichenbach couches this conception in *metaphorical* terms by speaking of "universal forces"[6] whose metrical "effects" on measuring rods are then said to be a matter of convention as follows: the customary definition of congruence in which the rod is held to be of equal length everywhere (*after* allowance for substance-specific thermal effects and the like) corresponds to equating the universal forces to zero; on the other hand, a non-customary definition of congruence, according to which the length of the rod varies with position or orientation (even *after* allowance for thermal effects, etc.), corresponds to assuming an appropriately specified non-vanishing universal force whose mathematical characterization will be explained below. Reichenbach did not anticipate that this metaphorical encumbrance of his formulation would mislead some people into making the ill-conceived charge that non-customary definitions of congruence

---

[3] *Ibid.*, pp. 33–34.

[4] H. Reichenbach: *The Philosophy of Space and Time, op. cit.*, p. 16.

[5] *Ibid.*, Sec. 8.

[6] *Ibid.*, Secs. 3, 6, 8.

are based on *ad hoc* invocations of universal forces. Inasmuch as this charge has been leveled against the conventionality of congruence, it is essential that we now divest Reichenbach's statement of its misleading potentialities.

Reichenbach[7] invites consideration of a large hemisphere made of glass which merges into a huge glass plane, as shown in cross-section by the surface G in the diagram, which consists of a plane with a hump. Using solid rods, human beings on this surface would readily determine it to be a Euclidean plane with a central hemispherical hump. He then supposes an opaque

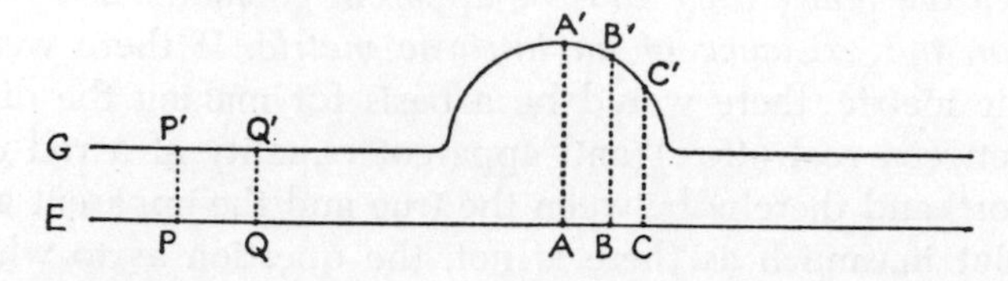

plane E to be located below the surface G as shown in the accompanying figure. Vertical light rays incident upon G will cast shadows of all objects on that glass surface onto E. As measured by *actual* solid rods, G-people will find A′B′ and B′C′ to be equal, while their projections AB and BC on the Euclidean plane E would be unequal. Reichenbach now wishes to prepare the reader for the recognition of the conventionality of congruence by having him deal with the following kind of question. Might it not be the case that:

(1) the *inequality* of AB and BC is only *apparent*, these intervals and other projections like them in the region R of E under the hemisphere being *really* equal, so that the *true* geometry of the surface E is *spherical* in R and Euclidean only outside it,

(2) the equality of A′B′ and B′C′ is only *apparent*, the *true* geometry of surface G being plane Euclidean *throughout*, since in the apparently hemispherical region R′ of G *real* equality obtains among those intervals which are the upward vertical projections of E-intervals in R that are equal in the *customary* sense of our daily life, and

(3) on each of the two surfaces, transported measuring rods respectively fail to coincide with really equal intervals in R and

[7] *Ibid.*, Sec. 3.

R′ respectively, because they do not remain truly congruent to themselves under transport, being deformed under the influence of undetectable forces which are "universal" in the sense that (a) they affect all materials alike, and (b) they penetrate all insulating walls?

On the basis of the conceptions presented in Chapter One, *which involve no kind of reference to universal forces,* one can fulfill Reichenbach's desire to utilize this question as a basis for expounding the conventionality of congruence by presenting the following considerations. The legitimacy of making a distinction between the real (true) and the apparent geometry of a surface *turns on the existence of an intrinsic metric.* If there were an intrinsic metric, there would be a basis for making the distinction between real (true) and apparent equality of a rod under transport, and thereby between the true and the apparent geometry. But inasmuch as there is not, the question as to whether a given surface is really a Euclidean plane with a hemispherical hump or only apparently so must be *replaced* by the following question: on a *particular convention of congruence* as specified by a choice of one of Carnap's functions f, does the coincidence behavior of the transported rod on the surface in question yield the geometry under discussion or not?

Thus the question as to the geometry of a surface is inherently ambiguous without the introduction of a congruence definition. And in view of the conventionality of spatial congruence, we are entitled to metrize G and E *either* in the customary way *or* in other ways so as to describe E as a Euclidean plane with a hemispherical hump R in the center and G as a Euclidean plane throughout. To assure the correctness of the latter *non*-customary descriptions, we need only *decree* the congruence of those respective intervals which our questioner called "really equal" as opposed to apparently equal in parts (1) and (2) in his question respectively. Accordingly, without the presupposition of an intrinsic metric there can be no question of an absolute or "real" deformation of *all* kinds of measuring rods alike under the action of universal forces, and, *mutatis mutandis,* the same statement applies to clocks. Since a rod undergoes no kind of objective physical change in the supposed "presence" of universal forces, that "presence" signifies no more than that we assign a different length to it in different positions or orientations by convention. Hence, just as the conversion of the length of a table from meters to feet does not involve the action of a force on the table as the "cause" of the change, so also reference to universal forces as "causes" of "changes" in the transported rod can have no literal but only *metaphorical* significance. Moreover, mention of universal forces is an entirely dispensable mode of speech in this context as is evident from the fact that the rule assigning to the transported rod lengths which vary with its position and orientation can be given by specifying Carnap's function f.

Reichenbach, on the other hand, chooses to formulate the conventionality of congruence by first distinguishing between what he calls "differential" and "universal" forces and then using "universal forces" *metaphorically* in his statement of the epistemological status of the metric. By "differential" forces[8] he means thermal and other influences which we called "perturbational" in Chapter One and whose presence is distorting by issuing in the dependence of the coincidence behavior of transported rods on the latter's chemical composition. Since we conceive of physical geometry as the system of metric relations which are *independent* of chemical composition, we correct for the substance-specific deformations induced by differential forces.[9] Reichenbach defines "universal forces" as having the twin properties of affecting all materials in the same way and as all-permeating because there are no walls capable of providing insulation against them. There is precedent for a literal rather than metaphorical use of universal forces to give a congruence definition: in order to provide a *physical realization* of a non-customary congruence definition which would metrize the interior of a sphere of radius R so as to be a model of an infinite 3-dimensional hyperbolic space, Poincaré[1] postulates that (a) each concentric sphere of radius $r < R$ is held at a constant

---

[8] *Ibid.*

[9] *Ibid.*, p. 26; H. P. Robertson: "Geometry as a Branch of Physics," *Albert Einstein: Philosopher-Scientist*, ed. by P. A. Schilpp (Evanston: Library of Living Philosophers; 1949), pp. 327–29, and "The Geometries of the Thermal and Gravitational Fields," *American Mathematical Monthly*, Vol. LVII (1950), pp. 232–45; and E. W. Barankin: "Heat Flow and Non-Euclidean Geometry," *American Mathematical Monthly*, Vol. XLIX (1942), pp. 4–14.

[1] H. Poincaré: *The Foundations of Science, op. cit.*, pp. 75–77.

absolute temperature $T \propto (R^2 - r^2)$, while the optical index of refraction is inversely proportional to $R^2 - r^2$, and (b) *contrary to actual fact,* all kinds of bodies within the sphere have the *same* coefficient of thermal expansion. It is essential to see that the expansions and contractions of these bodies under displacement have a *literal* meaning in this context, because they are *relative* to the actual displacement-behavior of our normally Euclidean kinds of bodies and are linked to thermal sources.[2]

But Reichenbach's *metaphorical* use of universal forces for giving the congruence definition and exhibiting the dependence of the geometry on that definition takes the following form: "Given a geometry G′ to which the measuring instruments conform [*after* allowance for the effects of thermal and other 'differential' influences], we can imagine a universal force F which affects the instruments in such a way that the actual geometry is an arbitrary geometry G, while the observed deviation from G is due to a universal deformation of the measuring instruments."[3] And he goes on to say that if g'_{ik} $(i = 1, 2, 3; k = 1, 2, 3)$ are the empirically obtained metrical coefficients of the geometry G′ and g_{ik} those of G, then the force tensor F is given mathematically by the tensor equation

$$g'_{ik} + F_{ik} = g_{ik},$$

where the g'_{ik}, which had yielded the observed geometry G′, are furnished experimentally by the measuring rods[4] and where the F_{ik} are the "correction factors" $g_{ik} - g'_{ik}$ which are added correctionally to the g'_{ik} so that the g_{ik} are obtained.[5] But since

[2] A precisely analogous *literal* use of universal forces is also made by Reichenbach [H. Reichenbach: *The Philosophy of Space and Time, op. cit.,* pp. 11–12] to convey pictorially a physical realization of a congruence definition which would confer a spherical geometry on the region R of the surface E discussed above.

[3] H. Reichenbach: *The Philosophy of Space and Time, op. cit.,* p. 33.

[4] For details on this experimental procedure, see, for example, *ibid.,* Secs. 39 and 40.

[5] We shall see in Section (ii) of this chapter that Reichenbach was *mistaken* in asserting [*ibid.,* pp. 33-34] that for a given surface or 3-space a particular metric geometry *determines* (1) a *unique* definition of congruence and, once a unit of length has been chosen, (2) a *unique* set of functions g_{ik} as the representations of the metric tensor in any particular coordinate system. It will turn out that there are infinitely many *incompatible* congruence definitions and as many correspondingly different metric tensors

Reichenbach emphasizes that it is a matter of *convention* whether we equate F_{ik} to zero or not,[6] this formulation is merely a *metaphorical* way of asserting that the following is a matter of convention: whether congruence is said to obtain among intervals having equal lengths ds given by the metric $ds^2 = g'_{ik}dx^i dx^k$ —which entails G′ as the geometric description of the observed coincidence relations—or among intervals having equal lengths ds given by the metric $ds^2 = g_{ik}dx^i dx^k$ which yields a different geometry G.[7] Clearly then, to equate the universal forces to zero is merely to choose the metric based on the tensor g'_{ik} which was obtained from measurements in which the rod was called congruent to itself everywhere. In other words, to stipulate $F_{ik} = 0$ is to choose the customary congruence standard based on the rigid body.

---

which impart the *same* geometry to physical space. Hence, while a given metric tensor yields a unique geometry, a geometry G does not determine a metric tensor *uniquely* to within a constant factor depending on the choice of unit. And thus it is incorrect for Reichenbach to speak here of the components g_{ik} of a particular metric tensor as "those of G" [*ibid.*, p. 33n] and to suppose that a unique F is specified by the requirement that a certain geometry G prevail in place of the observed geometry G′.

[6] H. Reichenbach: *The Philosophy of Space and Time, op. cit.*, pp. 16, 27–28, 33.

[7] It is to be clearly understood that the g_{ik} yield a congruence relation incompatible with the one furnished by the g'_{ik}, because in any given coordinate system they are different functions of the given coordinates and not proportional to one another. (A difference consisting in a mere proportionality could not involve a difference in the congruence classes but only in the unit of length used.) The incompatibility of the congruences furnished by the two sets of metric coefficients is a necessary though not a sufficient condition (cf. the preceding footnote 5) for the non-identity of the associated geometries G and G′.

The difference between the two metric tensors corresponding to *incompatible* congruences must not be confounded with a mere difference in the *representations* in different coordinate systems of the *one* metric tensor corresponding to a *single* congruence criterion (for a given choice of a unit of length): the former is illustrated by the *incompatible* metrizations $ds^2 = dx^2 + dy^2$ and $ds^2 = (dx^2 + dy^2)/y^2$ in which the corresponding metric coefficients are suitably *different* functions of the *same* rectangular coordinates, whereas the latter is illustrated by using first rectangular and then polar coordinates to express the *same metric* as follows: $ds^2 = dx^2 + dy^2$ and $ds^2 = d\rho^2 + \rho^2 d\theta^2$. In the latter case, we are not dealing with different *metrizations* of the space but only with different *coordinatizations* (parametrizations) of it, at least one pair of corresponding metric coefficients having members which are *different* functions of their respective coordinates but so chosen as to yield an *invariant* ds.

On the other hand, apart from one exception to be stated presently, to stipulate that the components F_{ik} may not all be zero is to adopt a non-customary metric given by a tensor g_{ik} corresponding to a specified variation of the length of the rod with position or orientation.

That there is one exception, which, incidentally, Reichenbach does not discuss, can be seen as follows: a given congruence determines the metric tensor up to a constant factor depending on the choice of unit and conversely. Hence two metric tensors will correspond to different congruences if and only if they differ other than by being proportional to one another. Thus, if g_{ik} and g'_{ik} are proportional by a factor different from 1, these two tensors furnish metrics differing only in the choice of unit and hence yield the same congruence. Yet in the case of such proportionality of g_{ik} and g'_{ik}, the F_{ik} *cannot* all be zero. For example, if we consider the line element

$$ds^2 = a^2 d\phi^2 + a^2 \sin^2\phi \, d\theta^2$$

on the surface of a sphere of radius a = 1 meter = 100 cm, the mere change of units from meters to centimeters would change the metric $ds^2 = d\phi^2 + \sin^2\phi \, d\theta^2$ into

$$ds^2 = 10000 \, d\phi^2 + 10000 \sin^2\phi \, d\theta^2.$$

And if these metrics are identified with g'_{ik} and g_{ik} respectively, we obtain

$$\begin{aligned} F_{11} &= g_{11} - g'_{11} = 10000 - 1 = 9999 \\ F_{12} &= F_{21} = g_{12} - g'_{12} = 0 \\ F_{22} &= g_{22} - g'_{22} = 10000 \sin^2\phi - \sin^2\phi = 9999 \sin^2\phi. \end{aligned}$$

It is now apparent that the F_{ik}, being given by the differences between the g_{ik} and the g'_{ik}, will not all vanish *and* that also these two metric tensors will yield the *same* congruence, if and only if these tensors are proportional by a factor different from 1. Therefore, a necessary and sufficient condition for obtaining *incompatible* congruences is that there be at least *one* non-vanishing component F_{ik} such that the metric tensors g_{ik} and g'_{ik} are not proportional to one another. The exception to our statement that a non-customary congruence definition is assured by the failure of at least one component of the F_{ik} to vanish is therefore given by the case of the proportionality of the metric tensors.[8]

Although Reichenbach's metaphorical use of "universal forces" issued in misleading and wholly unnecessary complexities which we shall point out presently, he himself was in no way victimized by them. Writing in 1951 concerning the invocation of universal forces in this context, he declared: "The assumption of such forces means merely a change in the coordinative definition of congruence."[9] It is therefore quite puzzling that, in 1956, Carnap, who had lucidly expounded the same ideas *non*-metaphorically as we saw, singled out Reichenbach's characterization and recommendation of the customary congruence definition in terms of equating the universal forces to zero as a praiseworthy part of Reichenbach's outstanding work *The Philosophy of Space and Time.* In his Preface to the latter work, Carnap says:

> Of the many fruitful ideas which Reichenbach contributed . . . I will mention only one, which seems to me of great interest for the methodology of physics but which has so far not found the attention it deserves. This is the principle of the elimination of universal forces. . . . Reichenbach proposes to accept as a general methodological principle that we choose that form of a theory among physically equivalent forms (or, in other words, that definition of "rigid body" or "measuring standard") with respect to which all universal forces disappear.[1]

The misleading potentialities of reference to metaphorical "universal forces" in the statement of the congruence definition and in related contexts manifest themselves in three ways as follows:

(1) The formulation of non-customary congruence definitions in terms of deformations by universal forces has inspired the erroneous charge that such congruences are *ad hoc* because they

[8] There is a most simple illustration of the fact that if the metric tensors are *not* proportional such that as few as *one* of the components F_{ik} is *non*-vanishing, the congruence associated with the g_{ik} will be incompatible with that of the g'_{ik} and will hence be non-customary. If we consider the metrics $ds^2 = dx^2 + dy^2$ and $ds^2 = 2dx^2 + dy^2$, then comparison of an interval for which $dx = 1$ and $dy = 0$ with one for which $dx = 0$ and $dy = 1$ will yield congruence on the first of these metrics but not on the second.

[9] H. Reichenbach: *The Rise of Scientific Philosophy* (Berkeley: University of California Press; 1951), p. 133.

[1] R. Carnap: Preface to H. Reichenbach's *The Philosophy of Space and Time* (New York: Dover Publications, Inc.; 1957), p. 7.

allegedly involve the *ad hoc* postulation of non-vanishing universal forces.

(2) In Reichenbach's statement of the congruence definition to be employed to explore the spatial geometry in a *gravitational* field, universal forces enter in both a *literal* and a *metaphorical* sense. The conflation of these two senses issues in a seemingly contradictory formulation of the customary congruence definition.

(3) Since the variability of the curvature of a space would manifest itself in alterations in the coincidence behavior of all kinds of solid bodies under displacement, Reichenbach speaks of bodies displaced in such a space as being subject to universal forces "destroying coincidences."[2] In a manner analogous to the gravitational case, the conflation of this literal sense with the metaphorical one renders the definition of rigidity for this context paradoxical.

We shall now briefly discuss these three sources of confusion in turn.

1. If a congruence definition itself had factual content, so that alternative congruences would differ in factual content, then it would be significant to say of a congruence definition that it is *ad hoc* in the sense of being an *evidentially-unwarranted* claim concerning facts. But inasmuch as ascriptions of spatial congruence to non-coinciding intervals are not factual but conventional, neither the customary nor any of the non-customary definitions of congruence can possibly be *ad hoc*. Hence the abandonment of the former in favor of the latter kind of definition can be no more *ad hoc* than the regraduation of a Centigrade thermometer into a Fahrenheit one or than the change from Cartesian to polar coordinates. By formulating non-customary congruence definitions in terms of the metaphor of universal forces, Reichenbach made it possible for his metaphorical sense to be misconstrued as literal. And once this error had been committed, its victims tacitly regarded the customary congruence definition as factually true and felt justified in dismissing other congruences as *ad hoc* on the grounds that they involved the *ad hoc* assumption of (literally conceived) universal forces. As well say that a change of the units of length is *ad hoc*.

---

[2] H. Reichenbach: *The Philosophy of Space and Time, op. cit.*, p. 27.

Thus we find that Ernest Nagel, for example, overlooks in writing about Poincaré that the invocation of universal forces to preserve Euclidean geometry can be no more *ad hoc* than a change from rectangular to polar coordinates in order to write the equation of a circle as $\rho = k$ instead of $\sqrt{x^2 + y^2} = k$. After granting that, if necessary, Euclidean geometry can be retained by an appeal to universal forces, Nagel writes: "Nevertheless, universal forces have the curious feature that their presence can be recognized *only* on the basis of geometrical considerations. The assumption of such forces thus has the appearance of an *ad hoc* hypothesis, adopted solely for the sake of salvaging Euclid."[3] But it is only Nagel's and not Poincaré's construal of the character of the relevant kind of universal forces which permits Nagel's conclusion that Poincaré must have recourse to some sort of *ad hoc* hypothesis in order to assure a Euclidean description of the posited observational facts. For within the framework of a physical theory which assumes space to be a mathematical continuum—an assumption on which Poincaré's entire thesis was predicated—there is no basis whatever for leveling an *ad hoc* charge against Poincaré. The invocation of the kind of universal force whose "presence can be recognized *only* on the basis of geometrical considerations" and which is introduced "solely for the sake of salvaging Euclid" can be no more *ad hoc* than the use of polar rather than rectangular coordinates. Nagel's concern that Poincaré's thesis can be upheld only on pain of an *ad hoc* assumption of universal forces is just as unwarranted as the following supposition: the numerical increase in the lengths of all objects produced by the conversion from meters to inches requires the *ad hoc* postulation of universal forces as the "physical cause" of the universal elongation.

2. Regarding the geometry in a gravitational field, Reichenbach says the following: "We have learned . . . about the difference between *universal* and *differential* forces. These concepts have a bearing upon this problem because we find that gravitation is a universal force. It does indeed affect all bodies in the same manner. This is the physical significance of the equality of gravitational and inertial mass."[4] It is entirely correct, of course,

---

[3] E. Nagel: *The Structure of Science* (New York: Harcourt, Brace and World; 1961), p. 264.

[4] H. Reichenbach: *The Philosophy of Space and Time, op. cit.*, p. 256.

that a uniform gravitational field (which has not been transformed away in a given space-time coordinate system) is a universal force in the *literal* sense *with respect to a large class of effects* such as the free fall of bodies. But there are other effects, such as the bending of elastic beams, with respect to which gravity is clearly a *differential* force in Reichenbach's sense: a wooden book shelf will sag more under gravity than a steel one. And this shows, incidentally, that *Reichenbach's classification of forces into universal and differential is not mutually exclusive.* Of course, just as in the case of any other force having differential effects on measuring rods, allowance is made for differential effects of gravitational origin in laying down the congruence definition.

The issue is therefore twofold: first, does the fact that gravitation is a universal force in the *literal* sense indicated above have a bearing on the spatial geometry, and second, in the presence of a gravitational field is the logic of the spatial congruence definition any different in regard to the role of *metaphorical* universal forces from what it is in the absence of a gravitational field? Within the particular context of the GTR, there is indeed a *literal* sense in which the gravitational field of the sun, for example, is causally relevant geometrically as a universal force. And the literal sense in which the coincidence behavior of the transported customary rigid body is objectively different in the vicinity of the sun, for example, from what it is in the absence of a gravitational field can be expressed in two ways as follows: (1) *relatively to the congruence defined by the customary rigid body,* the spatial geometry in the gravitational field is *not* Euclidean—contrary to pre-GTR physics—but is Euclidean in the absence of a gravitational field; (2) the geometry in the gravitational field is Euclidean if and only if the customary congruence definition is supplanted by one in which the length of the rod varies *suitably* with its position or orientation,[5] whereas it is Euclidean relatively to the customary congruence definition for a vanishing gravitational field. It will be noted, however, that formulation (1) makes no mention at all of any

[5] For the gravitational field of the sun, the function specifying a non-customary congruence definition issuing in a Euclidean geometry is given in R. Carnap: *Der Raum, op. cit.*, p. 58.

deformation of the rod by universal forces as that body is transported from place to place in a given gravitational field. Nor need there be any metaphorical reference to universal forces in the statement of the customary congruence definition ingredient in formulation (1). For that statement can be given as follows: in the presence no less than in the absence of a gravitational field, congruence is conventional, and hence we are free to adopt the customary congruence in a gravitational field as a basis for determining the spatial geometry. By encumbering his statement of a congruence definition with metaphorical use of "universal forces," Reichenbach enables the unwary to infer incorrectly that a rod subject to the universal force of gravitation in the specified *literal* sense *cannot* consistently be regarded as *free* from deforming universal forces in the *metaphorical* sense and hence cannot serve as the congruence standard. This conflation of the literal and metaphorical senses of "universal force" in the context of a theory which assumes the continuity of space thus issues in the mistaken belief that in the GTR the customary spatial congruence definition cannot be adopted consistently for the gravitational field. And those who were led to this misconception by Reichenbach's metaphor will therefore deem self-contradictory the following consistent assertion by him: "We do not speak of a change produced by the gravitational field in the measuring instruments, but regard the measuring instruments as 'free from deforming forces' in spite of the gravitational effects."[6] Moreover, those victimized by the metaphorical part of Reichenbach's language will be driven to reject as inconsistent Einstein's characterization of the geometry in the gravitational field in the GTR as given in formulation (1) above. And they will insist erroneously that formulations (1) and (2) are *not* equally acceptable alternatives on the grounds that formulation (2) is uniquely correct.

The confounding of the literal and metaphorical senses of "universal force" by reference to gravity in a theoretical context which presupposes the continuity of space is present, for example, in Ernest Nagel's treatment of universal forces with resulting potentialities of confusion. Thus, he incorrectly cites the force

---

[6] H. Reichenbach: *The Philosophy of Space and Time, op. cit.*, p. 256.

of gravitation in its role of being *literally* a "universal force" as a *species* of what are only *metaphorically* "universal forces" within a theory that assumes the continuity of physical space. Specifically, in speaking of Poincaré's assumption of universal forces whose "presence can be recognized *only* on the basis of geometrical considerations" because they are assumed "solely for the sake of salvaging Euclid"—i.e., "universal forces" in the *metaphorical* sense—Nagel says: " 'Universal force' is not to be counted as a 'meaningless' phrase, for it is evident that a procedure is indicated for ascertaining whether such forces are present or not. Indeed, gravitation in the Newtonian theory of mechanics is just such a universal force; it acts alike on all bodies and cannot be screened."[7] But Nagel's misidentification of Newtonian gravity as "just such a universal force" as the metaphorical kind of universal force whose "presence can be recognized *only* on the basis of geometrical considerations" can lead to the incorrect conclusion that a rod subject to a Newtonian gravitational field *cannot* be held to be "free" from "universal forces" in the sense of being self-congruent when performing its metrical function under transport.

Precisely the latter kind of error was committed by P. K. Feyerabend, who discusses the bearing of a hypothetical universal force of the *literal* kind on the metrical function of a transported rod. For Feyerabend supposes mistakenly that the rod must be held to be *distorted* in the presence of universal forces of the *literal* kind which "act upon all chemical substances in a similar way but which make themselves noticed in a slight change induced in the transitions probabilities of atoms radiating in that region."[8]

3. In a manner analogous to the gravitational case just discussed, we can assert the following: since congruence is conventional, we are at liberty to use the customary definition of it without regard to whether the geometry obtained by measurements expressed in terms of that definition is one of *variable*

[7] E. Nagel: *The Structure of Science* (New York: Harcourt, Brace and World; 1961), p. 264, n. 19.

[8] Cf. P. K. Feyerabend: "Comments on Grünbaum's 'Law and Convention in Physical Theory,' " in H. Feigl and G. Maxwell (eds.) *Current Issues in the Philosophy of Science* (New York: Holt, Rinehart and Winston; 1961), p. 157, and A. Grünbaum: "Rejoinder to Feyerabend," *ibid.*, pp. 164–67.

curvature or not. And thus we see that upon avoiding the intrusion of a metaphorical use of "universal forces," the statement of the congruence definition need not take any cognizance of whether the resulting geometry will be one of constant curvature or not. A geometry of constant curvature or so-called "congruence geometry" is characterized by the fact that the so-called "axiom of free mobility" holds in it: for example, on the surface of a sphere, a triangle having certain angles and sides in a given place can be moved about without any change in their magnitudes relatively to the customary standards of congruence for angles and intervals. By contrast, on the surface of an egg, the failure of the axiom of free mobility to hold can be easily seen from the following indicator of the variability of the curvature of that 2-space: a circle and its diameter made of any kind of wire are so constructed that one end P of the diameter is attached to the circle while the other end S is free *not* to coincide with the opposite point Q on the circle though coinciding with it in a given initial position on the surface of the egg. Since the ratio of the diameter and the circumference of a circle *varies* in a space of variable curvature such as that of the egg-surface, S will no longer coincide with Q if the circular wire and its attachment PS are moved about on the egg so as to preserve contact with the egg-surface everywhere. The indicator thus exhibits an objective destruction of the coincidence of S and Q which is wholly independent of the indicator's chemical composition (under uniform conditions of temperature, etc.).

One may therefore speak *literally* here, as Reichenbach does,[9] of universal forces acting on the indicator by *destroying coincidences*. And since the customary congruence definition is entirely permissible as a basis for geometries of variable curvature, there is, of course, no inconsistency in giving that congruence definition by equating universal forces to *zero* in the *metaphorical* sense, even though the destruction of coincidences attests to the quite *literal* presence of causally efficacious universal forces. But Reichenbach invokes universal forces *literally* without any warning of an impending metaphorical reference to them in the congruence definition. And the reader is therefore both startled and

[9] H. Reichenbach: *The Philosophy of Space and Time, op. cit.*, Sec. 6.

puzzled by the seeming paradox in Reichenbach's declaration that: "Forces *destroying coincidences* must also be set equal to zero, if they satisfy the properties of the universal forces mentioned on p. 13; only then is the problem of geometry uniquely determined."[1] Again the risk of confusion can be eliminated by dispensing with the metaphor in the congruence definition.

Although I believe that Reichenbach's *The Philosophy of Space and Time* is the most penetrating single book on its subject, the preceding analysis shows why I cannot share Nagel's judgment that, in that book, Reichenbach employs "The distinction between 'universal' and 'differential' forces . . . with great clarifying effect."[2]

Having divested Reichenbach's statements about universal forces of their misleading potentialities, we shall hereafter safely be able to discuss other issues raised by statements of his which are couched in terms of universal forces.

The first of these is posed by the following assertion by him: "We obtain a statement about physical reality only if in addition to the geometry G of the space its universal field of force F is specified. Only the combination

$$G + F$$

is a testable statement."[3] In order to be able to appraise this assertion, consider a surface on which some set of generalized curvilinear (or "Gaussian") coordinates has been introduced. The coordinatization of a space, whose purpose it is merely to number points so as to convey their topological neighborhood relations of betweenness, does not, *as such*, presuppose (entail) a metric. However, the statement of a *rule* assuring that different people will independently effect the *same* coordinatization of a given space may require reference to the use of a rod. But even in the case of coordinates such as rectangular (Cartesian) coordinates, whose assignment is carried out by the use of a rigid rod, it is quite possible to ignore the manner in which the coordinatization was effected and to regard the coordinates purely topologically, so that a metric very *different* from $ds^2 = dx^2 + dy^2$ can then still be introduced quite consistently. Accordingly, now put a metric

[1] *Ibid.*, p. 27.
[2] E. Nagel: *The Structure of Science, op. cit.*, p. 264, n. 18.
[3] H. Reichenbach: *The Philosophy of Space and Time, op. cit.*, p. 33.

$ds^2 = g_{ik}dx^idx^k$ onto the coordinated surface quite arbitrarily by a capricious choice of a suitable set of functions g_{ik} of the given coordinates. Assume that the latter specification of the geometry G is not coupled with any information regarding F.

Is it then correct to say that since this metrization provides no information at all about the coincidence behavior of a rod under transport on the surface, it conveys no factual information whatever about the surface or physical reality? That such an inference is mistaken can be seen from the following: depending upon whether the Gaussian curvature K associated with the stipulated g_{ik} is positive (spherical geometry), zero (Euclidean geometry), or negative (hyperbolic geometry), it is an objective fact about the surface that through a point outside a given geodesic of the chosen metric, there will be respectively 0, 1, or infinitely many other such geodesics which will not *intersect* the given geodesic. Whether or not certain lines on a surface intersect, is however merely a *topological* fact concerning it. And hence we can say that although an arbitrary metrization of a space without a specification of F is not altogether devoid of factual content pertaining to that space, such a metrization can yield no objective facts concerning the space not already included in the latter's topology.

We can therefore conclude the following: if the description of a space (surface) is to contain empirical information concerning the coincidence behavior of transported rods in that space and if a metric $ds^2 = g_{ik}dx^idx^k$ (and thereby a geometry G) is chosen whose congruences do not accord with those defined by the application of the transported rod, then indeed Reichenbach's assertion holds. Specifically, the chosen metric tensor g_{ik} and its associated geometry G must then be coupled with a specification of the different metric tensor g'_{ik} that would have been found experimentally, if the rod had actually been chosen as the congruence standard. But Reichenbach's provision of that specification via the universal force F is quite unnecessarily roundabout. For F is defined by $F_{ik} = g_{ik} - g'_{ik}$ and cannot be known without already knowing *both* metric tensors. Thus, there is a loss of clarity in pretending that the metric tensor g'_{ik}, which codifies the empirical information concerning the rod, is first obtained from the identity

$$g'_{ik} = g_{ik} - F_{ik}.$$

(B) THE "RELATIVITY OF GEOMETRY"

In order to emphasize the dependence of the metric geometry on the definition of congruence, Reichenbach speaks of "the relativity of geometry." But there is an important error in his characterization of this dependence, which takes the following forms: "If we change the coordinative definition of congruence, a different geometry will result. This fact is called the *relativity of geometry*,"[4] and, more explicitly, "There is nothing wrong with a coordinative definition established on the requirement that a certain kind of geometry is to result from the measurements. . . . A coordinative definition can also be introduced by the prescription what the result of the measurements is to be. 'The comparison of length is to be performed in such a way that Euclidean geometry will be the result'—this stipulation is a possible form of a coordinative definition."[5] Reichenbach's claim is that a given metric geometry *uniquely* determines a congruence class (or congruence definition) appropriate to it.

That this contention is mistaken will now be demonstrated: we shall show that besides the customary definition of congruence, which assigns the same length to the measuring rod everywhere and thereby confers a Euclidean geometry on an ordinary tabletop, there are infinitely many *other* definitions of congruence having the following property: they likewise yield a Euclidean geometry for that surface but are incompatible with the customary one by making the length of a rod depend on its orientation and/or position.

Thus, consider our horizontal tabletop equipped with a network of Cartesian coordinates x and y, but now metrize this surface by means of the *non*-standard metric

$$ds^2 = \sec^2\theta \, dx^2 + dy^2,$$

where $\sec^2\theta$ is a *constant greater than 1.* Unlike the standard metric, this metric assigns to an interval whose coordinates differ by dx *not* the length dx but the *greater* length $\sec\theta \, dx$ while continuing to assign the length dy to an interval whose coordinates differ only by dy. Although this metric thereby makes the

---

[4] H. Reichenbach: *The Rise of Scientific Philosophy, op. cit.*, p. 132.

[5] H. Reichenbach: *The Philosophy of Space and Time, op. cit.*, pp. 33–34.

length of a given rod dependent on its orientation, we shall show that the *infinitely* many different non-standard congruences generated by values of $\sec\theta$ which exceed 1 each impart a *Euclidean* geometry to the tabletop no less than does the standard congruence given by

$$ds^2 = dx^2 + dy^2.$$

Accordingly, our demonstration will show that the requirement of Euclideanism does *not* uniquely determine a congruence class of intervals but allows an *infinitude* of *incompatible* congruences. We shall therefore have established that there are infinitely many ways in which a measuring rod could squirm under transport on the tabletop as compared to its familiar de facto behavior while still yielding a Euclidean geometry for that surface.

To carry out the required demonstration, we first note the preliminary fact that the geometry yielded by a particular metrization is clearly independent of the particular coordinates in which that metrization is expressed. And hence if we expressed the standard metric

$$ds^2 = dx^2 + dy^2$$

in terms of the primed coordinates x′ and y′ given by the transformations

$$x = x' \sec\theta$$

$$y = y',$$

obtaining

$$ds^2 = \sec^2\theta \, dx'^2 + dy'^2$$

we would obtain a *Euclidean* geometry as before, since the latter equation would merely express the original standard metric in terms of the primed coordinates. Thus, when the same *invariant* ds of the standard metric is expressed in terms of both primed and unprimed coordinates, the metric coefficients g'_{ik} given by $\sec^2\theta$, 0 and 1 yield a Euclidean geometry no less than do the unprimed coefficients 1, 0, and 1.

This elementary ancillary conclusion now enables us to see that the following *non*-standard metrization (or remetrization) of the surface in terms of the *original*, unprimed rectangular coordinates must likewise give rise to a Euclidean geometry:

$$ds^2 = \sec^2\theta \, dx^2 + dy^2.$$

For the value of the Gaussian curvature and hence the prevailing geometry depends not on the particular coordinates (primed or unprimed) to which the metric coefficients g_{ik} pertain but only on the *functional form* of the g_{ik}[6] which is the same here as in the case of the g'_{ik} above.

More generally, therefore, the geometry resulting from the *standard* metrization is *also* furnished by the following kind of *non*-standard metrization of a space of points, expressed in terms of the same (unprimed) coordinates as the standard one: the *non*-standard metrization has unprimed metric coefficients g_{ik} which have the same *functional form* (to within an arbitrary constant arising from the choice of unit of length) as those primed coefficients g'_{ik} which are obtained by expressing the *standard* metric in some set or other of *primed* coordinates via a suitable coordinate transformation. In view of the large variety of allowable coordinate transformations, it follows at once that the class of *non*-standard metrizations yielding a Euclidean geometry for a tabletop is far wider than the already infinite class given by

$$ds^2 = \sec^2\theta \, dx^2 + dy^2, \text{ where } \sec^2\theta > 1.$$

Thus, for example, there is *identity* of *functional form* between the *standard* metric in *polar* coordinates, which is given by

$$ds^2 = d\rho^2 + \rho^2 d\theta^2,$$

and the *non*-standard metric in Cartesian coordinates given by

$$ds^2 = dx^2 + x^2 dy^2,$$

since x plays the same role *formally* as ρ, and similarly for y and θ. Consequently, the latter *non*-standard metric issues in a *Euclidean* geometry just as the former standard one does.

It is clear that the multiplicity of metrizations which we have proven for Euclidean geometry obtains as well for each of the non-Euclidean geometries. The failure of a geometry of two or more dimensions to determine a congruence definition uniquely does not, however, have a counterpart in the *one*-dimensional time-continuum: the demand that Newton's laws hold in their

[6] F. Klein: *Vorlesungen über Nicht-Euklidische Geometrie* (Berlin: Springer-Verlag; 1928), p. 281.

customary metrical form does determine a unique definition of temporal congruence. And hence it is feasible to rely on the law of translational or rotational inertia to define a time metric or "uniform time."

On the basis of this result, we can now show that a number of claims made by Reichenbach and Carnap respectively are false.

(1) In 1951, Reichenbach wrote, as will be recalled from the beginning of this Section: "If we change the coordinative definition of congruence, a different geometry will result. This fact is called the *relativity of geometry*."[7] That this statement is false is evident from the fact that if, in our example of the tabletop, we change our congruence definition from $ds^2 = dx^2 + dy^2$ to any one of the infinitely many definitions incompatible with it that are given by $ds^2 = \sec^2\theta\, dx^2 + dy^2$, precisely the same Euclidean geometry results. Thus, contrary to Reichenbach, the introduction of a non-vanishing universal force corresponding to an alternative congruence does not guarantee a change in the geometry. Instead, the correct formulation of the relativity of geometry is that in the form of the ds function the congruence definition uniquely determines the geometry, though not conversely, and that any one of the congruence definitions issuing in a geometry G′ can always be replaced by infinitely many *suitably* different congruences yielding a specified *different* geometry G. In view of the unique fixation of the geometry by the congruence definition in the context of the facts of coincidence, the repudiation of a given geometry in favor of a different one does indeed require a change in the definition of congruence. And the new congruence definition which is expected to furnish the new required geometry can do so in one of the following two ways:

(1) by determining a system of geodesics *different* from the one yielded by the original congruence definition, or

(2) if the geodesics determined by the new congruence definition are the *same* as those associated with the original definition, then the *angle* congruences must be different,[8] i.e., the new

[7] H. Reichenbach: *The Rise of Scientific Philosophy*, *op. cit.*, p. 132.

[8] The specification of the magnitudes assigned to angles by the components g_{ik} of the metric tensor was given in Chapter One, Section C.

congruence definition will have to require a different congruence class of angles.[9]

That (2) constitutes a genuine possibility for obtaining a different geometry will be evident from the following example of mapping the sphere geodesically on a plane, a mapping which issues in two *incompatible* definitions of congruence

$$ds^2_1 = g_{ik}dx^i dx^k, \text{ and}$$
$$ds^2_2 = g'_{ik}dx^i dx^k$$

that yield the *same* system of geodesics via the equations $\delta \int ds_1 = 0$ and $\delta \int ds_2 = 0$ and yet determine *different* geometries (Gaussian curvatures), because they require incompatible congruence classes of *angles* appropriate to these respective geometries. A horizontal surface which is a Euclidean plane on the *customary* metrization can alternatively be metrized to have the geometry of a hemisphere by projection from the *center* of a sphere through its *lower half* while the south pole is resting on that plane. Upon calling congruent on the horizontal surface segments and angles which are the projections of equal segments and angles respectively on the lower hemisphere, the great circle arcs of the hemisphere map into the Euclidean straight lines of the plane such that *every* straight of the Euclidean description is *also* a straight (geodesic) of the new hemispherical geometry conferred on the horizontal surface.[1] But the *angles* which are regarded as congruent on the horizontal surface in the new metrization are not congruent in the original metrization yielding a Euclidean description. The destruction by the new metrization of the angle congruences associated with the original metrization can be made apparent as follows.

Consider two triangles ABC and A′B′C′ which qualify as *similar* in the Euclidean geometry of the original metrization, so that $\measuredangle A = \measuredangle A'$, $\measuredangle B = \measuredangle B'$, and $\measuredangle C = \measuredangle C'$. Since the geodesics of

[9] For other general theorems governing the so-called "geodesic correspondence" or "geodesic mapping" relevant here, cf. L. P. Eisenhart: *An Introduction to Differential Geometry* (Princeton: Princeton University Press; 1947), Sec. 37, pp. 205–11, and D. J. Struik: *Differential Geometry, op. cit.*, pp. 177–80.

[1] The mathematical details can be found in D. J. Struik: *Differential Geometry, op. cit.*, p. 179.

the new metrization, i.e., of its associated non-Euclidean (spherical) geometry are the *same* as those of Euclidean geometry in this case, triangles ABC and A′B′C′ will still be *rectilinear* triangles on the new metrization. But since there are no similar triangles in the spherical geometry of constant positive Gaussian curvature which results from the new metrization, triangles ABC and A′B′C′ are no longer similar on the new metrization though still rectilinear. Hence the original angle congruences among these triangles can no longer obtain on the new metrization.

It must be pointed out, however, that if a change in the congruence definition *preserves* the geodesics, then its issuance in a *different* congruence class of *angles* is only a necessary and not a sufficient condition for imparting to the surface a metric geometry *different* from the one yielded by the original congruence definition. This fact becomes evident by reference to our earlier case of the tabletop's being a model of Euclidean geometry *both* on the customary metric $ds^2 = dx^2 + dy^2$ *and* on the different metric $d\bar{s}^2 = \sec^2\theta \, dx^2 + dy^2$: the geodesics as well as the geometries furnished by these incompatible metrics are the same, but the angles which are congruent in the new metric are generally not congruent in the original one. For these two metrics of the surface are not related by a conformal transformation as defined in Chapter One, Section C, and the conformal relation between the two metrics is a necessary and not only a sufficient condition for the sameness of their associated angle congruences.[2] That these two metrics issue in *incompatible* congruence classes of *angles* though in the *same* geometry can also be seen very simply as follows: a Euclidean triangle which is equilateral on the new metric $d\bar{s}$ will *not* be equilateral on the customary one ds, and hence the three angles of such a triangle will all be congruent to each other in the former metric but not in the latter.

It is clear now that an *arbitrary* change in the congruence definition for either line segments or angles or both cannot as such guarantee a different geometry.

(2) In reply to Hugo Dingler's contention[3] that the rigid body is uniquely specified by the geometry and only by the latter,

---

[2] Cf. D. J. Struik: *Differential Geometry, op. cit.*, pp. 169–70.

[3] H. Dingler: "Die Rolle der Konvention in der Physik," *Physikalische Zeitschrift*, Vol. XXIII (1922), p. 50.

Reichenbach mistakenly agrees[4] that the geometry is sufficient to define congruence and contests only Dingler's further claim that it is necessary.[5]

Carnap[6] discusses the dependencies obtaining between (a) the metric geometry, which he symbolizes by "R" in this German publication; (b) the topology of the space and the facts concerning the coincidences of the rod in it, symbolized by "T" for "Tatbestand"; and (c) the metric M ("Mass-setzung"), which entails a congruence definition and is given by the function f (and by the choice of a unit), as will be recalled from the beginning of this chapter.[7] And he concludes that the functional relations between R, M, and T are such "that if two of them are given, the third specification is thereby uniquely given as well."[8] Accordingly, he writes:

$$R = \Phi_1 (M, T)$$
$$M = \Phi_2 (R, T)$$
$$T = \Phi_3 (M, R).$$

While the first of these dependencies does hold, our example of imparting a Euclidean geometry to a tabletop by each of two incompatible congruence definitions shows that not only the second but also the third of Carnap's dependencies fails to hold. For the mere specification of M to the effect that the rod will be called congruent to itself everywhere and of R as Euclidean does not tell us whether the coincidence behavior T of the rod on the tabletop will be such as to coincide successively with those intervals that are equal according to the formula

---

[4] H. Reichenbach: "Discussion of Dingler's Paper," *Physikalische Zeitschrift*, Vol. XXIII (1922), p. 52, and H. Reichenbach: "Über die physikalischen Konsequenzen der relativistischen Axiomatik," *Zeitschrift für Physik*, Vol. XXXIV (1925), p. 35.

[5] In Chapter Four we shall assess the merits of Reichenbach's denial that the geometry is necessary, which he rests on the grounds that rigidity can be defined by the elimination of differential forces.

[6] R. Carnap: *Der Raum, op. cit.*, p. 54.

[7] Although both Carnap's metric M and the distance function

$$ds = \sqrt{g_{ik}dx^i dx^k}$$

can provide a congruence definition, they cannot be deduced from one another without information concerning the coincidence behavior of the rod in the space under consideration.

[8] R. Carnap: *Der Raum, op. cit.*, p. 54.

$ds^2 = dx^2 + dy^2$ or with the different intervals that are equal on the basis of one of the metrizations $ds^2 = \sec^2\theta\, dx^2 + dy^2$ (where $\sec^2\theta > 1$). In other words, the stated specifications of M and R do not tell us whether the rod behaves on the tabletop as we know it to behave in actuality or whether it squirms in any one of infinitely many different ways as compared to its actual behavior.

(3) As a corollary of our proof of the *non*-uniqueness of the congruence definition, we can show that the following statement by Reichenbach is false: "If we say: actually a geometry G applies but we measure a geometry G′, we define at the same time a force F which causes the difference between G and G′."[9] Using our previous notation, we note first that instead of determining a metric tensor g'_{ik} uniquely (up to an arbitrary constant), the geometry G determines an infinite class α of such tensors differing other than by being proportional to one another. But since $F_{ik} = g_{ik} - g'_{ik}$ (where the g'_{ik} are furnished by the rod prior to its being regarded as "deformed" by any universal forces), the failure of G to determine a tensor g_{ik} uniquely (up to an arbitrary constant) issues in there being as many *different* universal forces F_{ik} as there are different tensors g_{ik} in the class α determined by G. We see, therefore, that contrary to Reichenbach, there are infinitely many different ways in which the measuring rod can be held to be "deformed" while furnishing the *same* geometry G.

These criticisms of Reichenbach do not affect substantial portions of his immensely valuable contributions to the philosophy of geometry. But it is evident that it is less than discerning to draw the following conclusion with E. H. Hutten in regard to the logical analysis of the concept of space as understood since the advent of Einstein's relativity: "except for some change in terminology there is nothing to add to the exposition as found for instance, in Reichenbach's book on the Philosophy of Space and Time."[1]

---

[9] H. Reichenbach: *The Philosophy of Space and Time, op. cit.*, p. 27.

[1] E. H. Hutten: *The Language of Modern Physics* (London: George Allen and Unwin, Ltd., and New York: The Macmillan Company; 1956), p. 110.

Chapter 4

CRITIQUE OF EINSTEIN'S PHILOSOPHY OF GEOMETRY

(A) AN APPRAISAL OF DUHEM'S ACCOUNT OF THE FALSIFIABILITY OF ISOLATED EMPIRICAL HYPOTHESES IN ITS BEARING ON EINSTEIN'S CONCEPTION OF THE INTERDEPENDENCE OF GEOMETRY AND PHYSICS.

Since Einstein's central thesis concerning the epistemological status of physical geometry will be seen in Section C to be a geometrical version of Pierre Duhem's conception of the falsifiability of isolated empirical hypotheses, this first section will be devoted to a critical examination of Duhem's conception as articulated by W. V. O. Quine.

It has been maintained by writers other than Duhem and Quine that there is an important *asymmetry* between the *verification* and the *refutation* of a theory in empirical science. Refutation has been said to be conclusive or decisive while verification was claimed to be irremediably inconclusive in the following sense: If a theory T_1 entails observational consequences O, then the *truth* of T_1 does not, of course, follow deductively from the truth of the conjunction

$$(T_1 \rightarrow O) \cdot O.$$

On the other hand, the *falsity* of T_1 is indeed deductively inferable by *modus tollens* from the truth of the conjunction

$$(T_1 \rightarrow O) \cdot \sim O.$$

Thus, F. S. C. Northrop writes: "We find ourselves, therefore, in this somewhat shocking situation: the method which natural science uses to check the postulationally prescribed theories . . . is absolutely trustworthy when the proposed theory is not confirmed and logically inconclusive when the theory is experimentally confirmed."[1]

Under the influence of Duhem,[2] this thesis of asymmetry of conclusiveness between verification and refutation has been strongly denied as follows: *If* "T_1" denotes the kind of individual or *isolated* hypothesis H whose verification or refutation is at issue in the conduct of particular scientific experiments, then Northrop's formal schema is a misleading oversimplification. Upon taking cognizance of the fact that the observational consequences O are deduced *not* from H alone but rather from the conjunction of H and the relevant body of *auxiliary* assumptions A, the refutability of H is seen to be no more conclusive than its verifiability. For now it appears that Northrop's formal schema must be replaced by the following:

(i) $[(H \cdot A) \rightarrow O] \cdot O$ (verification)

and

(ii) $[(H \cdot A) \rightarrow O] \cdot \sim O$. (refutation).

The recognition of the presence of the auxiliary assumptions A in both the verification and refutation of H now makes apparent that the *refutation* of H *itself* by *adverse* empirical evidence $\sim O$ can be no more decisive than its *verification* (confirmation) by *favorable* evidence O. What can be inferred deductively from the refutational premise (ii) is *not* the falsity of H itself but only the much weaker conclusion that H and A cannot both be true. It is immaterial here that the *falsity* of the *conjunction* of H and A *can* be inferred *deductively* from the refutational premise (ii) while the truth of that conjunction can be inferred only *inductively* from the verificational premise (i). For this does not detract from the fact that there is parity of inconclusiveness between the refutation of H itself and the verification of H itself

[1] F. S. C. Northrop: *The Logic of the Sciences and the Humanities* (New York: The Macmillan Company; 1947), p. 146.

[2] Pierre Duhem: *The Aim and Structure of Physical Theory* (Princeton: Princeton University Press; 1954), Part II, Ch. vi, especially pp. 183–90.

in the following sense: (ii) does not entail (deductively) the falsity of H itself, just as (i) does not entail the truth of H by itself. In short, isolated component hypotheses of far-flung theoretical systems are not separately refutable but only contextually disconfirmable: *no one constituent hypothesis H* can ever be *extricated* from the ever-present web of collateral assumptions so as to be *open* to *separate refutation* by the evidence as part of an *explanans* of that evidence, just as no such isolation is achievable for purposes of verification. And Northrop's schema is an adequate representation of the actual logical situation only if "T_1" in his schema refers to the entire theoretical *system* of premises which enters into the deduction of O rather than to such mere *components* H as are at issue in specific scientific inquiries.

Under the influence of Duhem's emphasis on the confrontation of an entire theoretical system by the tribunal of evidence, writers such as W. V. O. Quine have made what I take to be the following claim: no matter what the specific content O′ of the prima facie adverse empirical evidence $\sim$O, we can always justifiably affirm the truth of H as part of the theoretical *explanans* of O′ by doing two things: First, blame the falsity of O on the falsity of A rather than on the falsity of H, and second, so modify A that the conjunction of H and the *revised* version A′ of A does entail (explain) the actual findings O′. Thus, in his "Two Dogmas of Empiricism," Quine writes: "Any statement can be held true come what may, if we make drastic enough adjustments elsewhere in the system."[3] And one of Quine's arguments in that provocative essay against the tenability of the analytic-synthetic distinction is that a supposedly synthetic statement, no less than a supposedly analytic one can be claimed to be true "come what may" on Duhemian grounds. Accordingly, we take the Duhem-Quine thesis—hereafter called the "D-thesis"—to involve the following set of contentions: *there is an inductive* (*epistemological*) *interdependence and inseparability between H and the auxiliary assumptions* A, and there is therefore an ingression of a kind of a priori choice into physical theory. For at the price of suitable compensatory modifications in the remainder of the theory, any one of its component hypotheses H may be retained in the face

[3] W. V. O. Quine: *From a Logical Point of View* (2nd ed.; Cambridge: Harvard University Press; 1961), p. 43. Cf. also p. 41, n. 17.

of seemingly contrary empirical findings as part of an *explanans* of these very findings. And this quasi a priori preservability of H is sanctioned by the far-reaching theoretical ambiguity and flexibility of the logical constraints imposed by the observational evidence.[4]

In the particular case of physical geometry, Duhem would point to the fact that in a sense to be specified *in detail* in Section C, the physical laws used to correct a measuring rod for substance-specific distortions presuppose a geometry and comprise the laws of optics. And hence he would *deny*, for example, that either of the following kinds of *independent* tests of geometry and optics are feasible:

1. Prior to and independently of knowing or presupposing the geometry, we find it to be a law of optics that the paths of light coincide with the geodesics of the congruence defined by rigid bodies.

Knowing this, we then use triangles consisting of a geodesic base line in the solar system and the stellar light rays connecting its extremities to various stars to determine the geometry of the system of rigid body geodesics: stellar parallax measurements will tell us whether the angle sums of the triangles are 180° (Euclidean geometry), less than 180° (hyperbolic geometry) or in excess of 180° (spherical geometry).

If we thus find that the angle sum is different from 180°, then we shall know that the geometry of the rigid body geodesics is not Euclidean. For in view of our prior independent ascertainment of the paths of light rays, such a non-Euclidean result could not be interpreted as due to the failure of optical paths to coincide with the rigid body geodesics.

2. Prior to and independently of knowing or presupposing the laws of optics, we ascertain what the geometry is relatively to the rigid body congruence.

Knowing this we then find out whether the paths of light rays coincide with the geodesics of the rigid body congruence by

---

[4] P. Duhem: *The Aim and Structure of Physical Theory, op. cit.* Duhem's explicit disavowal of both decisive falsifiability and crucial verifiability of an *explanans* will not bear K. R. Popper's reading of him [K. R. Popper: *The Logic of Scientific Discovery, op. cit.*, p. 78]: Popper, who is an exponent of decisive falsifiability [*ibid.*], misinterprets Duhem as allowing that tests of a hypothesis may be decisively falsifying and as denying only that they may be crucially verifying.

making a parallactic or some other determination of the angle sum of a light ray triangle.

Since we know the geometry of the rigid body geodesics independently of the optics, we know what the corresponding angle sum of a triangle whose sides are geodesics ought to be. And hence the determination of the angle sum of a light ray triangle is then decisive in regard to whether the paths of light rays coincide with the geodesics of the rigid body congruence.

In place of such independent confirmability and falsifiability of the geometry and the optics, Duhem affirms their *inductive* (epistemological) inseparability and interdependence.

In the present chapter, I shall endeavor, among other things, to establish two main conclusions:

(1) Quine's formulation of Duhem's thesis—which we call the "D-thesis"—is true *only* in various *trivial* senses of what Quine calls "drastic enough adjustments elsewhere in the system." And no one would wish to contest any of these thoroughly uninteresting versions of the D-thesis,

(2) in its *non*-trivial exciting form, the D-thesis is untenable in the following fundamental respects:

A. *Logically*, it is a *non-sequitur*. For *independently* of the particular empirical context to which the hypothesis H pertains, there is no logical guarantee at all of the existence of the *required kind* of revised set A′ of auxiliary assumptions such that

$$(H \cdot A') \rightarrow O'$$

for any one component hypothesis H and any O′. Instead of being guaranteed logically, the existence of the required set A′ needs *separate* and *concrete* demonstration for each particular context. In the absence of the latter kind of *empirical* support for Quine's unrestricted Duhemian claim, that claim is an unempirical dogma or article of faith which the pragmatist Quine is no more entitled to espouse than an empiricist would be.

B. The D-thesis is not only a *non-sequitur* but is actually *false*, as shown by an important counter-example, namely the separate falsifiability of a particular component hypothesis H.

Of these conclusions, (1) and (2A) will be defended in the present Section A, while arguments in support of (2B) will be deferred until Section C.

To forestall misunderstanding, let it be noted that my rejection of the very strong assertion made by Quine's D-thesis is not at all intended as a repudiation of the following far weaker contention, which I believe to be eminently sound: the logic of every disconfirmation, no less than of every confirmation of an isolated scientific hypothesis H is such as to *involve at some stage or other* an entire network of interwoven hypotheses in which H is ingredient rather than in every stage merely the separate hypothesis H. Furthermore, it is to be understood that the issue before us is the logical one whether in principle every component H is unrestrictedly preservable by a suitable A′, not the psychological one whether scientists possess sufficient ingenuity at every turn to propound the required set A′, *if it exists.* Of course, *if* there are cases in which the requisite A′ simply *does not even exist logically,* then surely no amount of ingenuity on the part of scientists will enable them to ferret out the non-existent required A′ in such cases.

I. *The Trivial Validity of the D-Thesis.*

It can be made evident at once that unless Quine restricts in very specific ways what he understands by "drastic enough adjustments elsewhere in the [theoretical] system," the D-thesis is a thoroughly unenlightening truism. For if someone were to put forward the false empirical hypothesis H that "Ordinary buttermilk is highly toxic to humans," this hypothesis could be saved from refutation in the face of the observed wholesomeness of ordinary buttermilk by making the following "drastic enough" adjustment in our system: changing the rules of English usage so that the intension of the term "ordinary buttermilk" is that of the term "arsenic" in its customary usage. Hence a *necessary* condition for the non-triviality of Duhem's thesis is that *the theoretical language* be *semantically stable* in the relevant respects.

Furthermore, it is clear that if one were to countenance that O′ itself qualifies as A′, Duhem's affirmation of the existence of an A′ such that

$$(H \cdot A') \rightarrow O'$$

would hold trivially, and H would not even be needed to deduce O′. Moreover, the D-thesis can hold trivially even in cases in which H is required in addition to A′ to deduce the *explanandum* O′: an A′ of the trivial form

$$\sim H \vee O'$$

requires H for the deduction of O′, but no one will find it enlightening to be told that the D-thesis can thus be sustained.

I am unable to give a formal *and* completely general *sufficient* condition for the *non*-triviality of A′. And, so far as I know, neither the originator nor any of the advocates of the D-thesis have even shown any awareness of the need to circumscribe the class of *non*-trivial revised auxiliary hypotheses A′ so as to render the D-thesis interesting. I shall therefore assume that the proponents of the D-thesis intend it to stand or fall on the kind of A′ which we would all recognize as *non*-trivial in *any given case*, a kind of A′ which I shall symbolize by A'_{nt}. And I shall endeavor to show that such a *non*-trivial form of the D-thesis is indeed untenable after first commenting on the attempt to sustain the D-thesis by resorting to the use of a *non-standard logic.*

The species of drastic adjustment consisting in recourse to a *non*-standard logic is specifically mentioned by Quine. Citing a hypothesis such as "there are brick houses on Elm Street," he claims that even a statement so "germane to sense experience . . . can be held true in the face of recalcitrant experience by pleading hallucination or by amending certain statements of the kind called logical laws."[5] I disregard for now the argument from hallucination. In the absence of *specifics* as to the ways in which alterations of logical laws will enable Quine to hold in the face of *recalcitrant* experience that a statement H like "there are brick houses on Elm Street" is *true,* I must conclude the following: the invocation of non-standard logics either makes the D-thesis *trivially* true or turns it into an interesting claim which is an unfounded dogma. For suppose that the non-standard logic used is a three-valued one. Then even if it were otherwise feasible to assert within the framework of such a logic that the particular statement H is "true," the term "true" would no longer have the meaning associated with the two-valued framework of logic within which the D-thesis was enunciated to begin with. It is not to be overlooked that a form of the D-thesis which allows it-

[5] W. V. O. Quine: *From a Logical Point of View, op. cit.,* p. 43.

self to be sustained by alterations in the meaning of "true" is no less trivial *in the context of the expectations raised by the D-thesis* than one which rests its case on calling arsenic "buttermilk." And this triviality obtains *in this context,* notwithstanding the fact that the two-valued and three-valued usages of the word "true" share what H. Putnam has usefully termed a common "core meaning."[6]

For suppose we had two particular substances I_1 and I_2 which are isomeric with each other. That is to say, these substances are composed of the same elements in the same proportions and with the same molecular weight but the arrangement of the atoms within the molecule is different. Suppose further that I_1 is not at all toxic while I_2 is highly toxic, as in the case of two isomers of trinitrobenzene.[7] Then if we were to call I_1 "duquine" and asserted that "duquine is highly toxic," this statement H could also be trivially saved from refutation in the face of the evidence of the wholesomeness of I_1 by the following device: only *partially* changing the meaning of "duquine" so that its intension is the second, highly toxic isomer I_2, thereby leaving the chemical "core meaning" of "duquine" intact. To forestall the *misunderstanding* that my charge of triviality here is guilty of the error committed by the Eddington-Putnam triviality-thesis (cf. Chapter One, Section D), let me point out precisely what I regard as trivial in this context. The preservation of H from refutation in the face of the evidence by a *partial* change in the meaning of "duquine" is trivial in the sense of being only a *trivial* fulfillment of *the expectations raised by the D-thesis.* But, in my view, the possibility as such of preserving H *by this particular kind of change in meaning* is not at all trivial. For this possibility as such reflects *a fact about the world:* the existence of isomeric substances of radically different degrees of toxicity (allergenicity)!

Even if one ignores the change in the meaning of "true"

---

[6] H. Putnam: "Three-Valued Logic," *Philosophical Studies,* Vol. VIII (1957), p. 74.

[7] Cf. H. L. Alexander: *Reactions With Drug Therapy* (Philadelphia: W. B. Saunders Co.; 1955), p. 14. Alexander writes: "It is true that drugs with closely related chemical structures do not always behave clinically in a similar manner, for antigenicity of simple chemical compounds may be changed by minor alterations of molecular structures. . . . 1,2,4-trinitrobenzene . . . is a highly antigenic compound. . . . 1,3,5 . . . trinitrobenzene is allergenically inert." (I am indebted to Dr. A. I. Braude for this reference.)

inherent in the resort to a three-valued logic, there is no reason to think that the D-thesis can be successfully upheld in such an altered logical framework: the arguments which I shall present against the *non*-trivial form of the D-thesis within the framework of the standard logic apply just as much, so far as I can see, in the three-valued and other non-standard logics of which I am aware. And if the reply is that there are other non-standard logics which are both viable for the purposes of science and in which my impending polemic against the non-trivial form of the D-thesis does not apply, then I retort: as it stands, Quine's assertion of the feasibility of a change in the laws of logic which would thus sustain the D-thesis is an unempirical dogma or at best a promissory note. And until the requisite collateral is supplied, it is not incumbent upon anyone to accept that promissory note.

II. *The Untenability of the Non-Trivial D-Thesis.*

The *non*-trivial D-thesis will now be seen to be a *non-sequitur.* The non-trivial D-thesis is that for every component hypothesis H of any domain of empirical knowledge and for any observational findings O′,

$$(\exists A'_{nt})\,[(H \cdot A'_{nt}) \rightarrow O'].$$

But this claim does not follow from the fact that the falsity of H is not deductively inferable from the premise

$$[(H \cdot A) \rightarrow O] \cdot \sim O,$$

which we shall call premise (ii) as at the beginning of this section. For the latter premise utilizes not the full empirical information given by O′ but only the part of that information which tells us that O′ is logically incompatible with O. Hence the *failure* of $\sim$O to permit the deduction of $\sim$H does *not* justify the assertion of the D-thesis that there always *exists* a non-trivial A′ such that the conjunction of H and that A′ entails O′. In other words, the fact that the falsity of H is *not* deducible (by *modus tollens*) from premise (ii) is quite insufficient to show that H can be preserved non-trivially as part of an *explanans* of *any* potential empirical findings O′. I conclude, therefore, from the analysis given so far that in its *non*-trivial form,

Quine's D-thesis is *gratuitous* and that the existence of the required non-trivial A′ would require *separate* demonstration for each particular case.

(B) THE INTERDEPENDENCE OF GEOMETRY AND PHYSICS IN POINCARÉ'S CONVENTIONALISM.

The literature treating of the interdependence of geometry and physics exhibits a pervasive confusion between two logically very different kinds of interdependence which can be respectively associated with Duhem's epistemological holism and Poincaré's conventionalism: an *inductive* (*epistemological*) interdependence and a *linguistic* one. The distinction between these two kinds of interdependence merits being drawn in its own right and is also of considerable importance for the critical estimate of Einstein's conception to be given in Section C.

In the present section, I shall therefore discuss the relevant difference between the views of Duhem and Poincaré as well as some related questions pertaining to the interpretation of Poincaré.

As we saw in Section C of Chapter One, the central theme of Poincaré's so-called conventionalism is essentially an elaboration of the thesis of alternative metrizability whose fundamental justification we owe to Riemann. Poincaré's much cited but widely misunderstood statement concerning the possibility of always giving a Euclidean description of any results of stellar parallax measurements reads as follows:

> If Lobachevski's geometry is true, the parallax of a very distant star will be finite; if Riemann's is true, it will be negative. These are results which seem within the reach of experiment, and there have been hopes that astronomical observations might enable us to decide between the three geometries.
>
> But in astronomy "straight line" means simply "path of a ray of light."
>
> If therefore negative parallaxes were found, or if it were demonstrated that all parallaxes are superior to a certain limit, two courses would be open to us; we might either renounce Euclidean geometry, or else modify the laws of optics and suppose that light does not travel rigorously in a straight line.

> It is needless to add that all the world would regard the latter solution as the more advantageous.
>
> The Euclidean geometry has, therefore, nothing to fear from fresh experiments.[8]

The context of this paragraph[9] makes it quite clear that Poincaré is using the observationally significant case of a stellar light ray triangle to explain that, if need be, the preservation of a Euclidean description by an alternative metrization is a genuinely live option. Hence, his account of the relevance of stellar parallax measurements to the determination of the metric geometry of physical space makes precisely the same point, albeit much less lucidly, as the following magisterially clear statement by him:

> In space we know rectilinear triangles the sum of whose angles is equal to two right angles; but equally we know curvilinear triangles the sum of whose angles is less than two right angles. . . . To give the name of straights to the sides of the first is to adopt Euclidean geometry; to give the name of straights to the sides of the latter is to adopt the non-Euclidean geometry. So that to ask what geometry it is proper to adopt is to ask, to what line is it proper to give the name straight? It is evident that experiment can not settle such a question.[1]

Now, the equivalence of this latter contention to Riemann's view of congruence becomes evident the moment we note that the legitimacy of identifying lines which are curvilinear in the usual geometrical parlance as "straights" is vouchsafed by the warrant for our choosing a new definition of congruence such that the previously curvilinear lines become geodesics of the new congruence.

And we note that whereas the *original* geodesics in space exemplified the formal relations obtaining between Euclidean "straight lines," the *different* geodesics associated with the new metrization embody the relations prescribed for straight lines by the formal postulates of *hyperbolic* geometry. Awareness of the fact that Poincaré begins the quoted passage with the words "In [physical] space" enables us to see that he is making the follow-

---

[8] H. Poincaré: *The Foundations of Science, op. cit.*, p. 81.
[9] *Ibid.*, pp. 81–86.
[1] *Ibid.*, p. 235.

ing assertion here: the same *physical* surface or region of three-dimensional *physical* space admits of *alternative* metrizations so as to constitute a physical realization of either the formal postulates of Euclidean geometry or of one of the non-Euclidean abstract calculi. To be sure, syntactically, this alternative metrizability involves a formal intertranslatability of the *relevant portions* of these incompatible geometrical calculi, the "intertranslatability" being guaranteed by a "dictionary" which pairs off with one another the *alternative names* (or descriptions) of each physical path or configuration. But the essential point made here by Poincaré is not that a purely formal translatability obtains; instead, Poincaré is emphasizing here that a given physical surface or region of physical 3-space can indeed be a model of one of the *non*-Euclidean geometrical calculi no less than of the Euclidean one. In this sense one can say, therefore, that Poincaré affirmed the conventional or definitional status of applied geometry.

Hence, we must reject the following wholly syntactical interpretation of the above citation from Poincaré, which is offered by Ernest Nagel, who writes: "The thesis he [Poincaré] establishes by this argument is simply the thesis that choice of notation in formulating a system of pure geometry is a convention."[2] Having thus misinterpreted Poincaré's conventionalist thesis as pertaining only to formal intertranslatability, Nagel fails to see that Poincaré's avowal of the conventionality of physical or *applied* geometry is none other than the assertion of the *alternative metrizability of physical space* (or of a portion thereof). And, in this way, Nagel is driven to give the following unfounded interpretation of Poincaré's conception of the status of applied (physical) geometry: "Poincaré also argued for the definitional status of *applied* as well as of *pure* geometry. He maintained that, even when an interpretation is given to the primitive terms of a pure geometry so that the system is then converted into statements about certain physical configurations (for example, interpreting 'straight line' to signify the path of a light ray), no experiment on physical geometry can ever decide against one of the alternative systems of physical geometry and in favor of another."[3] But far

[2] E. Nagel: *The Structure of Science, op. cit.*, p. 261.
[3] *Ibid.*

from having claimed that the geometry is still conventional even *after* the provision of a particular physical interpretation of a pure geometry, Poincaré merely reiterated the following thesis of alternative metrizability in the passages which Nagel[4] then goes on to quote from him: suitable alternative semantical interpretations of the term "congruent" (for line segments and/or for angles), and correlatively of "straight line," etc., can readily demonstrate that, subject to the restrictions imposed by the existing topology, it is always a live option to give *either* a Euclidean *or* a *non*-Euclidean description of any given set of *physico*-geometric facts. And since alternative metrizations are just as legitimate epistemologically as alternative systems of units of length or temperature, one can always, in principle, *reformulate* any physical theory based on a given metrization of space—or, as we saw in Chapter Two above, of time—so as to be based on an *alternative* metric.

There is therefore no warrant at all for the following caution expressed by Nagel in regard to the feasibility of what is merely a reformulation of physical theory on the basis of a new metrization: ". . . even if we admit universal forces in order to retain Euclid . . . we must incorporate the assumption of universal forces into the rest of our physical theory, rather than introduce such forces piecemeal subsequent to each observed 'deformation' in bodies. It is by no means self-evident, however, that physical theories can in fact always be devised that have built-in provisions for such universal forces."[5] Yet, precisely that fact is self-evident, and its self-evidence is obscured from view by the logical havoc created by the statement of a remetrization issuing in Euclidean geometry in terms of "universal forces." For that metaphor seems to have misled Nagel into imputing the status of an *empirical* hypothesis to the use of a *non*-standard spatial metric merely because the latter metric is described by saying that we "assume" appropriate universal forces. In fact, our discussion in Chapter Two has shown mathematically for the one-dimensional case of time how Newtonian mechanics is to be recast via suitable transformation equations, such as equation (4) there, so as to implement a remetrization given by $T = f(t)$, which can be

[4] *Ibid.*, pp. 261–62.
[5] *Ibid.*, pp. 264–65.

described metaphorically by saying that all clocks are "accelerated" by "universal forces."

Corresponding remarks apply to Poincaré's contention that we can always preserve Euclidean geometry in the face of any data obtained from stellar parallax measurements: if the paths of light rays are geodesics on the customary definition of congruence, as indeed they are in the Schwarzschild procedure cited by Robertson,[6] and if the paths of light rays are found parallactically to sustain non-Euclidean relations on that metrization, then we need only choose a different definition of congruence such that these same paths will no longer be geodesics and that the geodesics of the newly chosen congruence are Euclideanly related. From the standpoint of synthetic geometry, the latter choice effects a *renaming* of optical and other paths and thus is merely a *recasting of the same factual content in Euclidean language rather than a revision of the extra-linguistic content of optical and other laws.* The retainability of Euclideanism by remetrization, which is affirmed by Poincaré, therefore *involves a merely linguistic interdependence of the geometric theory of rigid solids and the optical theory of light rays.* And since Poincaré's claim here is a straightforward elaboration of the metric amorphousness of the continuous manifold of space, it is not clear how H. P. Robertson[7] can reject it as a "pontifical pronouncement" and even regard it as being in contrast with what he calls Schwarzschild's "sound operational approach to the problem of physical geometry." For Schwarzschild had rendered the question concerning the prevailing geometry *factual* only by the adoption of a particular spatial metrization based on the travel times of light, which does indeed turn the direct light paths of his astronomical triangle into geodesics.[8]

---

[6] H. P. Robertson: "Geometry as a Branch of Physics," *Albert Einstein: Philosopher-Scientist, op. cit.*, pp. 324–25.

[7] *Ibid.*, pp. 324–25.

[8] For very useful discussions of the actual astronomical methods used to determine the geometry of physical space in the large, see H. P. Robertson, *op. cit.*, pp. 323–25 and 330–32; Max Jammer's historical work *Concepts of Space* (Cambridge: Harvard University Press; 1954), pp. 147–48; William A. Baum: "Photoelectric Test of World Models," *Science*, Vol. CXXXIV (1961), p. 1426, and Allan Sandage: "Travel Time for Light from Distant Galaxies Related to the Riemannian Curvature of the Universe," *Science*, Vol. CXXXIV (1961), p. 1434.

Poincaré's interpretation of the parallactic determination of the geometry of a stellar triangle has also been obscured by Ernest Nagel's statement of it. Apart from encumbering that statement with the metaphorical use of "universal forces," Nagel fails to point out that the crux of the preservability (retainability) of Euclidean geometry lies in: First, the denial of the *geodesicity* (*straightness*) of optical paths which are found parallactically to sustain *non*-Euclidean relations on the *customary* metrization of line segments and angles, or at least in the rejection of the customary congruence for *angles* (cf. Chapter Three, Section B),[9] and second, the ability to guarantee the existence of a suitable new metrization whose associated geodesics are paths which do exhibit the formal relations of Euclidean straights. For Nagel characterizes the retainability of Euclidean geometry come what may by asserting that the latter's retention is effected "only by maintaining that the sides of the stellar triangles are not really Euclidean [*sic*] straight lines, and he [the Euclidean geometer] will therefore adopt the hypothesis that the optical paths are deformed by some fields of force."[1] But apart from the obscurity of the notion of the deformation of the optical *paths,* the unfortunate inclusion of the word "Euclidean" in this sentence of Nagel's obscures the very point which the advocate of Euclid is concerned to make in this context in the interests of his thesis. And this point is not, as Nagel would have it, that the optical paths are not really *Euclidean* straight lines, a fact whose admission (assuming the customary congruence for angles) provided the starting point of the discussion. Instead, what the proponent of Euclid is concerned to point out here is that the legitimacy of alternative metrizations enables him to offer a metric such that the optical paths do *not* qualify as *geodesics* (straights) from the

---

[9] Under the assumed conditions as to the parallactic findings, a new metrization may allow the optical paths to be interpretable as *Euclideanly*-related geodesics, but only if the customary *angle* congruences were abandoned and changed suitably as part of the remetrization (cf. Chapter Three, Section B). In that case, the paths of light rays would be straight lines *even in the Euclidean description* obtained by the new metrization, but the optical laws involving angles would have to be suitably restated. For theorems governing the so-called "geodesic mapping" or "geodesic correspondence" relevant here, cf. L. P. Eisenhart: *An Introduction to Differential Geometry, op. cit.,* Sec. 37, pp. 205–11, and D. J. Struik: *Differential Geometry, op. cit.,* pp. 177–80.

[1] E. Nagel: *The Structure of Science, op. cit.,* p. 263.

outset. For it is by denying altogether the *geodesicity* of the optical paths that the advocate of Euclid can uphold his thesis successfully in the face of the admitted prima facie *non*-Euclidean parallactic findings.

The invocation of the conventionality of congruence to carry out remetrizations is not at all peculiar to Poincaré. For F. Klein's relative consistency proof of hyperbolic geometry via a model furnished by the interior of a circle on the Euclidean plane,[2] for example, is based on one particular kind of possible remetrization of the circular portion of that plane, projective geometry having played the heuristic role of furnishing Klein with a suitable definition of congruence. Thus what from the point of view of synthetic geometry appears as intertranslatability via a dictionary, appears as alternative metrizability from the point of view of differential geometry.

There are two respects, however, in which Poincaré is open to criticism in this connection:

First, he maintained[3] that it would always be regarded as most convenient to preserve Euclidean geometry, even at the price of remetrization, on the grounds that this geometry is the simplest analytically.[4] As is well-known, precisely the opposite development materialized in the GTR: Einstein forsook the simplicity of the geometry itself in the interests of being able to maximize the simplicity of the definition of congruence. He makes clear in his fundamental paper of 1916 that had he insisted on the retention of Euclidean geometry in a gravitational field, then he could *not* have taken "one and the same rod, independently of its place and orientation, as a realization of the same interval."[5]

Second, even if the simplicity of the geometry itself were the sole determinant of its adoption, that simplicity might be judged by criteria other than Poincaré's analytical simplicity such as the simplicity of the undefined concepts used.

However, if Poincaré were alive today, he could point to an interesting recent illustration of the sacrifice of the simplicity and accessibility of the congruence standard on the altar of maximum

---

[2] R. Bonola: *Non-Euclidean Geometry, op. cit.*, pp. 164–75.

[3] H. Poincaré: *The Foundations of Science, op. cit.*, p. 81.

[4] *Ibid.*, p. 65.

[5] A. Einstein: "The Foundations of the General Theory of Relativity," in *The Principle of Relativity, op. cit.*, p. 161.

simplicity of the resulting theory. Astronomers have recently proposed to remetrize the time continuum for the following reason: as indicated in Chapter Two, when the mean solar second, which is a very precisely known fraction of the period of the earth's rotation on its axis, is used as a standard of temporal congruence, then there are three kinds of discrepancies between the actual observational findings and those predicted by the usual theory of celestial mechanics. The empirical facts thus present astronomers with the following choice: either they retain the rather natural standard of temporal congruence at the cost of having to bring the principles of celestial mechanics into conformity with observed fact by revising them appropriately, or they remetrize the time continuum, employing a less simple definition of congruence so as to preserve these principles intact. Decisions taken by astronomers in the last few years were exactly the reverse of Einstein's choice of 1916 as between the simplicity of the standard of congruence and that of the resulting theory. The mean solar second is to be supplanted by a unit to which it is non-linearly related: the sidereal year, which is the period of the earth's revolution around the sun, due account being taken of the irregularities produced by the gravitational influence of the other planets.[6]

We see that the implementation of the requirement of descriptive simplicity in theory-construction can take alternative forms, because agreement of astronomical theory with the evidence now available is achievable by revising either the definition of temporal congruence or the postulates of celestial mechanics. The existence of this alternative likewise illustrates that for an axiomatized physical theory containing a geo-chronometry, it is *gratuitous* to single out the postulates of the theory as having been prompted by *empirical* findings in contradistinction to deeming the *definitions of congruence* to be wholly a priori, or vice versa. This conclusion bears out geo-chronometrically Braithwaite's contention[7] that there is an important sense in which axiomatized physical theory does not lend itself to compliance with Heinrich

[6] G. M. Clemence: "Time and its Measurement," *op. cit.*

[7] R. B. Braithwaite: "Axiomatizing a Scientific System by Axioms in the Form of Identification," in *The Axiomatic Method* ed. by L. Henkin, P. Suppes, and A. Tarski (Amsterdam: North Holland Publishing Company; 1959), pp. 429–42.

Hertz's injunction to "distinguish thoroughly and sharply between the elements . . . which arise from the necessities of thought, from experience, and from arbitrary choice."[8] The same point is illustrated by the possibility of characterizing the factual innovation wrought by Einstein's abandonment of Euclidean geometry in favor of Riemannian geometry in the GTR in several ways as follows:

(1) *Upon using the customary definition of spatial congruence,* the geometry near the sun is *not* Euclidean, contrary to the claims of pre-GTR physics.

(2) The geometry near the sun is *not* Euclidean on the basis of the *customary* congruence, but it *is* Euclidean on a suitably modified congruence definition which makes the length of a rod a specified function of its position and orientation,[9]

(3) *within the confines* of the requirement of giving a *Euclidean* description of the non-classical facts postulated by the GTR, Einstein recognized the *factually dictated* need to *abandon* the *customary* definition of congruence, which had yielded a Euclidean description of the classically assumed facts. Thus, the revision of the Newtonian theory made necessary by the discovery of relativity can be formulated as either a change in the postulates of geometric theory or a change in the correspondence rule for congruence.

Having seen that Poincaré's remetrizational retainability of Euclidean geometry or of some other particular geometry involves a merely *linguistic* interdependence of the geometric theory of rigid solids and the optical theory of light rays, we are ready to contrast that interdependence with the quite different epistemological (inductive) interdependence affirmed by Duhem.

The Duhemian conception envisions scope for alternative geometric accounts of a given body of evidence to the extent that these geometries are associated with alternative, factually *non*-equivalent sets of physical laws which are used to compute *corrections* for substance-specific distortions.[1] Poincaré,[2] however, in-

---

[8] H. Hertz: *The Principles of Mechanics* (New York: Dover Publications, Inc.; 1956), p. 8.

[9] The function in question is given in R. Carnap: *Der Raum, op. cit.,* p. 58.

[1] For some details on just how factually non-equivalent correctional physical laws are associated with different metric geometries, see Section C of this chapter.

[2] H. Poincaré: *The Foundations of Science, op. cit.,* pp. 66–80.

stead of invoking the Duhemian *inductive* latitude, specifically bases the possibility of giving either a Euclidean or a non-Euclidean description of the same spatio-physical facts on alternative metrizability *quite apart from any considerations of substance-specific distorting influences and even after correcting for these in some way or other.* Says he: "No doubt, in our world, natural solids . . . undergo variations of form and volume due to warming or cooling. But we neglect these variations in laying the foundations of geometry, because, besides their being very slight, they are irregular and consequently seem to us accidental."[3] For the sake of specificity, our contrasting comparison of Poincaré and Duhem will focus on the feasibility of alternative geometric interpretations of stellar parallax data.

The attempt to explain parallactic data yielding an angle sum different from 180° for a stellar light ray triangle by different geometries which constitute live options *in the inductive sense of Duhem* would presumably issue in the following alternative between two theoretical systems. Each of these theoretical systems comprises a geometry G and an optics O which are epistemologically inseparable and which are inductively interdependent in the sense that the combination of G and O must yield the observed results:

(a) G_E: the geometry of the rigid body geodesics is Euclidean, and
O_1: the paths of light rays do *not* coincide with these geodesics but form a non-Euclidean system,

or

(b) $G_{non\text{-}E}$: the geodesics of the rigid body congruence are *not* a Euclidean system, and
O_2: the paths of light rays *do* coincide with these geodesics, and thus they form a non-Euclidean system.

To contrast this Duhemian conception of the feasibility of alternative geometric interpretations of the assumed parallactic data with that of Poincaré, we recall that the physically interpreted alternative geometries associated with two (or more) different metrizations *in the sense of Poincaré* have precisely the same

[3] *Ibid.*, p. 76.

total factual content, as do the corresponding two sets of optical laws. For an alternative metrization in the sense of Poincaré affects only the *language* in which the facts of optics and the coincidence behavior of a transported rod are described: the two geometric descriptions respectively associated with two alternative metrizations are *alternative representations of the same factual content,* and so are the two sets of optical laws corresponding to these geometries. Hence I maintain that Poincaré is affirming a *linguistic* interdependence of the geometric theory of rigid solids and the optical theory of light rays. By contrast, in the Duhemian account, G_E and $G_{non\text{-}E}$ not only differ in factual content but are logically incompatible, and so are O_1 and O_2. And on the latter conception, there is sameness of factual content *in regard to the assumed parallactic data* only between the *combined* systems formed by the two conjunctions (G_E and O_1) and ($G_{non\text{-}E}$ and O_2).[4] Thus, the need for the combined system of G and O to yield the empirical facts, coupled with the avowed epistemological (inductive) inseparability of G and O lead the Duhemian to conceive of the interdependence of geometry and optics as inductive (epistemological).

Hence whereas Duhem construes the interdependence of G and O inductively such that the geometry by itself is not accessible to empirical test, Poincaré's conception of their interdependence allows for an empirical determination of G by itself, if we have renounced recourse to an alternative metrization in which the length of the rod is held to vary with its position or orientation. This is not, of course, to say that Duhem regarded alternative metrizations as such to be illegitimate.

It would seem that it was Poincaré's discussion of the interdependence of optics and geometry by reference to stellar parallax measurements which has misled many writers such as Einstein,[5] Eddington,[6] and Nagel[7] into regarding him as a proponent of the Duhemian thesis. An illustration of the widespread conflation of

---

[4] These combined systems do not, however, have the same over-all factual content.

[5] A. Einstein: *Geometrie und Erfahrung* (Berlin: Julius Springer; 1921), p. 9, and A. Einstein: "Reply to Criticisms," in P. A. Schilpp (ed.) *Albert Einstein: Philosopher-Scientist* (Evanston: The Library of Living Philosophers; 1949), pp. 665–88.

[6] A. S. Eddington: *Space, Time and Gravitation, op. cit.,* p. 9.

[7] E. Nagel: *The Structure of Science, op. cit.,* p. 262.

the linguistic and inductive kinds of interdependence of geometry and physics (optics) is given by D. M. Y. Sommerville's discussion of what he calls "the inextricable entanglement of space and matter." He says:

> A . . . "vicious circle" . . . arises in connection with the astronomical attempts to determine the nature of space. These experiments are based upon the received laws of astronomy and optics, which are themselves based upon the euclidean assumption. It might well happen, then, that a discrepancy observed in the sum of the angles of a triangle could admit of an explanation by some modification of these laws, or that even the absence of any such discrepancy might still be compatible with the assumptions of non-euclidean geometry.

Sommerville then quotes the following assertion by C. D. Broad:

> All measurement involves both physical and geometrical assumptions, and the two things, space and matter, are not given separately, but analysed out of a common experience. Subject to the general condition that space is to be changeless and matter to move about in space, we can explain the same observed results in many different ways by making compensatory changes in the qualities that we assign to space and the qualities we assign to matter. Hence it seems theoretically impossible to decide by any experiment what are the qualities of one of them in distinction from the other.

And Sommerville's immediate comment on Broad's statement is the following:

> It was on such grounds that Poincaré maintained the essential impropriety of the question, "Which is the true geometry?" In his view it is merely a matter of convenience. Facts are and always will be most simply described on the euclidean hypothesis, but they can still be described on the non-euclidean hypothesis, with suitable modifications of the physical laws. To ask which is the true geometry is then just as unmeaning as to ask whether the old or the metric system is the true one.[8]

Having dealt with the misinterpretation of Poincaré as a Duhemian, it remains to remove the misunderstanding that Poincaré

[8] D. M. Y. Sommerville: *The Elements of Non-Euclidean Geometry* (New York: Dover Publications, Inc.; 1958), pp. 209–10. Reprinted through permission of the publisher.

intended his conventionalism as a *denial* of the following Carnap-Reichenbach thesis, which was discussed in Chapter Three: in principle, the question as to the geometry of physical space is *empirical* after the geometrical vocabulary (including the term "congruent" for intervals and angles) has been given a physical interpretation.

According to a widely accepted reading of Poincaré's writings, he is said to have maintained that even after a system of abstract geometry is given a semantical interpretation via a particular coordinative definition of congruence, no experiment can verify or falsify the resulting system of physical geometry, the choice of the particular metrical geometry being entirely a matter of convention.[9]

The principal basis for the belief that Poincaré took a stand in opposition to the kind of qualified metrical empiricism which is espoused by writers such as Reichenbach and Carnap seems to be Poincaré's treatment of "Experience and Geometry" in Chapter Five of his *Science and Hypothesis*,[1] the fifth section of which culminates in the statement that "whichever way we look at it, it is impossible to discover in geometric empiricism a rational meaning."[2] But there seems to be general unawareness of the fact that Poincaré lifted Sections 4 and 5 of this chapter verbatim out of the wider context of his earlier paper "Des Fondements de la Géométrie, à propos d'un Livre de M. Russell,"[3] which was fol-

[9] The view that Poincaré was an extreme geometric conventionalist, who rejected the *qualified* geometric empiricism of Carnap and Reichenbach, is expressed, for example, by the following authors: H. Reichenbach: *The Philosophy of Space and Time, op. cit.*, p. 36; *The Rise of Scientific Philosophy, op. cit.*, p. 133, and "The Philosophical Significance of the Theory of Relativity" in: P. A. Schilpp (ed.), *Albert Einstein: Philosopher-Scientist* (Evanston: Library of Living Philosophers; 1949), p. 297; E. Nagel: *The Structure of Science, op. cit.*, p. 261; "Einstein's Philosophy of Science," *The Kenyon Review*, Vol. XII (1950), p. 525, and "The Formation of Modern Conceptions of Formal Logic in the Development of Geometry," *Osiris*, Vol. VII (1939), pp. 212–16; H. Weyl: *Philosophy of Mathematics and Natural Science* (Princeton: Princeton University Press; 1949), p. 34, and O. Hölder: *Die Mathematische Methode* (Berlin: Julius Springer; 1924), p. 400, n. 2.

[1] Cf. H. Poincaré: *The Foundations of Science, op. cit.*, pp. 81–86.

[2] *Ibid.*, p. 86.

[3] This critique of Russell's *Foundations of Geometry* appeared in *Revue de Métaphysique et de Morale*, Vol. VII (1899), pp. 251–79; the transplanted excerpt is given in Sec. 12, pp. 265–67 of this paper.

lowed by his important rejoinder "Sur les Principes de la Géométrie, Réponse à M. Russell."[4] These neglected papers together with his posthumous *Dernières Pensées*[5] seem to me to show convincingly that Poincaré was not an opponent of the qualified kind of empiricist position taken by Reichenbach and Carnap. And I explain his apparent endorsement of unmitigated, anti-empiricist conventionalism in his more publicized writings on the basis of the historical context in which he wrote.

For at the turn of the century, the Riemannian kind of qualified empiricist conception of physical geometry, which takes full cognizance of the stipulational status of congruence and which we now associate with writers like Carnap and Reichenbach, had hardly secured a sufficient philosophical following to provide a stimulus and furnish a target for Poincaré's polemic. Instead, the then dominant philosophical interpretations of geometry were such aprioristic neo-Kantian ones as Couturat's and Russell's, on the one hand, and Helmholtz's type of empiricist interpretation, which made *inadequate* allowance for the stipulational character of congruence, on the other.[6] No wonder, therefore, that Poin-

---

[4] *Revue de Métaphysique et de Morale,* Vol. VIII (1900), pp. 73–86; the relevant paper by Russell is "Sur les Axiomes de la Géométrie," in Vol. VII (1899), pp. 684–707 of that same journal.

[5] H. Poincaré: *Dernières Pensées* (Paris: Flammarion; 1913), Chaps. 2 and 3.

[6] Cf. H. von Helmholtz: *Schriften zur Erkenntnistheorie,* ed. by P. Hertz and M. Schlick (Berlin: Julius Springer; 1921), pp. 15–20. H. Freudenthal has maintained (*Mathematical Reviews,* Vol. XXII [1961], p. 107) that instead of being a supporter of Riemann against Helmholtz, Poincaré was an exponent of Helmholtz's anti-Riemannian view that metric geometry presupposes a three-dimensional rather than a merely one-dimensional solid body as a congruence standard. Freudenthal backs that interpretation of Poincaré by the latter's declaration that "if then there were no solid bodies in nature, there would be no geometry" ("L'Espace et la Géométrie," *Revue de Métaphysique et de Morale,* Vol. III [1895], p. 638). According to Freudenthal ("Zur Geschichte der Grundlagen der Geometrie," *Nieuw Archief voor Wiskunde,* Vol. V [1957], p. 115), this declaration shows that "Poincaré still thinks quite in the empiricist spirit of Helmholtz's space problem and has not even penetrated to Riemann's conception, which is aware of a metric without rigid bodies." But, contrary to Freudenthal, it seems clear from the context of Poincaré's declaration that his mention of the role of solid bodies pertains not at all to a Helmholtzian insistence on a three-dimensional congruence standard as against Riemann's one-dimensional one; instead it concerns the role of solids in the genesis of the notion of mere change of position as against other changes of state, solids being distinguished from liquids and gases by the fact that their displace-

caré's conventionalist emphasis in his better known but incomplete writings seems, in the contemporary context, to place him into the ranks of such extreme conventionalists as H. Dingler.[7]

As evidence for my non-standard interpretation of Poincaré as being a *qualified* geometric empiricist rather than an extreme conventionalist, I cite the following crucial and unequivocal concluding passage from Poincaré's rejoinder to Russell, who had maintained that the "axiom of free mobility" furnishes a uniquely true criterion of congruence as an a priori condition for the possibility of metric geometry in the Kantian presuppositional sense and not in the sense of a coordinative definition. Poincaré writes:

> Finally, I have never said that one can *ascertain by experiment whether certain bodies preserve their form.* I have said just the contrary. The term "to preserve one's form" has no meaning by itself. But I confer a meaning on it by *stipulating* that certain bodies will be said to preserve their form. These bodies, thus chosen, can henceforth serve as instruments of measurement. But if I say that these bodies preserve their form, it is because *I choose to do so* and not because experience obliges me to do so.
>
> In the present context I choose to do so, because by a series of *observations* ("constatations") analogous to those which were under discussion in the previous section [i.e., observations showing the coincidence of certain points with others in the course of the movements of bodies] *experience has proven* to me that their movements form a Euclidean group. I have been able to make these observations in the manner just indicated

---

ments lend themselves to compensation by a corresponding movement of our own bodies, which issues in the restoration of the set of sense impressions we had of the solids prior to their displacement.

But this view is, of course, entirely consonant both with Riemann's conception of the congruence standard as one-dimensional and with his claim that, being continuous, physical space has no intrinsic metric, the latter having to be brought in from elsewhere, as is done by the use of the rigid body. In fact, how except by embracing precisely this view could Poincaré have espoused the conventionality of congruence and the resulting alternative metrizability of physical space on which he founded his thesis of the feasibility of either a Euclidean or a non-Euclidean description?

[7] Poincaré himself deplored the widespread misunderstandings of his philosophical work and its misappropriation by "all the reactionary French journals." Cf. his *La Mécanique Nouvelle,* cited in R. Dugas: "Henri Poincaré devant les Principes de la Mécanique," *Revue Scientifique,* Vol. LXXXIX (1951), p. 81.

> *without having any preconceived idea concerning metric geometry.* And, having made them, I judge that the convention will be convenient and I adopt it.[8]

It must also be remembered that Poincaré's declaration that "no geometry is either true or false"[9] was made by him as part of a discussion in which he contrasted his endorsement of this proposition with his complete rejection of the following two others: First, the truth of Euclidean geometry is known to us a priori independently of all experience, and second, one of the geometries is true and the others false, but we can never know which one is true. The entire tenor of this discussion makes it clear that Poincaré is concerned there with abstract, uninterpreted geometries whose relations to physical facts are as yet indeterminate by virtue of the absence of coordinative definitions. It is because he is directing his critique against those who fail to grasp that the identification of the equality predicate "congruent" with its denotata is not a matter of factual truth but of coordinative definition that he asks in *Science and Hypothesis*: "how shall one know [without circularity] that any concrete magnitude which I have measured with my material instrument really represents the abstract distance?"[1]

But one might either contest my interpretation here or conclude that Poincaré was inconsistent by pointing to the following passage by him:

> Should we . . . conclude that the axioms of geometry are experimental verities? . . . If geometry were an experimental science, it would not be an exact science, it would be subject to a continual revision. Nay, it would from this very day be convicted of error, since we know that there is no rigorously rigid solid.
>
> *The axioms of geometry therefore are . . . conventions* . . . Thus it is that the postulates can remain rigorously true even

[8] H. Poincaré: "Sur les Principes de la Géométrie, Réponse à M. Russell," *op. cit.*, pp. 85–86, italics in the latter paragraph are mine.

[9] *Ibid.*, pp. 73–74.

[1] H. Poincaré: *Foundations of Science, op. cit.*, p. 82. Cf. also the paper cited in the preceding footnote, where he writes (p. 77): "One would thus have to define distance by measurement" and (p. 78): "The geometric [abstract] distance is thus in need of being defined; and it can be defined only by means of measurement."

> though the experimental laws which have determined their adoption are only approximative.[2]

The only way in which I can construe the latter passage and others like them in the face of our earlier citations from him is by assuming that Poincaré maintained the following: there are practical rather than logical obstacles which frustrate the complete elimination of perturbational distortions, and the resulting vagueness (spread) as well as the finitude of the empirical data provide scope for the exercise of a certain measure of convention in the determination of a metric tensor.

This reading of Poincaré accords with the interpretation of him in L. Rougier's *La Philosophie Géométrique de Henri Poincaré*. Rougier writes:

> The conventions fix the language of science which can be indefinitely varied: once these conventions are accepted, the facts expressed by science necessarily are either true or false. . . . Other conventions remain possible, leading to other modes of expressing oneself; but the truth, thus diversely translated, remains the same. One can pass from one system of conventions to another, from one language to another, by means of an appropriate dictionary. The very possibility of a translation shows here the existence of an invariant. . . . Conventions relate to the variable language of science, not to the invariant reality which they express.[3]

(C) CRITICAL EVALUATION OF EINSTEIN'S CONCEPTION OF THE INTERDEPENDENCE OF GEOMETRY AND PHYSICS: PHYSICAL GEOMETRY AS A COUNTER-EXAMPLE TO THE NON-TRIVIAL D-THESIS.

Einstein has articulated and endorsed Duhem's claim by reference to the special case of testing a hypothesis of physical geometry. In opposition to the Carnap-Reichenbach conception, Einstein maintains[4] that no hypothesis of physical geometry is *separately* falsifiable, i.e., in isolation from the remainder of

[2] *Ibid.*, pp. 64–65. Similar statements are found on pp. 79 and 240.

[3] L. Rougier: *La Philosophie Géométrique de Henri Poincaré* (Paris: F. Alcan; 1920), pp. 200–201.

[4] Cf. A. Einstein: "Reply to Criticisms," *op. cit.*, pp. 676–78.

physics, even though all of the terms in the vocabulary of the geometrical theory, including the term "congruent" for line segments and angles, have been given a specific physical interpretation. And the substance of his argument is briefly the following: In order to follow the practice of ordinary physics and use rigid solid rods as the physical standard of congruence in the determination of the geometry, it is essential to make computational allowances for the thermal, elastic, electromagnetic, and other deformations exhibited by solid rods. The introduction of these corrections is an essential part of the logic of testing a physical geometry.[5] For the presence of inhomogeneous thermal and other such influences issues in a dependence of the coincidence behavior of transported solid rods on the latter's *chemical composition*, whereas physical geometry is conceived as the system of metric relations exhibited by transported solid bodies independently of their particular chemical composition. The demand for the computational *elimination* of such substance-specific distortions as a prerequisite to the experimental determination of the geometry has a thermodynamic counterpart: the requirement of a means for measuring temperature which does not yield the discordant results produced by expansion thermometers at other than fixed points when different thermometric substances are employed. This thermometric need is fulfilled successfully by Kelvin's thermodynamic scale of temperature.

But Einstein argues that the geometry itself can never be accessible to experimental falsification in isolation from those other laws of physics which enter into the calculation of the corrections compensating for the distortions of the rod. And from this he then concludes that you can always preserve any geometry you like by suitable adjustments in the associated correctional physical laws. Specifically, he states his case in the form of a dialogue in which he attributes his own Duhemian view to Poincaré and offers that view in opposition to Hans Reichenbach's

---

[5] For a very detailed treatment of the relevant computations, see B. Weinstein: *Handbuch der Physikalischen Massbestimmungen* (Berlin: Julius Springer), Vol. I (1886), and Vol. II (1888); A. Pérard: *Les Mesures Physiques* (Paris: Presses Universitaires de France; 1955); U. Stille: *Messen und Rechnen in der Physik* (Braunschweig: Vieweg; 1955), and R. Leclercq: *Guide Théorique et Pratique de la Recherche Expérimentale* (Paris: Gauthier-Villars; 1958).

conception, which was discussed in Chapter Three. But I submit that Poincaré's text will *not* bear Einstein's interpretation. For, as we saw in Section B, when speaking of the variations which solids exhibit under distorting influences, Poincaré says "we neglect these variations in laying the foundations of geometry, because, besides their being very slight, they are irregular and consequently seem to us accidental."[6] I am therefore taking the liberty of replacing the name "Poincaré" in Einstein's dialogue by the term "Duhem and Einstein." With this modification, the dialogue reads as follows:

> Duhem and Einstein: The empirically given bodies are not rigid, and consequently can not be used for the embodiment of geometric intervals. Therefore, the theorems of geometry are not verifiable.
>
> Reichenbach: I admit that there are no bodies which can be *immediately* adduced for the "real definition" [i.e., physical definition] of the interval. Nevertheless, this real definition can be achieved by taking the thermal volume-dependence, elasticity, electro- and magneto-striction, etc., into consideration. That this is really and without contradiction possible, classical physics has surely demonstrated.
>
> Duhem and Einstein: In gaining the real definition improved by yourself you have made use of physical laws, the formulation of which presupposes (in this case) Euclidean geometry. The verification, of which you have spoken, refers, therefore, not merely to geometry but to the entire system of physical laws which constitute its foundation. An examination of geometry by itself is consequently not thinkable. —Why should it consequently not be entirely up to me to choose geometry according to my own convenience (i.e., Euclidean) and to fit the remaining (in the usual sense "physical") laws to this choice in such manner that there can arise no contradiction of the whole with experience?[7]

By speaking here of the "real definition" (i.e., the coordinative definition) of "congruent intervals" by the corrected transported rod, Einstein is ignoring that the actual and potential physical meaning of congruence in physics *cannot* be given exhaustively

[6] H. Poincaré: *The Foundations of Science, op. cit.*, p. 76.
[7] A. Einstein: "Reply to Criticisms," *op. cit.*, pp. 676–78 as modified.

by any *one* physical criterion or test condition. But here as elsewhere in this book, we can safely ignore this open cluster character of the concept of congruence. Our concern as well as Einstein's is merely to single out *one* particular congruence class from among an infinitude of such alternative classes. And as long as our specification of that one chosen class is unambiguous, it is wholly immaterial that there are also other physical criteria (or test conditions) by which it could be specified.

Einstein is making two major points here: First, in obtaining a physical geometry by giving a physical interpretation of the postulates of a formal geometric axiom system, the specification of the physical meaning of such theoretical terms as "congruent," "length," or "distance" is not at all simply a matter of giving an operational definition in the strict sense. Instead, what has been variously called a "rule of correspondence" (Margenau and Carnap), a "coordinative definition" (Reichenbach), an "epistemic correlation" (Northrop), or a "dictionary" (N. R. Campbell) is provided here through the mediation of hypotheses and laws which are collateral to the geometric theory whose physical meaning is being specified. Einstein's point that the physical meaning of congruence is given by the transported rod *as corrected theoretically* for idiosyncratic distortions is an illuminating one and has an abundance of analogues throughout physical theory, thus showing, incidentally, that strictly operational definitions are a rather simplified and limiting species of rules of correspondence. In particular, we see that the physical interpretation of the term "length," which is often adduced as the prototype of all "operational" definitions in Bridgman's sense, is not given operationally in any distinctive sense of that ritually invoked term. Einstein's second claim, which is the cardinal one for our purposes, is that the role of collateral theory in the physical definition of congruence is such as to issue in the following circularity, from which there is no escape, he maintains, short of acknowledging the existence of an a priori element in the sense of the Duhemian ambiguity: the rigid body is not even defined without first decreeing the validity of Euclidean geometry (or of some other particular geometry). For before the corrected rod can be used to make an empirical determination of the de facto geometry, the required corrections must be computed via laws, such as those of elasticity, which involve Euclideanly calculated areas and volumes.[8] But clearly the warrant for thus introducing Euclidean geometry *at this stage* cannot be empirical.

In the same vein, H. Weyl endorses Duhem's position as follows:

> Geometry, Mechanics, and Physics form an inseparable theoretical whole. . . .[9] Philosophers have put forward the thesis that the validity or non-validity of Euclidean geometry cannot be proved by empirical observations. It must in fact be granted that in all such observations essentially physical assumptions, such as the statement that the path of a ray of light is a straight line and other similar statements, play a prominent part. This merely bears out the remark already made above that it is only the whole composed of geometry and physics that may be tested empirically.[1]

If Einstein's and Weyl's Duhemian thesis were to prove correct, then it would have to be acknowledged that there is a sense in which physical geometry itself does not provide a geometric characterization of physical reality. For by this characterization we understand the articulation of the system of relations obtaining between bodies and transported solid rods quite apart from their substance-specific distortions. And to the extent to which physical geometry is a priori in the sense of the Duhemian ambiguity, there is an ingression of a priori elements into physical theory to take the place of distinctively geometric gaps in our knowledge of the physical world.

I now wish to set forth my doubts regarding the soundness of Einstein's geometrical form of the D-thesis by demonstrating the separate falsifiability of the geometric hypothesis H. And I shall do so in two parts, the first of which deals with the simplified case in which effectively no deforming influences are present in a certain region whose geometry is to be ascertained. Having argued in Section (A) that the non-trivial D-thesis is a *non-sequitur,* my present aim is to show by geometric counter-example that it is also false.

---

[8] Cf. I. S. Sokolnikoff: *Mathematical Theory of Elasticity* (New York: McGraw-Hill Book Company; 1946), and S. Timoshenko and J. N. Goodier: *Theory of Elasticity* (New York: McGraw-Hill Book Company; 1951).

[9] H. Weyl: *Space-Time-Matter* (New York: Dover Publications, Inc.; 1950), p. 67.

[1] *Ibid.,* p. 93.

If we are confronted with the problem of the falsifiability of the geometry ascribed to a region which is effectively free from deforming influences, then the correctional physical laws play no role as auxiliary assumptions, and the latter reduce to the claim that the region in question is, in fact, effectively free from deforming influences. And if such freedom can be affirmed without presupposing collateral theory, then the geometry alone rather than only a wider theory in which it is ingredient will be falsifiable. By contrast, if collateral theory were presupposed here, then Duhem and Einstein might be able to adduce its modifiability to support their claim that the geometry itself is not separately falsifiable.

Specifically, they might argue then that the collateral theory could be modified such that the region then turns out not to be free from deforming influences with resulting inconclusive falsifiability of the geometry. The question is therefore whether freedom from deforming influences can be asserted and ascertained independently of (sophisticated) collateral theory. My answer to this question is "Yes." For quite independently of the conceptual elaboration of such physical magnitudes as temperature, whose constancy would characterize a region free from deforming influences, the absence of perturbations is certifiable for the region as follows: two solid rods of very different chemical constitution which coincide at one place in the region will also coincide everywhere else in it (independently of their paths of transport). It would not do for the Duhemian to object here that the certification of two solids as quite different chemically is theory-laden to an extent permitting him to uphold his thesis of the inconclusive falsifiability of the geometry. For suppose that observations were so ambiguous as to permit us to assume that two solids which appear strongly to be chemically different are, in fact, chemically identical in all relevant respects. If so rudimentary an observation were thus ambiguous, then no observation could ever possess the required univocity to be incompatible with an observational consequence of a total theoretical system. And if that were the case, Duhem could hardly avoid the following conclusion: "observational findings are always so unrestrictedly ambiguous as not to permit even the refutation of any given total theoretical system." But such a result would be tantamount to the absurdity that any total theoretical system can be espoused as true a priori.

By the same token, incidentally, I cannot see what methodological safeguards would prevent Quine from having to countenance such an outcome within the framework of his D-thesis. In view of his avowed willingness to "plead hallucination" to deal with observations not conforming to the hypothesis that "there are brick houses on Elm Street," one wonders whether he would be prepared to say that *all* human observers who make disconfirming observations on Elm Street are hallucinating. And, if so, why not discount all observations incompatible with an *arbitrary* total theoretical system as hallucinatory? Thus, it would seem that if Duhem is to maintain, as he does, that a total theoretical *system* is refutable by confrontation with observational results, then he must allow that the coincidence of diverse kinds of rods at different places in the region (independently of their paths of transport) is certifiable observationally. Accordingly, the absence of deforming influences is ascertainable *independently* of any assumptions as to the geometry and of other (sophisticated) collateral theory.

Let us now employ our earlier notation and denote the geometry by "H" and the assertion concerning the freedom from perturbations by "A." Then, once we have laid down the congruence definition and the remaining semantical rules, the physical geometry H becomes *separately* falsifiable as an *explanans* of the posited empirical findings O′. It is true, of course, that A is only more or less highly confirmed by the ubiquitous coincidence of chemically different kinds of solid rods. But the inductive risk thus inherent in affirming A does not arise from the alleged inseparability of H and A, and that risk can be made exceedingly small without any involvement of H. Accordingly, the actual logical situation is characterized not by the Duhemian schema but instead by the schema

$$[\{(H \cdot A) \rightarrow O\} \cdot \sim O \cdot A] \rightarrow \sim H.$$

It will be noted that we identified the H of the Duhemian schema with the geometry. But since a geometric theory, at least in its synthetic form, can be axiomatized as a conjunction of logically independent postulates, a particular axiomatization of

H could be decomposed logically into various sets of component subhypotheses. Thus, for example, the hypothesis of Euclidean geometry could be stated, if we wished, as the conjunction of two parts consisting respectively of the Euclidean parallel postulate and the postulates of absolute geometry. And the hypothesis of hyperbolic geometry could be stated in the form of a conjunction of absolute geometry and the hyperbolic parallel postulate.

In view of the logically compounded character of a geometric hypothesis, Professor Grover Maxwell has suggested that the Duhemian thesis may be tenable in this context if we construe it as pertaining not to the falsifiability of a geometry as a whole but to the falsifiability of its component subhypotheses in any given axiomatization. There are two ways in which this proposed interpretation might be understood: first, as an assertion that *any one component sub*hypothesis eludes conclusive refutation on the grounds that the empirical findings can falsify the set of axioms only as a whole, or second, in any given axiomatization of a physical geometry there exists *at least one component sub*-hypothesis which eludes conclusive refutation.

The first version of the proposed interpretation will not bear examination. For suppose that H is the hypothesis of Euclidean geometry and that we consider absolute geometry as one of its subhypotheses and the Euclidean parallel postulate as the other. If now the empirical findings were to show, on the one hand, that the geometry is hyperbolic, then indeed absolute geometry would have eluded refutation; but if, on the other hand, the prevailing geometry were to turn out to be spherical, then the mere replacement of the Euclidean parallel postulate by the spherical one could not possibly save absolute geometry from refutation. For absolute geometry alone is logically incompatible with spherical geometry and hence with the posited empirical findings.

If one were to read Duhem as per the very cautious *second* version of Maxwell's proposed interpretation, then our analysis of the logic of testing the geometry of a *perturbation-free* region could not be adduced as having furnished a counter-example to so mild a form of Duhemism. And the question of the validity of this highly attenuated version is thus left open by our analysis without any detriment to that analysis.

We now turn to the critique of Einstein's Duhemian argument

as applied to the empirical determination of the geometry of a region which *is* subject to deforming influences.

There can be no question that when deforming influences are present, the laws used to make the corrections for deformations involve areas and volumes in a fundamental way (e.g., in the definitions of the elastic stresses and strains) and that this involvement presupposes a geometry, as is evident from the area and volume formulae of differential geometry, which contain the square root of the determinant of the components g_{ik} of the metric tensor.[2] Thus, the empirical determination of the geometry involves the joint assumption of a geometry and of certain collateral hypotheses. But we see already that this assumption cannot be adequately represented by the conjunction H • A of the Duhemian schema, where H represents the geometry.

Now suppose that we begin with a set of Euclideanly formulated physical laws P_0 in correcting for the distortions induced by perturbations and then use the thus Euclideanly corrected congruence standard for empirically exploring the geometry of space by determining the metric tensor. *The initial stipulational affirmation of the Euclidean geometry G_0 in the physical laws P_0 used to compute the corrections in no way assures that the geometry obtained by the corrected rods will be Euclidean!* If it is non-Euclidean, then the question is: What will be required by Einstein's fitting of the physical laws to preserve Euclideanism and avoid a contradiction of the theoretical system with experience? Will the adjustments in P_0 necessitated by the retention of Euclidean geometry entail merely a change in the dependence of the length assigned to the transported rod on such nonpositional parameters as temperature, pressure, and magnetic field? Or could the putative empirical findings compel that the length of the transported rod be likewise made a nonconstant function of its position and orientation as independent variables in order to square the coincidence findings with the requirement of Euclideanism? The possibility of obtaining non-Euclidean results by measurements carried out in a spatial region uniformly characterized by standard conditions of temperature, pressure, electric and magnetic field strength, etc., shows it to be extremely

[2] L. P. Eisenhart: *Riemannian Geometry* (Princeton: Princeton University Press; 1949), p. 177.

doubtful, as we shall now show, that the preservation of Euclideanism could always be accomplished short of introducing *the dependence of the rod's length on the independent variables of position or orientation.*

But the introduction of the latter dependence is none other than so radical a change in the meaning of the word "congruent" that this term now denotes a class of intervals different from the original congruence class denoted by it. And such tampering with the semantical anchorage of the word "congruent" violates the requirement of semantical stability, which is a necessary condition for the *non*-triviality of the D-thesis, as we saw in Section A.

Suppose that, relatively to the customary congruence standard, the geometry prevailing in a given region when free from perturbational influences is that of a strongly non-Euclidean space of spatially and temporally constant curvature. Then what would be the character of the alterations in the customary correctional laws which Einstein's thesis would require to assure the Euclideanism of that region relatively to the customary congruence standard under *perturbational* conditions? The required alterations would be *independently falsifiable,* as will now be demonstrated, because they would involve affirming that such coefficients as those of linear thermal expansion *depend on the independent variables of spatial position.* That such a space dependence of the correctional coefficients might well be necessitated by the exigencies of Einstein's Duhemian thesis can be seen as follows by reference to the law of linear thermal expansion. In the usual version of physical theory, the first approximation of that law[3] is given in terms of the deviation $\triangle T$ from the standard temperature by

$$L = L_0 (1 + \alpha \cdot \triangle T).$$

[3] This law is only the first approximation, because the rate of thermal expansion varies with the temperature. The general equation giving the magnitude m_t (length or volume) at a temperature t, where m_o is the magnitude at 0°C, is

$$m_t = m_o (1 + \alpha t + \beta t^2 + \gamma t^3 + \ldots),$$

where α, β, γ, etc., are empirically determined coefficients [Cf. *Handbook of Chemistry and Physics* (Cleveland: Chemical Rubber Publishing Company; 1941), p. 2194]. The argument which is about to be given by reference to the approximate form of the law can be readily generalized to forms of the law involving more than one coefficient of expansion.

If Einstein is to guarantee the Euclideanism of the region under discussion by means of logical devices that are consonant with his thesis, and if our region is subject only to thermal perturbations for some time, then we are confronted with the following situation: unlike the customary law of linear thermal expansion, the revised form of that law needed by Einstein will have to bear the twin burden of effecting both of the following two kinds of superposed corrections: first, the *changes* in the lengths ascribed to the transported rod in different positions or orientations which would be required even if our region were everywhere at the standard temperature, merely for the sake of rendering Euclidean its otherwise non-Euclidean geometry, and second, corrections compensating for the effects of the de facto deviations from the standard temperature, these corrections being the sole onus of the usual version of the law of linear thermal expansion. What will be the consequences of requiring the revised version of the law of thermal elongation to implement the *first* of these two kinds of corrections in a context in which the deviation $\triangle T$ from the standard temperature is the *same* at some *different* points of the region (e.g., $\triangle T = 0$), that temperature deviation having been measured in the manner chosen by the Duhemian? Specifically, what will be the character of the coefficients α of the revised law of thermal elongation under the posited circumstances, if Einstein's thesis is to be implemented by effecting the *first* set of corrections? Since the new version of the law of thermal expansion will then have to guarantee that the lengths L assigned to the rod at the various points of *equal* temperature T *differ* appropriately, it would seem clear that logically possible empirical findings could compel Einstein to make the coefficients α of solids *depend* on the *space coordinates.*

But such a spatial dependence is *independently falsifiable*: comparison of the thermal elongations of an aluminum rod, for example, with an invar rod of essentially zero α by, say, the Fizeau method might well show that the α of the aluminum rod is a characteristic of aluminum which is not dependent on the space coordinates. And even if it were the case that the α's are found to be space dependent, how could Duhem and Einstein assure that this space dependence would have the particular functional form required for the success of their thesis?

We see that the required resort to the introduction of a spatial dependence of the thermal coefficients might well not be open to Einstein. Hence, in order to retain Euclideanism, it would then be necessary to remetrize the space in the sense of abandoning the customary definition of congruence, entirely apart from any consideration of idiosyncratic distortions and even after correcting for these in some way or other. But this kind of remetrization, though entirely admissible in other contexts, does not provide the requisite support for Einstein's Duhemian thesis! For Einstein offered it as a criticism of Reichenbach's conception. And hence it is the avowed onus of that thesis to show that the geometry by itself cannot be held to be empirical, i.e., separately falsifiable, even when, with Reichenbach, we have sought to assure its empirical character by choosing and then adhering to the usual (standard) definition of spatial congruence, which excludes resorting to such remetrization.

Thus, there may well obtain observational findings O′, expressed in terms of a particular definition of congruence (e.g., the customary one), which are such that there does not exist any non-trivial set A′ of auxiliary assumptions capable of preserving the Euclidean H in the face of O′. And this result alone suffices to invalidate the Einsteinian version of Duhem's thesis to the effect that any geometry, such as Euclid's, can be preserved in the face of any experimental findings which are expressed in terms of the customary definition of congruence.

It might appear that my geometric counterexample to the Duhemian thesis of unavoidably contextual falsifiability of an *explanans* is vulnerable to the following criticism: "To be sure, Einstein's geometric articulation of that thesis does not leave room for saving it by resorting to a remetrization in the sense of making the length of the rod vary with position or orientation even *after* it has been corrected for idiosyncratic distortions. But why saddle the Duhemian thesis as such with a restriction peculiar to Einstein's particular version of it? And thus why not allow Duhem to save his thesis by countenancing those alterations in the congruence definition which are remetrizations?"

My reply is that to deny the Duhemian the invocation of such an alteration of the congruence definition in this context is not a matter of gratuitously requiring him to justify his thesis within the confines of Einstein's particular version of that thesis; instead, the imposition of this restriction is entirely legitimate here, and the Duhemian could hardly wish to reject it as unwarranted. For it is of the essence of Duhem's contention that H (in this case, Euclidean geometry) can always be preserved not by tampering with the principal semantical rules (interpretive sentences) linking H to the observational base (i.e., the rules specifying a particular congruence class of intervals, etc.), but rather by availing oneself of the alleged *inductive latitude* afforded by the ambiguity of the experimental evidence to do the following: (a) leave the factual commitments of H essentially *unaltered* by retaining both the statement of H and the principal semantical rules linking its terms to the observational base, and (b) replace the set A by A′ such that A and A′ are logically incompatible under the hypothesis H. *The qualifying words "principal" and "essential" are needed here in order to obviate the possible objection that it may not be logically possible to supplant the auxiliary assumptions A by A′ without also changing the factual content of H in some respect.* Suppose, for example, that one were to abandon the optical hypothesis A that light will require equal times to traverse congruent closed paths in an inertial system in favor of some rival hypothesis. Then the semantical linkage of the term "congruent space intervals" to the observational base is changed to the extent that this term no longer denotes intervals traversed by light in equal round-trip times. But such a change in the semantics of the word "congruent" is innocuous in this context, since *it leaves wholly intact the membership of the class of spatial intervals that is referred to as a "congruence class."* In this sense, then, the modification of the optical hypothesis leaves intact both the *"principal"* semantical rule governing the term "congruent" *and* the *"essential"* factual content of the geometric hypothesis H, which is predicated on a particular congruence class of intervals. That "essential" factual content is the following: relatively to the congruence specified by unperturbed transported rods—among other things—the geometry is Euclidean.

Now, the essential factual content of a geometrical hypothesis can be changed either by preserving the original statement of the hypothesis while changing one or more of the principal semantical rules or by keeping all of the semantical rules intact and suitably changing the statement of the hypothesis. We can see, therefore, that the retention of a Euclidean H by the device of changing through remetrization the semantical rule governing the meaning of "congruent" (for line segments) effects a retention not of the essential *factual commitments* of the original Euclidean H but only of its *linguistic trappings*. That the thus "preserved" Euclidean H actually repudiates the essential factual commitments of the *original* one is clear from the following: the *original* Euclidean H had asserted that the coincidence behavior common to all kinds of solid rods is Euclidean, *if* such transported rods are taken as the physical realization of congruent intervals; but the Euclidean H which survived the confrontation with the posited empirical findings only by dint of a *remetrization* is predicated on a denial of the very assertion that was made by the original Euclidean H, which it was to "preserve." It is as if a physician were to endeavor to "preserve" an a priori diagnosis that a patient has acute appendicitis in the face of a negative finding (yielded by an exploratory operation) as follows: he would redefine "acute appendicitis" to denote the healthy state of the appendix!

Hence, the confines within which the Duhemian must make good his claim of the preservability of a Euclidean H do not admit of the kind of change in the congruence definition which alone would render his claim tenable under the assumed empirical conditions. Accordingly, the geometrical critique of Duhem's thesis given here does not depend for its validity on restrictions peculiar to Einstein's version of it.

Even apart from the fact that Duhem's thesis precludes resorting to an alternative metrization to save it from refutation in our geometrical context, the very feasibility of alternative metrizations is vouchsafed not by any general Duhemian considerations pertaining to the logic of falsifiability but by a property peculiar to the subject matter of geometry (and chronometry): the latitude for convention in the ascription of the spatial (or temporal) equality relation to intervals in the continuous manifolds of physical space (or time).

But what of the possibility of actually extricating the unique underlying geometry (to within experimental accuracy) from the network of hypotheses which enter into the testing procedure?

That contrary to Duhem and Einstein, the geometry itself may well be empirical, once we have renounced the kinds of alternative congruence definitions employed by Poincaré is seen from the following possibilities of its successful empirical determination. After assumedly obtaining a non-Euclidean geometry G_1 from measurements with a rod corrected on the basis of Euclideanly formulated physical laws P_o, we can revise P_o so as to conform to the non-Euclidean geometry G_1 just obtained by measurement. This retroactive revision of P_o would be effected by recalculating such quantities as areas and volumes on the basis of G_1 and changing the functional dependencies relating them to temperatures and other physical parameters. Let us denote by "P'_1" the set of physical laws P resulting from thus revising P_o to incorporate the geometry G_1. Since various physical magnitudes ingredient in P'_1 involve lengths and durations, we now use the set P'_1 to correct the rods (and clocks) with a view to seeing whether rods and clocks thus corrected will reconfirm the set P'_1. If not, modifications need to be made in this set of laws so that the functional dependencies between the magnitudes ingredient in them reflect the new standards of spatial and temporal congruence defined by P'_1-corrected rods and clocks. We may thus obtain a new set of physical laws P_1. Now we employ this set P_1 of laws to correct the rods for perturbational influences and then determine the geometry with the thus corrected rods. Suppose the result is a geometry G_2 different from G_1. Then *if*, upon repeating this two step process several more times, there is convergence to a geometry of constant curvature we must continue to repeat the two step process an additional finite number of times until we find the following: the geometry G_n ingredient in the laws P_n providing the basis for perturbation-corrections is indeed the same (to within experimental accuracy) as the geometry obtained by measurements with rods that have been corrected via the set P_n.

If there is such convergence at all, it will be to the same geometry G_n even if the physical laws used in making the *initial* corrections are not the set P_o, which presupposes Euclidean geometry, but a different set P based on some non-Euclidean geometry or other. That there can exist only one such geometry of constant curvature G_n would seem to be guaranteed by the identity of G_n with the unique underlying geometry G_t characterized by the following properties: (1) G_t would be exhibited by the coincidence behavior of a transported rod if the *whole* of the space were actually free of deforming influences, (2) G_t would be obtained by measurements with rods corrected for distortions on the basis of physical laws P_t presupposing G_t, and (3) G_t would be found to prevail in a given relatively small, perturbation-free region of the space quite independently of the assumed geometry ingredient in the correctional physical laws. Hence, *if* our method of successive approximation does converge to a geometry G_n of *constant* curvature, then G_n would be this unique underlying geometry G_t. And, in that event, we can claim to have found empirically that G_t is indeed the geometry prevailing in the entire space which we have explored.

But what if there is no convergence? It might happen that whereas convergence would obtain by starting out with corrections based on the set P_0 of physical laws, it would *not* obtain by beginning instead with corrections presupposing some particular non-Euclidean set P or vice versa: just as in the case of Newton's method of successive approximation,[4] there are conditions, as A. Suna has pointed out to me, under which there would be no convergence. We might then nonetheless succeed as follows in finding the geometry G_t empirically, *if* our space is one of constant curvature.

The geometry G_r resulting from measurements by means of a corrected rod is a single-valued function of the geometry G_a assumed in the correctional physical laws, and a Laplacian demon having sufficient knowledge of the facts of the world would know this function $G_r = f(G_a)$. Accordingly, we can formulate the problem of determining the geometry empirically as the problem of finding the point of intersection between the curve representing this function and the straight line $G_r = G_a$. That there exists one and only one such point of intersection follows from the existence of the geometry G_t defined above, provided that our space is one of constant curvature. Thus, what is now needed is to make determinations of the G_r corresponding to a number of geometrically different sets of correctional physical laws P_a, to draw the most reasonable curve $G_r = f(G_a)$ through this finite number of points (G_a, G_r), and then to find the point of intersection of this curve and the straight line $G_r = G_a$.

Whether this point of intersection turns out to be the one repre-

[4] R. Courant: *Vorlesungen über Differential und Integralrechnung* (Berlin: Julius Springer; 1927), Vol. I, p. 286.

senting Euclidean geometry or not is beyond the reach of our conventions, barring a remetrization. And thus the least that we can conclude is that since empirical findings can greatly narrow down the range of uncertainty as to the prevailing geometry, there is no assurance of the latitude for the choice of a geometry which Einstein takes for granted in the manner of the D-thesis. Einstein's Duhemian position would appear to be safe from this further criticism only if our proposed method of determining the geometry by itself empirically cannot be generalized in some way to cover the general relativity case of a space of variable curvature and if the latter kind of theory turns out to be true. The extension of our method to the case of a geometry of variable curvature is not simple. For in that case, the geometry G is no longer represented by a single scalar given by a Gaussian curvature, and our graphical method breaks down.

If my proposed method of escaping from the web of the Duhemian ambiguity were shown to be unsuccessful, and if there should happen to be no other scientifically viable means of escape, then, it seems to me, we would unflinchingly have to resign ourselves to this relatively unmitigable type of uncertainty. No, says the philosopher Jacques Maritain, who enticingly beckons us to take heart. The *scientific* elusiveness of the correct geometric description of external reality must *not* lead us to suppose, he tells us, that philosophy, when divorced from mathematical physics, cannot rescue us from the labyrinth of the Duhemian perplexity and unveil for us the structure of what he calls *ens geometricum reale* (real geometrical being).[5] As against Maritain's conception of the capabilities of philosophy as an avenue of cognition, I wish to uphold the following excellent declaration by Professor Bridgman: "The physicist emphatically would *not* say that his knowledge presumptively will not lead to a full understanding of reality for the reason that there are other kinds of knowledge than the knowledge in which he deals."[6] To justify my endorsement of Professor Bridgman's statement in this

[5] J. Maritain: *The Degrees of Knowledge* (London: G. Bles Company; 1937), p. 207. A new translation of this work was published in New York in 1959 by Charles Scribner's Sons; all page references here are to the 1937 translation.

[6] P. W. Bridgman: "The Nature of Physical 'Knowledge,'" in: L. W. Friedrich (ed.) *The Nature of Physical Knowledge* (Bloomington: Indiana University Press; 1960), p. 21.

context, I shall give a brief critique of Maritain's philosophy of geometry as presented in his book *The Degrees of Knowledge.*[7]

I have selected Maritain's views for rebuttal because they typify the conception of those who believe that the philosopher as such has at his disposal means for fathoming the structure of external reality which are not available to the scientist. In outline, Maritain endeavors to justify this idea in regard to geometry along the following lines. Says he: "There is no clearer word than the word reality, which means *that which is.* . . . What is meant when it is asked whether real space is euclidean or non-euclidean . . . ?"[8] To prepare for his answer to this question, he explains the following: "The word *real* has not the same meaning for the philosopher, the mathematician and the physicist.[9] . . . For the physicist a space is 'real' when the geometry to which it corresponds permits of the construction of a physico-mathematical universe which coherently and completely symbolizes physical phenomena, and where all our graduated readings find themselves 'explained.' And it is obvious that from this point of view no space of any kind holds any sort of privileged position.[1] But . . . the question is to know what is real space in the philosophical meaning of the word, i.e., as a 'real' entity . . . designating an object of thought capable of an extra-mental existence. . . ."[2] One is immediately puzzled as to how Maritain conceives that his distinction between physically real space and the philosophically real space which is avowedly extra-mental is not an empty distinction without a difference. And, instead of being resolved, this puzzlement only deepens when he tells us that by the *extra-mental geometric* features of existing bodies he understands "those properties which the mind recognizes in the elimination of all the physical."[3] But let us suspend judgment concerning this difficulty and see whether it is not cleared up by his treatment of the following question posed by him: How are we to know whether it is Euclidean geometry or one of the non-Euclidean geometries that represents the structure of the philosophically real, i.e.,

[7] J. Maritain: *The Degrees of Knowledge, op. cit.*, pp. 201–12.
[8] *Ibid.*, p. 201.
[9] *Ibid.*
[1] *Ibid.*, p. 202.
[2] *Ibid.*, p. 203.
[3] *Ibid.*, p. 207.

extra-mental or external space?[4] In regard to this question, he makes the following assertions:

First, the capabilities of physical measurements to yield the answer to the question are nil,[5] because a geometry is presupposed in the theory of our measuring instruments which forms the basis of corrections for "accessory variations due to various physical circumstances."[6] Of course, we recognize this contention to be a strong form of the Duhemian one, although Maritain does not refer to Duhem.

Second, the several non-Euclidean geometries depend for their consistency on their formal translatability into Euclidean geometry. This translation is effected by providing a Euclidean model of the particular non-Euclidean geometry in the sense of embedding an appropriately curved non-Euclidean surface in the three-dimensional Euclidean space. And the privileged position which Euclidean geometry enjoys as the underwriter of the consistency of the non-Euclidean geometries thus issues in a correlative dependence of the intuitability of the non-Euclidean geometries on the primary intuitability of the Euclideanism of the embedding three-dimensional hyperspace.[7]

Using the twin arguments from consistency and intuitability, Maritain then reaches the following final conclusion:

> The non-Euclidean spaces can then without the least intrinsic contradiction be the object of consideration by the mind, but there would be a contradiction in supposing their existence outside the mind, and thereby suppressing, for their benefit, the existence of the foundation on which the notion of them is based.
>
> Either way we are thus led to admit, despite the use which astronomy makes of them, that these non-euclidean spaces are rational [i.e., *purely mental*] beings; and that the *geometric* properties of existing bodies, those properties which the mind recognizes in the elimination of all the physical, are those which characterize euclidean space. For philosophy it is euclidean space which appears as an *ens geometricum reale*.[8]

---

[4] Cf. *ibid.*, p. 204.
[5] Cf. *ibid.*, p. 204.
[6] *Ibid.*, p. 205.
[7] Cf. *ibid.*, pp. 202, 205–206.
[8] *Ibid.*, pp. 206–7.

I submit that Maritain's thesis is unsound in its entirety and can be completely refuted as follows:

First, as Hilbert and Bernays have explained,[9] the consistency of the Euclidean axiom system is not vouchsafed by its intuitive plausibility as an adequate description of the space of our immediate physical environment. Instead, we establish the consistency of Euclidean geometry by providing a model of the formal Euclidean postulates in the domain of real numbers in the manner of analytic geometry.[1] Now, Maritain overlooks that precisely the same procedure of providing a real number model can be used to establish the internal consistency of the various non-Euclidean geometries without the mediation of a prior translation into Euclidean geometry (except possibly in an irrelevant heuristic sense). And he is misled by the fact that, historically, the consistency of the several non-Euclidean geometries was established by means of a translation into Euclidean geometry, as for example in Klein's relative consistency proof of hyperbolic geometry via a model furnished by the interior of a circle in the Euclidean plane. For surely the temporal priority of Euclidean geometry inherent in the historical circumstances of our discovery of the internal consistency of the various non-Euclidean geometries hardly serves to establish the logical primacy of Euclidean geometry as the sole guarantee of their consistency. And Maritain's error on this count is only compounded by his intuitability argument for the uniqueness of Euclideanism as the only possible structure of extra-mental reality. The latter argument is vitiated by the inveterate error of being victimized by the misleading connotation of *embedding in a Euclidean hyperspace,* which is possessed by the terms "curved space" and "curvature of a surface." This connotation springs from unawareness that the Gaussian curvature of a 2-space and the Riemannian curvatures for the various orientations at points of a 3-space are intrinsically definable and discernible properties of these spaces, requiring no embedding. Moreover, Maritain overlooks here that even when the consistency proof of hyperbolic geometry, for example, is

---

[9] D. Hilbert and P. Bernays: *Grundlagen der Mathematik* (Berlin: Julius Springer; 1934), Vol. I, §1, section A, pp. 2–3.

[1] Cf. L. P. Eisenhart: *Coordinate Geometry* (New York: Dover Publications, Inc.; 1960), Appendix to Chap. 1, pp. 279–92.

given on the basis of Euclidean geometry—which we saw is quite unnecessary—this can be accomplished without embedding, as in the case of the aforementioned *two*-dimensional Klein model, just as readily as by Beltrami's procedure of embedding a surface of constant negative Gaussian curvature (containing singular lines) in Euclidean 3-space.

Lastly, it can surely not be maintained that "the *geometric* properties of existing bodies" are "those properties which the mind recognizes in the elimination of all the physical." For, in that case, geometry would be the study of purely imagined thought-objects, which will, of course, turn out to have Euclidean properties, if Maritain's imagination thus endows them. And the geometry of such an imagined space could then not qualify as the geometry of Maritain's real or extra-mental space. The geometric theory of external reality does indeed abstract from a large class of physical properties in the sense of being the metrical study of the coincidence behavior of transported solids independently of the solids' substance-specific physical properties. But this kind of abstracting does not deprive metrical coincidence behavior of its physicality. And if the methods of the physicist cannot fathom the laws of that behavior, then certainly no other kind of intellectual endeavor will succeed in doing so.

It is true, of course, that even apart from experimental errors, not to speak of quantum limitations on the accuracy with which the metric tensor *of space-time* can be meaningfully ascertained by measurement,[2] no *finite* number of data can uniquely determine the functions constituting the representations g_{ik} of the metric tensor in any given coordinate system. But the criterion of *inductive* simplicity which governs the free creativity of the geometer's imagination in his choice of a particular metric tensor here is the same as the one employed in theory formation in any of the non-geometrical portions of empirical science. And choices made on the basis of such inductive simplicity are in principle true or false, unlike those springing from considerations of descriptive simplicity, which merely reflect conventions.

---

[2] E. P. Wigner: "Relativistic Invariance and Quantum Phenomena," *Reviews of Modern Physics*, Vol. XXIX (1957), pp. 255–68, and H. Salecker and E. P. Wigner: "Quantum Limitations of the Measurement of Space-Time Distances," *The Physical Review*, Vol. CIX (1958), pp. 571–77.

Chapter 5

EMPIRICISM AND THE GEOMETRY OF VISUAL SPACE

A very brief review of the account of our knowledge of visual space given by Carnap, Helmholtz, and Reichenbach will precede the discussion of some problems posed by very recent *experimental* studies of the geometry of *visual* space.

Distinguishing the space of physical objects from the space of visual experience ("Anschauungsraum"), Carnap sided with empiricism even in his earliest work to the extent of maintaining that the topology of *physical* space is known a posteriori and that the coincidence relations among points disclosed by experience yield a unique metrization for that space once a specific coordinative definition of congruence has been chosen freely.[1] But the neo-Kantian *parti pris* of that period enters in his epistemological interpretation of the axioms governing the topology of *visual* space: "Experience does not provide the justification for them, the axioms are . . . independent of the 'quantity of experience,' i.e., knowledge of them does not, as in the case of a posteriori propositions, become ever more reliable through multiply repeated experience. For, as Husserl has shown, we are dealing here not with facts in the sense of empirically ascertained realities but rather with the essence ("eidos") of certain presentations whose special nature can be grasped in a single immediate ex-

---

[1] R. Carnap: *Der Raum, op. cit.*, pp. 39, 45, 54, 63.

perience."[2] Reminding us of Kant's distinction between knowledge acquired "with" experience, on the one hand, and "from" experience, on the other, the early Carnap classified these axioms as synthetic a priori propositions in that philosopher's sense.

This theory of the phenomenological a priori was a stronger version of Helmholtz's claim that "space can be transcendental [a priori] while its axioms are not."[3] For Helmholtz's concession to Kantianism was merely to regard an *amorphous* visual extendedness as an a priori condition of spatial experience[4] while proclaiming the a posteriori character of the topological and metrical articulations of that extendedness on the basis of his pioneering method of imagining ("sich ausmalen") the specific sensory contents we would have in worlds having alternative spatial structures.[5]

The phenomenological a priori will not do, however, as an account of our knowledge of the properties of visual space. For it is an *empirical* fact that the experiences resulting from ocular activity have the indefinable attribute which is characteristic of visual extendedness rather than that belonging to tactile explorations or to those experiences that would issue from our possession

---

[2] *Ibid.*, p. 22. Cf. also p. 62. For a more recent defense of the thesis that "there are synthetic a priori judgments of spatial intuition," cf. K. Reidemeister: "Zur Logik der Lehre vom Raum," *Dialectica,* Vol. VI (1952), p. 342. For a discussion of related questions, see P. Bernays: "Die Grundbegriffe der reinen Geometrie in ihrem Verhältnis zur Anschauung," *Naturwissenschaften,* Vol. XVI (1928).

[3] H. von Helmholtz: *Schriften zur Erkenntnistheorie,* P. Hertz and M. Schlick (eds.), (Berlin: Julius Springer; 1921), p. 140.

[4] *Ibid.*, pp. 2, 70, 121–22, 140–42, 144–45, 147–48, 152, 158, 161–62, 163, 168, 172, 174. Helmholtz attempts to characterize the distinctive attribute of space, not possessed by other tri-dimensional manifolds, in the following way: "in space, the distance between two points on a vertical can be compared to the horizontal distance between two points on the floor, because a measuring device can be applied successively to these two pairs of points. But we cannot compare the distance between two tones of equal pitch and differing intensity with that between two tones of equal intensity and differing pitch" (*ibid.*, p. 12). Schlick, however, properly notes in his commentary (*ibid.*, p. 28) that this attribute is necessary but not sufficient to render the distinctive character of space.

[5] *Ibid.*, pp. 5, 22, 164–65. Cf. also K. Gerhardt's papers: "Nichteuklidische Kinematographie," *Naturwissenschaften,* Vol. XX (1932), p. 925, and "Nichteuklidische Anschauung und optische Täuschungen," *Naturwissenschaften,* Vol. XXIV (1936), p. 437.

of a sense organ responding to magnetic disturbances. In the class of all logically possible experiences, the "Wesensschau" provided by our ocular activity must be held to give rise to empirical knowledge. For the only way to assure a priori that all future deliverances of our eyes will possess the characteristic attribute which Husserl would have us ascertain in a single glance is by resorting to a covert tautology via refusing to call the resulting knowledge "knowledge of visual space," unless it possesses that attribute.

Reichenbach made a particularly telling contribution to the disintegration of the Kantian *metrical* a priori of *visual* space by showing that such intuitive compulsion as inheres in the Euclideanism of that space derives from facts of logic in which the Kantian interpretation cannot find a last refuge and that the counter-intuitiveness of non-Euclidean relations is merely the result of both ontogenetic and phylogenetic adaptation to the Euclidicity of the physical space of ordinary life.[6]

In very recent years, experimental mathematico-optical researches by R. K. Luneburg[7] and A. A. Blank[8] have even led these authors to contend that although the *physical* space in which sensory depth perception by binocular vision is effective is Euclidean, the binocular *visual* space resulting from psychometric coordination possesses a Lobatchevskian *hyperbolic* geometry of constant curvature. This contention suggests several questions.

[6] H. Reichenbach: *The Philosophy of Space and Time, op. cit.*, pp. 32–34 and 37–43.

[7] R. K. Luneburg: *Mathematical Analysis of Binocular Vision* (Princeton: Princeton University Press; 1947), and "Metric Methods in Binocular Visual Perception," in *Studies and Essays*, Courant Anniversary Volume (New York: Interscience Publishers, Inc.; 1948), pp. 215–39.

[8] A. A. Blank: "The Luneburg Theory of Binocular Visual Space," *Journal of the Optical Society of America*, Vol. XLIII (1953), p. 717; "The non-Euclidean Geometry of Binocular Visual Space," *Bulletin of the American Mathematical Society*, Vol. LX (1954), p. 376; "The Geometry of Vision," *The British Journal of Physiological Optics*, Vol. XIV (1957), p. 154; "The Luneburg Theory of Binocular Perception," in S. Koch (ed.) *Psychology, A Study of a Science* (New York: McGraw-Hill Book Company, Inc.; 1958), Study I, Vol. I, Part III, Sec. A. 2; "Axiomatics of Binocular Vision. The Foundations of Metric Geometry in Relation to Space Perception," *Journal of the Optical Society of America*, Vol. XLVIII (1958), p. 328, and "Analysis of Experiments in Binocular Space Perception," *Journal of the Optical Society of America*, Vol. XLVIII (1958), p. 911.

The first of these is how human beings manage to get about so easily in a Euclidean physical environment even though the geometry of visual space is presumably hyperbolic. Blank suggests the following as a possible answer to this question: First, man's motor adjustment to his physical environment does not draw on *visual* data alone; moreover, these do contribute physically true information, since they supply a good approximation to the relative directions of objects and since the mapping of physical onto visual space preserves the topology (though not the metric) of physical space, thereby enabling man to control his motor responses by feedback, as in the parking of a car or threading the eye of a needle; and second, the thesis of the hyperbolicity of visual space rests on data obtained under experimental conditions which are far more restrictive than those accompanying ordinary visual experience. Under ordinary conditions, we secure depth perception by relying on the coordination of our two ocular images which we have learned in the past in the usual contexts. But in order to ascertain the laws of merely *one* of the sources of spatial information—stereoscopic depth perception alone—the experimenters of the Luneburg-Blank theory endeavored to deny their subjects precisely that contextual reliance: there were no guideposts of perspectives and familiar objects whose positions the subject had determined by tactile means, the only visible objects being isolated point lights in an otherwise completely dark room; in fact, the subject was not even allowed to move his head to make judgments by parallax. Since these contextual guideposts are also available in monocular vision, the experimenters assumed that they play no part in the innate physiological processes governing the distinctive sensations of three-dimensional space which are obtained binocularly.

Several additional questions arise in regard to the Luneburg theory upon going beyond its own restricted objectives of furnishing an account of binocular visual perception and attempting to incorporate its thesis of the non-Euclidean structure of visual space in a comprehensive theory of spatial *learning*: (1) how is man able to arrive at a rather correct apprehension of the Euclidean metric relations of his environment by the use of a physiological instrument whose deliverances are claimed to be non-

Euclidean? (2) how can students be taught *Euclidean* geometry by *visual* methods, methods which certainly convey more than the topology of Euclidean space and whose success is therefore not explained by the fact that the purportedly hyperbolic visual space preserves the topology of Euclidean physical space? (3) if we have literally been seeing one of the non-Euclidean geometries of constant negative Gaussian curvature all along, why did it require two thousand years of research in axiomatics even to *conceive* these geometries, the Euclideanism of physical space being affirmed throughout this period? (4) why did such thinkers as Helmholtz and Poincaré first have to retrain their *Anschauung* conceptually in a counterintuitive direction before achieving a ready pictorialization of the Lobatchevski-Bolyai world, a feat which very few can duplicate even now? (5) if we took two groups of school children of equal intelligence and without prior formal geometrical education and taught Euclid to one group while teaching Lobatchevski-Bolyai to the other, why is it the case (if indeed that is the case!) that, in all probability, the first group would exhibit a far better mastery of their material?

The need to answer these questions becomes even greater, if we assume that our ideas concerning the geometry of our immediate physical environment are formed, in the first instance, not by the physical geometry of yardsticks or by the formal study of Euclidean geometry but rather by the psychometry of our visual sense data.

A. A. Blank, to whom the writer submitted these questions, has suggested that these questions may have answers which lie in part along the following lines: First, man has to learn the significance of ever-changing patterns of visual sensations for the metric of *physical* space by *discounting* much of the psychometry of visual sensation, thereby developing the habit of not being very perceptive of the metrical details of his visual experience. Thus, we learn before adulthood to associate with the non-rigid sequence of visual sensations corresponding to viewing a chair in various positions and contexts the attribute of physical rigidity, generally ignoring all but those aspects of the changing appearances that can serve as a basis for action. In fact, laboratory findings show that for any physical configuration whatever, there are an infinity of others which give the same binocular clues.[9] Since we retain those aspects of visual experience which enable us to place objects in the contexts useful for action, Euclidean relations can be more readily pictured (though not actually seen or made visible) than those of Lobatchevski; second, those geometrical judgments disclosed by binocular perception which are *common* to both Euclidean and hyperbolic geometry[1] will be true physically as well.

Moreover, there are certain small two-dimensional elements of visual space which are essentially isometric with the corresponding elements of the Euclidean space of physical stimuli. For example, in a plane parallel to the line joining the rotation centers of the eyes, physical metric relations are seen undistorted in the vicinity of a point at the base of the perpendicular to the plane from a point located half way between the eyes. We can therefore obtain first-order visual approximations to the physical Euclidean geometry from viewing small diagrams frontally in this way. In a like manner, we can understand how the concept of similar figures, which is uniquely characteristic of Euclidean geometry among spaces of *constant* curvature, can be conveyed in the context of a non-Euclidean visual geometry: all Riemannian geometries are locally Euclidean, thus possessing a group of similarity transformations in the small; third, the presumed greater ease with which students would master Euclid than Lobatchevski is due to the greater analytical simplicity of the numerical relations of Euclidean geometry.

---

[9] Cf. A. A. Blank: "The Luneburg Theory of Binocular Visual Space," *op. cit*, pp. 721–22, and L. H. Hardy, G. Rand, M. C. Rittler, A. A. Blank, and P. Boeder: *The Geometry of Binocular Space Perception* (New York: Columbia University College of Physicians and Surgeons; 1953), pp. 15ff. and 39ff.

[1] For the axioms of the so-called "absolute" geometry relevant here, see R. Baldus: *Nichteuklidische Geometrie*, edited by F. Löbell (3rd revised edition; Berlin: Walter de Gruyter and Company; 1953), Sammlung Göschen, Vol. CMLXX, Chap. ii.

Chapter 6

THE RESOLUTION OF ZENO'S METRICAL PARADOX OF EXTENSION FOR THE MATHEMATICAL CONTINUA OF SPACE AND TIME

It is a commonplace in the analytic geometry of physical space and time that an *extended* straight line segment, having positive length, is treated as "consisting of" *unextended* points, each of which has zero length. Analogously, time intervals of positive duration are resolved into instants, each of which has zero duration.

Ever since some of the Greeks defined a point as "that which has no part,"[1] philosophers and mathematicians have questioned the consistency of conceiving of an extended continuum as an aggregate of unextended elements. On the long list of investigators who have examined this question in the context of the specific mathematical and philosophical theories of their time, we find such names as Zeno of Elea,[2] Aristotle,[3] Cavalieri,[4] Tacquet,[5]

[1] This definition is given in Euclid: *The Thirteen Books of Euclid's Elements,* trans. T. L. Heath (Cambridge: Cambridge University Press; 1926), p. 153.

[2] S. Luria: "Die Infinitesimaltheorie der antiken Atomisten," *Quellen und Studien zur Geschichte der Mathematik, Astronomie, und Physik* (Berlin, 1933), Abteilung B: Studien, II, p. 106.

[3] Aristotle: *On Generation and Corruption,* Book I, Chap. ii, 316^{a}15-317^{a}17; A. Edel: *Aristotle's Theory of the Infinite* (New York: Columbia University Press; 1934), pp. 48–49, 76–78; T. L. Heath: *Mathematics in Aristotle* (Oxford: Oxford University Press; 1949), pp. 90, 117.

Pascal,[6] Bolzano,[7] Leibniz,[8] Paul du Bois-Reymond,[9] and Georg Cantor[1] to mention but a few. Writing on this issue more recently, P. W. Bridgman declared:

> with regard to the paradoxes of Zeno . . . if I literally thought of a line as consisting of an assemblage of points of zero length and of an interval of time as the sum of moments without duration, paradox would then present itself.[2]

This criticism of the mathematical theory of physical space and time is a challenge to the basic Cantorean conceptions underlying analytic geometry and the mathematical theory of motion.[3] Bridgman's view also calls into question such philosophies of science as rely on these conceptions for their interpretations of our mathematical knowledge of nature. Accordingly, it is essential that we inquire whether contemporary point-set theory succeeds in avoiding an inconsistency upon resolving positive linear intervals into extensionless point-elements.

In the present chapter, I shall endeavor to exhibit those features of present mathematical theory which do indeed preclude the existence of such an inconsistency. It will then be clear what kind of mathematical and philosophical theory does succeed in avoiding Zeno's mathematical (metrical) paradoxes of plurality, paradoxes that should be distinguished from his paradoxes of *motion*, on which I shall comment in Chapter Seven. My concern with the views which various writers have attributed to Zeno is exclusively systematic, and I make no claims whatever regarding

---

[4] C. B. Boyer: *The Concepts of the Calculus* (New York: Hafner Publishing Company; 1949), p. 140.

[5] *Ibid.*

[6] *Ibid.*, p. 152.

[7] *Ibid.*, p. 270, and B. Bolzano: *Paradoxes of the Infinite*, ed. D. A. Steele (New Haven: Yale University Press; 1951).

[8] B. Russell: *The Philosophy of Leibniz* (London: G. Allen & Unwin, Ltd.; 1937), p. 114.

[9] P. du Bois-Reymond: *Die Allgemeine Funktionentheorie* (Tübingen: Lauppische Buchhandlung; 1882), Vol. I, p. 66.

[1] G. Cantor: *Gesammelte Abhandlungen*, ed. E. Zermelo (Berlin: Julius Springer; 1932), pp. 275, 374.

[2] P. W. Bridgman: "Some Implications of Recent Points of View in Physics," *Revue Internationale de Philosophie*, Vol. III, No. 10 (1949), p. 490.

[3] G. Cantor: *Gesammelte Abhandlungen*, *op. cit.*, p. 275.

the historicity of Zeno's arguments or concerning the authenticity of views which I shall associate with his name. According to S. Luria,[4] Zeno invokes two basic axioms in his mathematical paradoxes of plurality. Having divided all magnitudes into positive and "dimensionless," i.e., unextended magnitudes, Zeno assumed that (1) the sum of an infinite number of equal positive magnitudes of arbitrary smallness must necessarily be infinite, (2) the sum of any finite or *infinite* number of "dimensionless" magnitudes must necessarily be *zero.*

The second of these axioms seems to command the assent of P. W. Bridgman and was also enunciated by the mathematician Paul du Bois-Reymond[5] who then inferred that we cannot regard a line as an aggregate of "dimensionless" points, however dense an order we postulate for this aggregate. Zeno himself is presumed to have used these axioms as a basis for the following dilemma:[6] If a line segment is resolved into an aggregate of infinitely many *like* elements, then two and only two cases are possible. Either these elements are of equal positive length and the aggregate of them is of infinite length (by Axiom 1) or the elements are of zero length and then their aggregate is necessarily of zero length (by Axiom 2). The first horn of this dilemma is valid but does not have relevance to the modern analytic geometry of space and time. It is the second horn that we must refute in the context of present mathematical theory if we are to solve the problem which we have posed.

To carry out this refutation, we must first ascertain the logical relationships between the modern concepts of metric, length, measure, and cardinality, when applied to (infinite) point-sets. For in the second horn of his dilemma, Zeno avers that a line cannot be regarded as an aggregate of points no matter what cardinality we postulate for the aggregate. And du Bois-Reymond endorsed this contention by reminding us that points are "dimensionless," i.e., unextended and by maintaining that if we conceive the line to be "merely an aggregate of points" then we are *eo*

---

[4] S. Luria: "Die Infinitesimaltheorie der antiken Atomisten," *op. cit.*, p. 106.

[5] P. du Bois-Reymond: *Die Allgemeine Funktionentheorie, op. cit.*, p. 66.

[6] H. Hasse and H. Scholz: *Die Grundlagenkrisis der griechischen Mathematik* (Charlottenburg: Pan-Verlag; 1928), p. 11.

ipso abandoning the view that "a line and a point are entirely different things."[7]

We see that du Bois-Reymond is conforming to the long intuitive tradition of using the concepts of length and dimensionality interchangeably to characterize (sensed) extension. It will therefore be best to begin our analysis by noting that we must distinguish the traditional *metrical* usage of the term "dimensionless" from the contemporary *topological* meaning of "zero dimension." This distinction has become necessary by virtue of the autonomous development of the topological theory of dimension apart from metrical geometry. Prior to this development, any positive interval of Cartesian *n*-space was simply called "*n*-dimensional" by definition. Thus line segments having length were called "one-dimensional" and surfaces having area "two-dimensional." By contrast, in the topological theory of dimension developed in the present century, it is a non-trivial *theorem* that lines are topologically one-dimensional, surfaces two-dimensional, and, generally, that Cartesian *n*-space is *n*-dimensional. In fact, it is this theorem which warrants the use of the name "dimension theory" for the branch of topology dealing with such non-metrical properties of point-sets as make for the validity of this theorem.[8] By contrast, the traditional *metrical* sense of dimensionality identifies dimensionality with length or measure of extendedness. It is only the latter sense of "dimension" and "dimensionless" which is relevant to the metrical problem of this chapter. Hence I refer the reader to another publication of mine[9] for an account of how the twentieth century theory of dimension can *consistently* affirm the following *additivity* properties for dimension in the *topological* sense of "zero-dimensional" and "one-dimensional": The point-set constituting the number-axis or any finite interval in it (e.g., an infinite straight line or a finite line segment respectively) is *one*-dimensional even though it is the set-theoretic sum of *zero*-dimensional subsets. The zero-dimensional subsets are: (a) Any unit point-set (such a set has a single point as its only member

---

[7] P. du Bois-Reymond: *Die Allgemeine Funktionentheorie, op. cit.*, p. 65.

[8] K. Menger: *Dimensionstheorie* (Leipzig: B. G. Teubner; 1928), p. 244.

[9] A. Grünbaum: "A Consistent Conception of the Extended Linear Continuum as an Aggregate of Unextended Elements," *Philosophy of Science*, Vol. XIX (1952), pp. 290–95.

and hence can be loosely referred to as a "point," whenever such usage does not permit ambiguities), (b) Any *finite* collection of one or more points, (c) Any denumerable set (in particular the set of rational real points), (d) The set of irrational real points, which is non-denumerably infinite.

Accordingly, we must now deal with the *metrical* question of how the definition of length can *consistently* assign zero length to *unit* point-sets or individual points while assigning positive *finite* lengths to such unions (sums) of these unit point-sets as constitute a finite interval. To furnish an answer to the latter question will be to refute the second horn of Zeno's dilemma. We shall furnish an analysis satisfying these requirements after devoting some attention to the consideration of prior related problems.

Length or extension is defined as a property of point-sets rather than of individual points, and zero length is assigned to the *unit set,* i.e., to a set containing only a single point. While it is both logically correct and even of central importance to our problem that we treat a line interval of geometry as a set of point-elements, strictly speaking the definition of "length" renders it incorrect to refer to such an interval as an "aggregate of unextended points." For the property of being unextended characterizes unit point-*sets* but is not possessed by their respective individual point-*elements,* just as temperature is a property only of aggregates of molecules and not of individual molecules. The entities which can therefore be properly said to be unextended are *included in* but are not *members of* the aggregate of points constituting a line interval. Accordingly, the line interval is a union of unextended unit point-sets and, strictly, not an "aggregate of unextended points." Though strictly incorrect, I wish to use the latter designation in order to avoid the more cumbersome expression "union of unextended unit point-sets."

I shall now present such portions of the theory of metric spaces as bear immediately on our problem.

The structure characterizing the class of all real numbers (positive, negative, and zero) arranged in order of magnitude is that of a linear Cantorean continuum.[1]

---

[1] E. V. Huntington: *The Continuum and Other Types of Serial Order* (2nd ed.; Cambridge: Harvard University Press; 1942), pp. 10, 44.

The Euclidean point-sets or "spaces" which we shall have occasion to consider are "metric" in the following complex sense:[2]

1. There is a one-one correspondence between the points of an *n*-dimensional Euclidean space E^n and a certain real coordinate system $(x_1, \ldots, x_n)$
2. If the points x, y have the coordinates x_i, y_i, then there is a real function d(x, y) called their (Euclidean) *distance* given by

$$d(x, y) = \left\{ \sum_{1}^{n} (x_i - y_i)^2 \right\}^{1/2}.$$

The basic properties of this function are given by certain distance *axioms.*[3]

A finite interval on a straight line is the (ordered) set of all real points between (and sometimes including one or both of) two fixed points called the "endpoints" of the interval. Since the points constituting an interval satisfy condition 1 above in the definition of "metric," it is possible to define the "distance" between the fixed endpoints of a given interval. The *number* representing this distance is the *length* of the point-*set* constituting the interval. Let "a" and "b" denote respectively the points a and b *or* their respective real number coordinates, depending upon the context. We then define the length of a finite interval (a, b) as the non-negative quantity b–a regardless of whether the interval {x} is closed ($a \leqq x \leqq b$), open ($a < x < b$) or half-open ($a \leqq x < b$ or $a < x \leqq b$). (It is understood that the symbols "<" and "=" have a purely ordinal meaning here.) Therefore, the set-theoretic addition of a single point to an open interval (or to a half-open interval at the open end) has no effect at all on the *length* of the resulting interval as compared with the length of the original interval. In the limiting case of a = b, the interval is called "degenerate," and here the closed interval reduces to a set containing the single point x = a, while each of the other three intervals is empty. *It follows that the length of a degen-*

[2] S. Lefschetz: *Introduction to Topology* (Princeton: Princeton University Press; 1949), p. 28.

[3] *Ibid.*

erate interval is zero. Loosely speaking, a single point has 0 length.[4]

Zeno is challenging us to obtain a result differing from zero when determining the length of a finite interval on the basis of the known zero lengths of its degenerate subintervals, each of which has a single point as its only member. But since each positive interval has a *non-denumerable* infinity of degenerate subintervals, we see already that the result of determining the length of that interval by "compounding," in some unspecified way, the zero lengths of its degenerate subintervals is far less obvious than it must have seemed to Zeno, who did not distinguish between countably and non-countably infinite sets!

Although length is similar to cardinality in being a property of *sets* and not of the elements of these, it is essential to realize that the cardinality of an interval is not a function of the length of that interval. The independence of cardinality and length becomes demonstrable by combining our definition of length with Cantor's proof of the equivalence of the set of all real points between 0 and 1 with the set of all real points between *any* two fixed points on the number axis. It is therefore not the case that the longer of two positive intervals has "more" points. Once the independence of cardinality and length of such point-sets is established, it is possible to eliminate several of the confusions which have vitiated certain treatments of the infinite divisibility of intervals, as we shall see below. Thus, it will become impossible to infer in finitist manner that the division of an interval into two or more subintervals imparts to each of the resulting subintervals a cardinality lower than the cardinality of the original interval.

An interesting illustration of the independence of cardinality and length is provided by the so-called "ternary set" (Cantor discontinuum). This set has zero length (and zero dimension) while having the cardinality of the continuum.[5]

We shall be concerned with ascertaining why Zeno's *paradoxical* result that the length of a given *positive* interval (a, b) is

---

[4] H. Cramér: *Mathematical Methods of Statistics* (Princeton: Princeton University Press; 1946), pp. 11, 19.

[5] R. Courant and H. Robbins: *What is Mathematics?* (New York: Oxford University Press; 1941), p. 249.

zero is *not* deducible from the following two propositions in our geometry: (1) Any positive or non-degenerate interval is the union of a continuum of degenerate subintervals, and (2) The length of a degenerate (sub)interval is zero. It is obvious that *if the theory is consistent,* Zeno's result cannot be deducible. Such a result would contradict the proposition that the length of the interval (a, b) is b–a $(a \neq b)$. Furthermore, this result would be incompatible with Cantor's theorem that all positive intervals have the same cardinality regardless of length, for this theorem shows that no inference regarding the length of a non-degenerate interval can be drawn from propositions (1) and (2). In order to show later that our theory does have the required consistency, i.e., that it does not lend itself to the deduction of Zeno's paradoxical result, we must now consider the determination of (a) the length of the union of a *finite* number of non-overlapping intervals of known lengths, and (b) the length of the union of a *denumerable* infinity of such intervals.

If an interval "i" is the union of a *finite* number of intervals, no two of which have a common point, i.e., if

$$i = i_1 + i_2 + i_3 + \ldots + i_n, \quad (i_p i_q = 0 \text{ for } p \neq q),$$

then it follows readily from the theory previously developed that the length of the total interval is equal to the *arithmetic sum* of the individual lengths of the subintervals. We therefore write

$$L(i) = L(i_1) + L(i_2) + L(i_3) + \ldots + L(i_n).$$

If now we *define* the arithmetic sum of a progression of finite cardinal numbers as the limit of a sequence of partial arithmetic sums of members of the sequence, then a non-trivial proof can be given[6] that the following theorem holds: The length of an interval which is subdivided into an enumerable number of subintervals without common points is equal to the arithmetic sum of the lengths of these subintervals.

Thus, both for a finite number and for a *countably* infinite number of non-overlapping subintervals, the length L(i) of the total interval is an additive function of the interval i. The length of an interval is a numerical measure of the comprehensiveness

[6] Cf. H. Cramér: *Mathematical Methods of Statistics, op. cit.,* pp. 19–21.

(extension) of that interval's *membership* but not of its cardinality. The latter does not depend upon the comprehensiveness of the membership of an interval.

It will be recalled that "length" was defined only for intervals. So far, we have not assigned any property akin to length to such sets as the set of all rational points between and including 0 and 1. There are many occasions, however, when it is desirable to have some kind of *measure* of the extensiveness, as it were, of point-sets quite different from intervals. Problems of this kind as well as problems encountered in the theory of (Lebesgue) integration have prompted the introduction of the generalized metrical concept of "measure" to deal with sets other than intervals as well. For details on the definition of "measure" for various kinds of point-sets, the reader is referred to Cramér[7] and Halmos.[8]

Since the theory of infinite divisibility has been used fallaciously in an attempt to deduce Zeno's metrical paradox, we shall now point out the relevant fallacies before dealing with the crux of our problem to refute the second horn of Zeno's metrical dilemma.

In an exchange of views with Leibniz, Johann Bernoulli committed an important fallacy: he treated the actually infinite set of natural numbers as having a *last* or "∞^{th}" term which can be "reached" in the manner in which an inductive cardinal can be reached by starting from zero.[9] Bernoulli's view is clearly self-contradictory, since no discrete denumerable infinity of terms could possibly have a last term.

When giving arguments in behalf of his theory of infinitesimals, C. S. Peirce[1] committed the same Bernoullian fallacy by reasoning as follows: (1) the decimal expansion of an irrational number has an infinite number of terms, (2) the infinite decimal expansion has a *last* element at the "infinitieth place," and since the latter is "infinitely far out" in the decimal expansion, this element

---

[7] H. Cramér: *Mathematical Methods of Statistics, op. cit.*, pp. 22ff.

[8] P. R. Halmos: *Measure Theory* (New York: D. Van Nostrand Company, Inc.; 1950.)

[9] H. Weyl: *Philosophy of Mathematics and Natural Science, op. cit.*, p. 44.

[1] C. Hartshorne and P. Weiss (eds.): *The Collected Papers of Charles Sanders Peirce* (Cambridge: Harvard University Press; 1935), Vol. VI, para. 125.

is infinitely small or infinitesimal in comparison to finite magnitudes, (3) since continuity requires irrationals, continuity presupposes infinitesimals. Furthermore, the method of defining irrational points by nested intervals[2] was *misconstrued* by Du Bois-Reymond[3] such that he was then able to charge it with committing the Bernoullian fallacy.[4]

We are now concerned with this fallacy, because it is always committed when the attempt is made to use the *infinite divisibility* of positive intervals as a basis for deducing Zeno's metrical paradox and for then denying that a positive interval can be an aggregate of unextended points. Precisely this kind of deduction of the paradox is attributed to Zeno by H. D. P. Lee[5] and P. Tannery,[6] both of whom seem to be unaware of the fallacy involved.

The following basic assumptions are involved in their version of Zeno's arguments:

1. Infinite *divisibility* guarantees the possibility of a *completable* process of "infinite division," i.e., of a completable yet infinite set of division operations.

2. The completion of this process of "infinite division" is achieved by the last operation in the series and terminates in

---

[2] R. Courant and H. Robbins: *What is Mathematics?*, *op. cit.*, pp. 68–69.

[3] P. du Bois-Reymond: *Die Allgemeine Funktionentheorie*, Vol. I, *op. cit.*, pp. 58–67.

[4] Du Bois Reymond's fundamental error lies in supposing that the method of nested intervals allows and requires the "coalescing" of the end-points of a supposedly "next-to-the-last" interval into a single point such that this "coalescing" is the *last step* in an infinite progression of nested interval formations. If the method in question did require such a coalescing, then it would indeed be as objectionable logically as is the Bernoullian conception of the ∞^{th} or last natural number. This is not the case, however, for while the method does indeed make reference to a progression of intervals, it neither allows nor requires that the irrational point is the "last" or "∞^{th}" such "contracted" interval. Instead of appealing to "coalescence," the method specifies the irrational point *by the mode of variation* of the intervals in the *entire sequence*. It is therefore a property of the entire sequence which enables us to define the kind of point which is being asserted to exist. It would seem that Du Bois-Reymond permitted himself to be misled by pictorial language like "the interval contracts into a point."

[5] H. D. P. Lee: *Zeno of Elea* (Cambridge: Luzac and Company; 1936), p. 23.

[6] P. Tannery: "Le Concept Scientifique du Continu: Zénon d'Élée et Georg Cantor," *Revue Philosophique*, Vol. XX, No. 2 (1885), pp. 391–92.

"reaching" a last product of division: a mathematical point of zero extension.[7]

3. An *actual infinity* of distinct elements is generated by such an allegedly completable process of "infinite division."

4. Since the divisions begin with a first operation, each have an immediate successor, and each, except the first, have a specific predecessor, they jointly constitute a progression.

By assumptions 3 and 4, the "final elements" to which Zeno's metrical argument is to be applied are presumed to have been generated by *a completed progression of division operations*. This consequence, however, is absurd. For it is the very essence of a progression not to have a last term and not to be completable! To maintain the self-contradictory proposition that in such an actually infinite aggregate there is a "last" division which insures the completability of the process of "infinite division" by "reaching" a "final" product of division is indeed to commit the Bernoullian fallacy.

Several consequences follow at once:

(1) We do not ever "arrive" by "infinite division" at an actual infinity of mathematical points, and therefore there is no question at all of generating an *actual* infinity of unextended elements by "infinite division."

(2) *The facts of infinite divisibility do not by themselves legitimately give rise to the metrical paradoxes of* Zeno, for these *may* arise only if we postulate an actual infinity of point-elements *ab initio*. It is because Cantor's theory rests on this latter postulate and not because his number-axis is infinitely divisible that we must inquire whether the line as conceived by Cantor is beset by the metrical difficulties pointed out by Zeno.

To show that this latter assertion is justified within the context of point-set theory, we shall now construct on the foundations of that theory a treatment of infinite divisibility consistent with it.

No clear meaning can be assigned to the "division" of a line unless we specify whether we understand by "line" an entity like a *sensed* "continuous" chalk mark on the blackboard or the very differently continuous line of Cantor's theory. The "continuity"

[7] For a discussion of the temporal rather than ordinal aspects of the concept of a discrete denumerable set of operations, see my "Messrs. Black and Taylor on Temporal Paradoxes," *Analysis*, Vol. XII (1952), pp. 144–48.

of the *sensed* linear expanse consists essentially in its failure to exhibit visually noticeable gaps as the eye scans it from one of its extremities to the other. There are no distinct *elements* in the sensed "continuum" of which the seen line can be said to be a structured aggregate. By contrast, the continuity of the Cantorean line consists precisely in the complicated structural relatedness of (point) *elements* which is specified by the postulates for real numbers. We *cannot* always perceive a distinct third gap between any two visually discernible gaps (sections) in the sensed line. Thus the visually discernible gaps (sections) in that line do not constitute a discernibly dense set. This means that any significant assertion concerning possible divisibility of a *sensed* line must be compatible with the existence of thresholds of perception. Division of the *sensed* line will mean the creation of one or more perceptible gaps in it. Contrariwise, any attribution of (infinite) "divisibility" to a Cantorean line must be based on the fact that *ab initio* that line and its intervals are already "divided" into an actual dense infinity of point-elements of which the line (interval) is the aggregate. Accordingly, the Cantorean line can be said to be already actually *infinitely divided.* "Division" of the line can therefore mean neither the creation of visual gaps in it nor the "separation" of the point elements from one another to make them distinct. What we will mean in speaking of the "division" of the Cantorean line is the formation of non-overlapping subintervals from intervals of the line, subject to the following three conditions:

1. For purposes of division, the degenerate interval is not a subinterval of any interval.
2. The empty set is not a subinterval of any interval.
3. In the case of finite point-sets in general and of the degenerate interval in particular, "division" will mean the formation of proper non-empty *subsets.*

Note that division is an *operation* while divisibility is a *property* of certain point-sets in the case of the Cantorean line. This distinction enabled us to reject the fallacious argument according to which the infinite divisibility of a given point-set insures that a progression of $\aleph_0$ division operations on that set will "terminate" in a single point as the "product" of "infinite *division.*" We saw that "infinite *division*" *of this kind* is self-contradictory. Therefore, confusion is avoided by basing our analysis only on infinite *divisibility* and on the actual infinite "dividedness" of sets.[8]

It follows from our definition of division and from the properties of finite sets that the division of a *finite* point-set of two or more members necessarily effects a reduction in its cardinality. This reduction is in marked contrast to the behavior of *intervals* since their cardinality remains unaffected by division. Since the degenerate interval has no proper non-empty subset, that unique kind of interval is *indivisible*. We see that on our theory, *indivisibility is not a metrical property* at all but a set-theoretic one. This theory has enabled us to assign a precise meaning to the indivisibility of a unit point-set by (1) defining division as an operation on sets only and not on their elements, (2) defining divisibility of finite sets as the formation of proper non-empty subsets of these, (3) showing that the degenerate interval is indivisible by virtue of its lack of a subset of the required kind.

It is of importance to realize that our analysis has shown how we can assert the following *two* propositions *perfectly consistently:*

1. The line and intervals in it are *infinitely divisible.*
2. The line and intervals in it are each a union of *indivisible* degenerate intervals.

We are now prepared to deal with the crux of our problem by using point-set theory to refute the second horn of Zeno's metrical dilemma.

Since a positive interval is the union of a continuum of degenerate intervals,[9] we must now determine what meaning, *if any,* we can assign to "summing" the lengths of these degenerate intervals so as to obtain the length of the total interval. The answer which we shall give to this problem will not be *ad hoc,* since the reasoning on which it is based will not depend upon

[8] Nevertheless it is often convenient *by way of elliptic parlance* to designate the membership of a set through mention of an actual infinity of operations which, as it were, "generate" the elements of the set in question.

[9] The word "continuum" can designate either the ordering structure of the real numbers or their cardinality. The context will indicate which of these meanings is intended or whether both are jointly involved.

the particular lengths which Zenonians wish us to "compound" but rather on the fact that the number of lengths to be "added" is *not denumerable*.

Earlier, we determined the length of the union of a *finite* number of non-overlapping intervals of known lengths on the basis of these latter lengths. In addition, we made a corresponding determination of the length of the union of a *denumerable* infinity of non-overlapping intervals. If now we attempt to subdivide an interval into a *non-denumerable* infinity of such intervals, we find that they *cannot be non-degenerate*. For Cantor has shown that any collection of positive non-overlapping intervals on a line is at most denumerably infinite.[1] It follows that the degenerate subintervals which are at the focus of our interest are the only kind of (non-overlapping) subintervals of which there are non-denumerably many in a given interval. Quite naturally, therefore, they create a special situation. The latter is due to the fact that our theory does *not* assign any meaning to "forming the *arithmetic* sum," when we are attempting to "sum" a *super-denumerable* infinity of individual numbers (lengths)! This fact is independent of whether the individual numbers in such a non-denumerable set of numbers are zeros or finite cardinal numbers differing from zero.

Consequently, the theory under discussion cannot be deemed to be *ad hoc* for precluding the possibility of "adding," in Zenonian fashion, the zero lengths of the continuum of points which "compose" the interval (a, b) to obtain zero as the length of this interval. Though the finite interval (a, b) is the union of a continuum of degenerate subintervals, *we cannot meaningfully determine its length in our theory by "adding" the individual zero lengths of the degenerate subintervals.* We are here confronted with an instance in which set-theoretic addition is possible while arithmetic addition is not.

We have shown that geometrical theory as here presented does *not* have the paradoxical feature of both assigning the non-zero length b–a to the interval (a, b) *and* permitting the inference that (a, b) must have zero length on the grounds that its points each have zero length. More precisely, we have shown that geo-

---

[1] G. Cantor: *Gesammelte Abhandlungen, op. cit.*, p. 153.

metrical theory can consistently affirm the following four propositions simultaneously:

1. The finite interval (a, b) is the union of a continuum of degenerate subintervals.
2. The length of each degenerate (sub)interval is 0.
3. The length of the interval (a, b) is given by the number b–a.
4. The length of an interval is not a function of its cardinality.

Our analysis has manifestly refuted the Zenonian allegation of inconsistency if made against set-theoretical geometry.

The set-theoretical analysis of the various issues raised or suggested by Zeno's paradoxes of plurality has enabled me to give a *consistent* metrical account of an extended line segment as an aggregate of unextended points. Thus, Zeno's mathematical paradoxes are avoided in the formal part of a geometry built on Cantorean foundations. The consistency of the metrical analysis which I have given depends crucially on the *non-denumerability* of the infinite point-sets constituting the intervals on the line. For the length (measure) of an *enumerably* infinite point-set (like the set of rational points between and including 0 and 1) is *zero* (upon denumeration of the set), as can readily be inferred from our analysis above.[2] Thus, if any set of rational points were regarded as constituting an extended line segment, then the customary mathematical theory under consideration could assert the length of that segment to be greater than zero only at the cost of permitting itself to become self-contradictory!

It might seem that my conclusion concerning the fundamental logical importance of non-denumerability could be criticized in the following way: The need for non-denumerably infinite point-sets to avoid metrical contradictions derives from defining the arithmetic sum of an infinite series of numbers as the limit of its partial sums as we have done.[3] Without this definition, it would not have been possible to infer that the sum of the individual zero lengths of the points (unit point-sets) in an enumerable point-set turns out to be *zero* upon denumeration of that set. Consequently, by omitting this definition, it would presum-

[2] Cf. also H. Cramér: *Mathematical Methods of Statistics, op. cit.*, p. 257.

[3] Cf. also G. H. Hardy: *A Course of Pure Mathematics* (9th ed.; New York: The Macmillan Company; 1945), pp. 145–47.

ably have been possible to assign a finite length to certain enumerable sets without contradiction. Thus it might be argued that a non-denumerably infinite point-set is indispensable for consistency only *relatively* to a formulation of the theory containing the definition that the arithmetic "sum" of an infinite series of numbers is the limit of its partial sums.

My reply to this objection is that the omission of the latter definition from the system would entail incurring the loss of the theory of infinite convergent series in analysis and geometry and of whatever part of the theory depends in part or whole upon the presence of this definition in the foundations. It follows that instead of being a merely incidental feature of the theory, the introduction of the definition in question was dictated by important theoretical considerations. The requirement that the points on the line be non-denumerably infinite which must then be satisfied to insure metrical consistency therefore has a corresponding significance.

Consequently, it is not clear how theorists whose philosophical commitments do not allow them to avail themselves of a super-denumerable set would avoid Zeno's mathematical paradoxes in keeping with their theory.

Proponents of Zeno's view might still argue that this *arithmetical* rebuttal is unconvincing on purely geometric grounds, maintaining that if extension (space) is to be composed of elements, these must themselves be extended. Specifically, geometers like Veronese objected[4] to Cantor that in the array of points on the line, their extensions are all, as it were, "summed geometrically" before us. And from this geometric perspective, it is not cogent, in their view, to suppose that even a super-denumerable infinity of unextended points would be able to sustain a positive interval, especially since the Cantorean theory can claim arithmetical consistency here only because of the obscurities that obligingly surround the meaning of the arithmetic "sum" of a super-denumerable infinity of numbers.

Is this objection to Cantor conclusive? I think not. Whence does it derive its plausibility? It would seem that it achieves persuasiveness via a tacit appeal to a *pictorial* representation of the points of mathematical physics in which they are arrayed in

[4] See E. W. Hobson: *The Theory of Functions of a Real Variable* (2nd ed.; Cambridge: Cambridge University Press; 1921) Vol. I, pp. 56–57.

the consecutive manner of beads on a string to form a line. But the properties that any such representation imaginatively attributes to points are not even allowed, let alone prescribed, by the formal postulates of geometric theory. The spuriousness of the difficulties adduced against the Cantorean conception of the line becomes apparent upon noting that not only the cardinality of its constituent points altogether eludes pictorialization but also their dense ordering: between any two points, there is an infinitude of others. Thus, in complete contrast to the discrete order of the beads on a string, *no* point is immediately adjacent to any other.

These considerations show that from a genuinely geometric point of view, a physical interpretation of the formal postulates of geometry cannot be obtained by the inevitably misleading pictorialization of *individual* points of the theory. Instead, we can provide a physical interpretation quite unencumbered by the intrusion of the irrelevancies of *visual* space, if we associate *not* the term *"point"* but the term *"linear continuum of points"* of our theory with an appropriate body in nature. By a point of this body we then mean nothing more or less than an element of it possessing the formal properties prescribed for points by the postulates of geometry. And, on this interpretation, the ground is then cut from under the geometric *parti pris* against Cantor by the modern legatees of Zeno.

I can best illustrate the importance of being mindful of the need for a super-denumerable point-set to avoid Zeno's metrical contradictions by showing how (1) Russell, and (2) the proponents of a strict operationism in geometry were led into error by failing to be aware of this need.

1. Russell neglected the essential contribution made by the cardinality and ordinal structure of the continuum toward the avoidance of Zeno's mathematical paradoxes in the mathematical theory of motion. This philosophical neglect of his is clear in the following passages:

> Mathematicians have distinguished different degrees of continuity, and have confined the word "continuous," for technical purposes, to series having a certain high degree of continuity. But for *philosophical* purposes, all that is important in continuity is introduced by the lowest degree of continuity, which is called "compactness" [i.e., denseness].

> . . . What do we mean by saying that the motion is continuous? It is not necessary for our purposes to consider the whole of what the mathematician means by this statement: *only part of what he means is philosophically important.* One part of what he means is that, if we consider any two positions of the speck occupied at any two instants, there will be other intermediate positions occupied at intermediate instants. . . .[5]

We know that the mere existence of the denseness property guarantees only a denumerably infinite point-set. Since a superdenumerably infinite point-set is required by the demands of metrical consistency, it follows that there are *philosophical* reasons for requiring a higher degree of continuity than is insured by the denseness property alone.

My analysis has shown that within the framework of the mathematics used in physical theory we cannot postulate the line to consist of the rational points alone, since the latter constitute a *denumerable* set. The choice between postulating the line to consist of the real points and resolving it into the rational points alone is therefore *not* a matter of mere arithmetical convenience in the analytic part of geometry or of mere aesthetic satisfaction to the scientist. Russell is unaware of this fact, for he asserts that a "rational space" may be actual. Says he: "a space . . . in which the points of a line form a series ordinally similar to the rationals, will, with suitable axioms, be empirically indistinguishable from a continuous space, *and may be actual.*"[6]

2. The Greeks certainly were *not* led to the discovery of incommensurable magnitudes by merely *operationally* carrying out

[5] B. Russell: *Our Knowledge of the External World* (London: George Allen & Unwin, Ltd.; 1926), pp. 138, 139–40; my italics.

[6] B. Russell: *The Principles of Mathematics* (Cambridge: Cambridge University Press; 1903), p. 444; my italics. A similar criticism applies to Dedekind. He maintains that if we postulated a discontinuous space consisting of the algebraic points alone, then "the discontinuity of this space would not be noticed in Euclid's science, would not be felt at all" (R. Dedekind: *Essays on the Theory of Number* [Chicago: Open Court Publishing Company; 1901], pp. 37–38). Since the set of algebraic points is still denumerable (A. Fraenkel: *Einleitung in die Mengenlehre* [New York: Dover Publications, Inc.; 1946], p. 40), the length (measure) of a segment consisting entirely of such points is paradoxically both zero and positive. I therefore cannot follow F. Waismann, who comments approvingly on Dedekind's statement by saying: "As to physical space one has become accustomed to conceding the justification of this concept" (F. Waismann: *Introduction to Mathematical Thinking* [New York: F. Ungar Publishing Company; 1951], p. 212).

the iterative transport of measuring sticks.[7] And it is impossible to show by direct physical operations alone that there are hypotenuses whose length cannot be represented by any rational number. For the denseness of the rational points guarantees that we can never claim anything but a rational result on the strength of operational accuracy alone. A radical operationist approach to geometry might therefore suggest that this science be constructed so as to use only the system of rational points.[8] The analysis given in this chapter has shown that in the absence of an alternative mathematical theory which is demonstrably viable for the purposes of physics, such an operationist approach to geometry and to the theoretical measurables of physics must be rejected on logical grounds alone.

Though I am claiming to have shown that set-theoretical geometry can successfully meet the challenge of Zeno's mathematical paradoxes *if it is otherwise consistent,* I am *not* asserting that Zeno and his followers were either mistaken within the context of the mathematical knowledge of their time or that they were deficient in philosophical acumen.

---

[7] For the historical details, see K. von Fritz: "The Discovery of Incommensurability by Hippasus of Metapontum," *Annals of Mathematics,* Vol. XLVI (1945).

[8] Cf. the approximative geometry of J. Hjelmslev (J. Hjelmslev: "Die natürliche Geometrie," *Abhandlungen aus dem mathematischen Seminar der Hamburger Universität,* Vol. II (1923), p. 1 ff.) and Weyl's comments on it (H. Weyl: *Philosophy of Mathematics and Natural Science, op. cit.,* pp. 143–44).

PART II

Philosophical Problems of the Topology of Time and Space

Chapter 7

THE CAUSAL THEORY OF TIME

The causal theory of time, which had occupied an important place in the thought of Leibniz and of Kant, again became a subject of central philosophic interest during the current century after its detailed elaboration and logical refinement at the hands of G. Lechalas,[1] H. Reichenbach,[2] K. Lewin,[3] R. Carnap,[4] and H. Mehlberg.[5] Specifically, it earned its new prominence in recent decades by its role in the magisterial and beautiful constructions of the relativistic topology of both time and space by Reichenbach[6] and Carnap.[7] More recently, I used the causal

[1] G. Lechalas: *Étude sur l'espace et le temps* (Paris: Alcan Publishing Company; 1896).

[2] H. Reichenbach: *Axiomatik der relativistischen Raum-Zeit-Lehre* (Braunschweig: Friedrich Vieweg & Sons; 1924).

[3] K. Lewin: "Die zeitliche Geneseordnung," *Zeitschrift für Physik,* Vol. XIII (1923), p. 16.

[4] R. Carnap: "Über die Abhängigkeit der Eigenschaften des Raumes von denen der Zeit," *Kantstudien,* Vol. XXX (1925), p. 331.

[5] H. Mehlberg: "Essai sur la théorie causale du temps," *Studia Philosophica,* Vol. I (1935), and Vol. II (1937).

[6] H. Reichenbach: *Philosophie der Raum-Zeit-Lehre* (Berlin: Walter de Gruyter & Company; 1928), esp. pp. 307–308. The English translation entitled *The Philosophy of Space and Time, op. cit.,* will hereafter be cited as "*PST.*"

[7] R. Carnap: *Abriss der Logistik* (Vienna: Julius Springer: 1929), Sec. 36, pp. 80–85. Cf. also his *Symbolische Logik* (Vienna: Julius Springer; 1954), Secs. 48–50, pp. 169–81; an English translation, *Introduction to Symbolic Logic and Its Applications* was published by Dover Publications, Inc., New York, in 1958. For an interesting comparison of Kant's version of the theory with the conceptions propounded by Carnap, Reichenbach, and Mehlberg, see H. Scholz: "Eine Topologie der Zeit im Kantischen Sinne," *Dialectica,* Vol. IX (1955), p. 66.

theory of time to show semantically that, with respect to the relation *"later than,"* the events of physics can meaningfully possess the seemingly counter-intuitive *denseness* property of the linear Cantorean continuum. And, in this way, I was able to supply the semantical *nervus probandi* which had been lacking in Russell's mathematical refutation of Zeno's paradoxes of motion.[8]

In order to assess the causal theory of time philosophically, we must inquire whether the physical meaning of the primitive *asymmetric* causal relation employed in most versions of it can be understood without possessing a *prior* understanding of the *temporal* terms which it is intended to define. I shall therefore devote the present chapter to an examination of the major defenses and criticisms of the causal theory of time, beginning with the Reichenbachian version of the theory, which is based on the mark method.[9]

My aim will be to show that (1) Reichenbach's mark method fails to define a serial temporal order within the class of physical events, being vitiated by circularity in its attempt to define the required asymmetric relation; (2) although the classical Leibniz-Reichenbach version of the causal theory of time is vulnerable to these criticisms, it is possible to define *temporal betweenness* on the basis of the postulate of causal continuity for reversible mechanical processes; but nomologically contingent boundary conditions must prevail, if the betweenness defined by causal continuity is to have the formal properties of the triadic relation ordering the points on an *open* (straight) *undirected* line, hereafter called "o-betweenness"; alternatively, the boundary conditions may issue in a temporal order exhibiting the formal properties possessed by the order of points on a *closed* (circular) *undirected* line with respect to a tetradic relation of separation, hereafter called "separation closure"; (3) as distinct from Reichenbach's mark method, a significant *modification* of his most recent account of the *anisotropy* of time as a statistical

[8] A. Grünbaum: "Relativity and the Atomicity of Becoming," *op. cit.*, esp. pp. 168–69. Much material concerning time in this article, especially the endorsement of Reichenbach's version of the causal theory of time, is superseded by the treatment given in this book. In addition, the article contains distorting misprints.

[9] Cf. H. Reichenbach: *PST, op. cit.*, pp. 135–38.

property of the *entropic* behavior of *space ensembles* of "branch systems,"[1] is successful; but (4) his conception of becoming as the transience of a physically-distinguished "now" along one of the two opposite directions of time is untenable. Of these four contentions, only (1) and (2) will be defended in the present chapter, (3) and (4) being deferred to Chapter Eight (Part A, II) and Chapter Ten, respectively.

Reichenbach introduces his mark method by giving the following topological coordinative definition of temporal sequence:[2] "If E_2 is the effect of E_1, then E_2 is said to be later than E_1." To show that causality defines an asymmetric temporal relation without circularity, he invites attention to the fact that when E_1 is the cause of E_2, small variations in E_1 in the form of the addition of a marking event e to E_1 will be connected with corresponding variations in E_2, but not conversely. Thus, if we denote an event E that is slightly varied (marked) by "E^e," we shall find, he tells us, that we observe only the combinations

$$E_1E_2 \qquad E^e{}_1E^e{}_2 \qquad E_1E^e{}_2$$

but *never* the combination

$$E^e{}_1E_2.$$

In the observed combinations, the events E_1 and E_2 play an asymmetric role, thereby defining an order, and it is clear that this order would be unaffected by interchanging the subscripts in the symbols naming the events involved. The event whose name does not have an "e" in the non-occurring combination is called the effect and the later event.[3]

---

[1] H. Reichenbach: "Les fondements logiques de la mécanique des quanta," *Annales de l'Institut Henri Poincaré*, Vol. XIII (1953), pp. 140–58, and "La signification philosophique du dualisme ondes-corpuscules," in A. George (ed.) *Louis de Broglie, Physicien et Penseur* (Paris: Editions Albin Michel; 1953), pp. 126–34. The most detailed account is given in his book *The Direction of Time* (Berkeley: University of California Press; 1956). Hereafter this book will be cited as "DT."

[2] H. Reichenbach: *PST, op. cit.*, p. 136. Cf. also G. W. Leibniz: "Initia rerum mathematicorum metaphysica," *Mathematische Schriften*, ed. Gerhardt (Berlin: Schmidt's Verlag; 1863), Vol. VII, p. 18; H. Weyl: *Philosophy of Mathematics and Natural Science, op. cit.*, pp. 101, 204, and "50 Jahre Relativitätstheorie," *Naturwissenschaften*, Vol. XXXVIII (1951), p. 74; H. Poincaré: *Letzte Gedanken*, trans. Lichtenecker (Leipzig: Akademische Verlagsgesellschaft; 1913), p. 54.

[3] H. Reichenbach: *PST, op. cit.*, p. 137.

Reichenbach's formulation of his principle contains no restriction to causal chains which are either materially genidentical, such as stones, or possess the quasi-material genidentity of individual light rays. Thus, although the illustrations he gives of the principle do involve stones and light rays, it has been widely interpreted as not requiring this kind of restriction. Moreover, it is not clear even from his illustration of the stone whether the members of his three observed pairs of events are to belong to *three* rather than to only one or two different genidentical causal chains, a specification which is of decisive importance if such a restriction is to obviate certain of the criticisms of his method which are about to follow. Accordingly, those objections which might thus have been obviated will also be stated.

An illustration of the use of Reichenbach's method given by W. B. Taylor in an attempt to demonstrate its independence of prior knowledge of temporal order[4] will serve to let me set forth the objections to it.

An otherwise dark room has two holes in opposite walls such that a single light ray traverses the room. Since the light source is hidden in the wall behind one of the two holes, we are not able to tell, as we face the light ray from a perpendicular direction, whether it travels across the room from left to right or right to left and thus do not know the location of the light source. In order to ascertain the direction of this causal process, let us isolate three events E_L in it which occur at the left end of the beam and also three events E_R of the process occurring at the right end. And now suppose that *one* of the three events at *each* end is *marked*, say, by means of the presence of transparent colored glass at the point and instant of its occurrence. The crux of the matter lies in Reichenbach's claim that if the events E_L are the respective (partial) causes of the events E_R, we shall observe only the combinations

$$E_L E_R \qquad E_L^e E_R^e \qquad E_L E_R^e$$

and *never* the combination

$$E_L^e E_R.$$

[4] W. B. Taylor: *The Meaning of Time in Science and Daily Life*, doctoral dissertation (Los Angeles: University of California at Los Angeles; 1953), pp. 37–39.

We shall now see that in order to obtain these particular kinds and pairs of events, on which the method relies to define an asymmetric relation, Reichenbach must either make tacit and inadmissible use of prior temporal knowledge or invoke the special requirement of *irreversible* marking processes. For in the absence of information concerning the temporal order within either triplet of events at the left end or at the right end, all that we can say is that our observations at the two ends can be represented by *temporally neutral* triangular arrays:

$$\begin{array}{ccc} & E_L & \\ E_L & & E_L^e \end{array} \qquad \begin{array}{ccc} & E_R^e & \\ E_R & & E_R^e \end{array}$$

And, again not presupposing temporal information, we are entitled to *interpret* our data to the effect that we observed the three combinations

$$E_L E_R^e \qquad E_L E_R^e \qquad E_L^e E_R$$

$$\text{but } \textit{not } E_L^e E_R^e.$$

In the context of Reichenbach's program, this latter interpretation seems to be fully as legitimate as the formation of his own particular grouping, which likewise constitutes a mere interpretation. But the legitimacy of this alternate interpretation is fatal to Reichenbach's claim that E_L and E_R play an asymmetric role, since the alternate interpretation contains the combination $E_L^e E_R$, which is the very combination that he had to rule out in order to show that E_R is later than E_L!

Even if we know the temporal sequence *within* each of the two triplets of events to be, say,

$$\begin{array}{c} E_L \\ E_L^e \\ E_L \end{array} \quad \text{and} \quad \begin{array}{c} E_R \\ E_R^e \\ E_R^e \end{array}$$

where the *downward* direction is the direction of increasing time, it is still not clear which particular event at the right end is to be associated with a particular event at the left end to form a pair. Hence, Reichenbach's causal theory allows us to form event-pairs so as to obtain both his own asymmetric interpretation *and* the alternate interpretation above even from the two internally *ordered* triplets of events. The alternate interpretation is obtainable by combining the first event in the E_L column with the second in the E_R column, the respective third events in the two columns and the second in the E_L column with the first in the E_R column.

Nor is this legitimate alternate interpretation the only source of difficulties for the mark-method. For without a criterion for uniting the spatially separated events under consideration in pairs, we might unwittingly mark the pulse that actually emanated from E_L^e upon its arrival at the right, instead of marking one of the other two pulses, which emanate from *unmarked* events at the left. Since the pulse coming from E_L^e will already be bearing a mark, the marking of that particular pulse upon arrival at the right will be *redundant.* We could then interpret our data as forming the combinations

$$E_L E_R \qquad E_L^e E_R^e \qquad E_L E_R$$

in which E_L and E_R occur with complete *symmetry.*[5] The proponent of the mark method cannot, of course, avert this embarrassing consequence by requiring that we mark at the right end only a pulse not already bearing a mark from before.[6]

Equally unavailing is the following attempt by W. B. Taylor to justify the mark method:

> In the example concerning the light ray, it was said that we first mark E_L and *then* (a moment later) see if the mark appears on E_R. As it stands, this procedure employs time order, which would be undesirable for the purposes at hand. But this way of stating the procedure can be eliminated by saying that we mark E_L and observe (tenseless form of the verb "observe")

---

[5] Using the term "partial cause" so as to allow some of its designata to be *non*-simultaneous, Hugo Bergmann offers another argument for the variational *symmetry* of cause and effect (*Der Kampf um das Kausalgesetz in der jüngsten Physik* [Braunschweig: F. Vieweg and Sons; 1929], pp. 16–19) by showing that Reichenbach succeeds merely in proving that the variation of one of the *partial* causes does not vary the other, but *not* that the effect can be varied without varying the cause. Reichenbach's retort ("Das Kausalproblem in der Physik," *Naturwissenschaften,* Vol. XIX [1931], p. 719) to Bergmann is superseded by his subsequent acknowledgement (*DT, op. cit.,* pp. 198–99) of an inadequacy here which we shall soon consider.

[6] For a similar objection, cf. H. Mehlberg: "Essai sur la théorie causale du temps," *op. cit.,* Vol. I, pp. 214–15.

> whether that same mark appears (again tenseless) on the E_R which is causally connected with E_L. The reason that time order appears to be presupposed is perhaps that we must use verbs (e.g., observe, appear) to describe the procedure of marking and because English verbs in their usual usage are in a tensed, and hence time-referent form. If the observer does not himself mark the events, but instead relies on other agencies to provide the marks, then this apparent circularity does not arise. He, in this case, simply observes the event pairs E_LE_R to see in what arrangement the mark e appears.[7]

This argument will not do, since we saw that it is not possible, without knowledge of the temporal order, to "simply observe" particular, observationally *given* event-pairs E_LE_R to see in what arrangement the mark e appears. The issue is not that time is presupposed in having to mention in temporal succession the names of the members of already-given event-pairs or in stating first the result of our observation on one event in such a pair and then on the other. Instead, the difficulty is that time order is presupposed in assuring and singling out the membership of Reichenbach's specific three event-pairs to begin with, though not in the internal arrangement of their members after the pairs themselves are already chosen. Once we grant the uniqueness of the Reichenbachian choice of event-pairs, it is quite true that temporal order is only apparently presupposed in the description of the experiment, since E_L and E_R do occur asymmetrically in those pairs, independently of the order in which they are named within the pairs.

A further difficulty, which will turn out to have an important bearing on the relation of causality to the anisotropy of time, lies in the fact that in the case of *reversible* marking processes (if there be such), the mark method must make illicit use of temporal betweenness to preclude failure of its experiments. For, as Mehlberg has rightly observed in his searching paper on the causal theory of time,[8] if the mark e were removed, in some way or other, from a signal originating at E^L_e *while that signal is in transit*, the experiment would yield the combination $E^e_LE_R$, which is precisely the one disallowed by Reichenbach. To prevent such

[7] W. B. Taylor: *The Meaning of Time in Science and Daily Life, op. cit.*, pp. 38–39.

[8] H. Mehlberg: "Essai sur la théorie causale du temps," *op. cit.*, Vol. I, p. 214; also pp. 207–57.

an eventuality, the mark-method must either incur failure by requiring that the physical system under consideration be *closed* during the *time interval between* E_L^e and E_R^e[9] or it must have recourse to an irreversible marking process such as passing white light through a color filter.

The question arises, therefore, whether the requirement that the marking processes be *irreversible* does not constitute an invocation of a new criterion of temporal order, since it is based on a much more restricted class of kinds of occurrences than the one to which causality is held to apply, the latter including *reversible* processes. Indeed, if the meaning of causality is correctly explicated by the mark method, and if that method—its other difficulties aside—can hope to be successful in defining a serial temporal order via an *asymmetric* causal relation for only those causal processes which are irreversible or which are rendered irreversible by its application, then two conclusions follow: first, causality as such is *not sufficient* to define those topological properties which are conferred on physical time by irreversible processes, and second, Reichenbach's causal criterion cannot be logically *independent* of a criterion grounded on such processes. Though declining to discuss the issue in his book of 1924, Reichenbach admitted then that the independence of these two criteria is open to question while nevertheless affirming the autonomy of the causal criterion and characterizing the concordance of the temporal orders based on causality, on the one hand, and irreversibility, on the other, as an empirical fact.[1] In his last paper on the subject, published just before his death, and in his

[9] In another connection, Reichenbach defines a closed system as a system "not subject to differential forces" (*PST, op. cit.*, pp. 22–23 and 118), differential forces being forces whose presence is correlated with *changes* of varying degree in different kinds of materials. But the absence of a change at any given instant t means the constancy of a certain value (or values) *between* the termini of a time interval containing the instant t (no *anisotropy* of time being assumed). Thus Reichenbach recognizes that the concept of closed system presupposes the ordinal concept of temporal betweenness. We see incidentally that the meaning of temporal betweenness is presupposed by the statement of the second law of thermodynamics, which concerns *closed* systems.

[1] H. Reichenbach: *Axiomatik der relativistischen Raum-Zeit-Lehre, op. cit.*, pp. 21–22. For critical comments on Reichenbach's early views concerning this issue, see E. Zilsel: "Über die Asymmetrie der Kausalität und die Einsinnigkeit der Zeit," *Naturwissenschaften,* Vol. XV (1927), p. 282.

posthumous book *DT*, he abandoned the ambitious program of the mark method to define an anisotropic serial time at one stroke. Instead, he offered a construction in which he relied on a certain kind of thermodynamic irreversibility and not merely on causality in an attempt to achieve this purpose.[2] In the next chapter we shall examine this construction critically and see in what specific sense irreversible processes can be held to define the anisotropy of time.

Before proceeding to a consideration of other issues, a problem encountered by the causal theory of time merits being stated.

On the causal theory of time, the existence of an *actual* causal chain linking two events is only a sufficient and not a necessary condition for their sustaining a relation of being temporally apart. Such a relation is likewise held to obtain between any two events for which it is merely physically *possible* to be the termini of a causal chain even if no actual chain connects them. Thus, causal connectibility rather than actual causal connectedness is the defining relation of being temporally apart. And causal non-connectibility rather than non-connectedness is the defining relation of topological simultaneity. But as Mehlberg has noted,[3] physical possibility, in turn, must then be definable or understood in such a manner as not to presuppose the ordinal concepts of time which enter into the laws that tell us what causal processes are physically possible. It is not clear that Mehlberg's theory of causal decompositions[4] provides a basis on which the difficulty might be circumvented. For he maintains that *not*-E must be held to be a physical event if E is such an event. And on the strength of this contention, Mehlberg thinks that any two events which are the termini of physically possible causal chains must therefore be held to be actually causally connected. Mehlberg thus proposes to guarantee the actual rather than merely poten-

---

[2] H. Reichenbach: "Les fondements logiques de la mécanique des quanta," *op. cit.*, pp. 137–38; *DT*, *op. cit.*, p. 198n. For his earlier views, see *PST*, *op. cit.*, pp. 138–39; "The Philosophical Significance of the Theory of Relativity," *op. cit.*, pp. 304–306, and "Ziele und Wege der physikalischen Erkenntnis," *Handbuch der Physik* (Berlin: Julius Springer; 1929), Vol. IV, pp. 53, 59–60, 64, 65.

[3] H. Mehlberg: "Essai sur la théorie causale du temps," *op. cit.*, Vol. I, pp. 191, 195, and Vol. II, p. 143.

[4] *Ibid.*, Vol. I, pp. 165–66, 240–41, and Vol. II, pp. 145–46, 169–72.

tial existence of all the required causal chains by asserting[5] that for any event not belonging to a class of simultaneous events, there is at least one event in that class with which the first event is causally connected.

We saw that the mark method is vitiated by circularity, because it explicates the causal relation in such a manner that: first, the method is then required to provide a criterion for designating, within a pair of causally connected events, the *one* event which is *the* cause of the other by virtue of being the earlier of the two in anisotropic time, and second, the method leaves itself vulnerable to the charge of having tacitly employed temporal criteria to secure the particular pairs of events which are essential to its success. Despite the failure of the variational conception of causality offered by the mark method, the kind of causality exhibited by reversible processes is competent, as we shall see shortly, to define some of the topological features of time. Although Leibniz and more recent proponents of his causal definition of the relation "later than"[6] were mistaken concerning the character and extent of the logical connection between the topology of time and the structure of causal chains, their affirmation of the existence of such a connection was sound. We shall see now that reversible genidentical causal processes do indeed define an order of temporal *betweenness* (albeit only in part) and also relations of *simultaneity* in the class of physical events.

By contrast to the mark method, our construction will have the following features: First, we consider a kind of causal relation between two different events which is *symmetric* and involves no reference at all to one of the two events being *the* cause of the other, the criterion for the latter characterization to be supplied subsequently in Chapter Eight; second, we shall eschew resting the asymmetry, and thereby, the seriality of the relation of "earlier than" on the causal relation itself; in fact, the latter relation will be seen to be neutral with respect to whether the

---

[5] *Ibid.*, Vol. II, p. 169.

[6] For a discussion of the ancestral role of the Leibnizian conception, see H. Reichenbach: "Die Bewegungslehre bei Newton, Leibniz und Huyghens," *Kantstudien*, Vol. XXIX (1924), p. 416, and the English translation of this paper by Maria Reichenbach in the posthumous volume of Reichenbach's selected essays entitled *Modern Philosophy of Science* (London: Routledge and Kegan Paul; 1959), pp. 46–66.

temporal order has the formal properties of the "o-betweenness" exhibited by the points on an undirected straight line or those of "separation-closure" relating the points on an undirected circle;[7] instead of being made to depend on causality itself, the o-betweenness of time will depend for its existence on the *boundary* conditions, which determine the relations of the various causal chains to one another; and third, instead of construing causality variationally in the manner of the mark method, we shall begin with material objects each of which possesses genidentity (i.e., the kind of sameness that arises from the persistence of an object for a period of time) and whose behavior therefore provides us with genidentical causal chains. And we shall then consider the causal relation between any pair of events belonging to a genidentical causal chain.

Consider any (ideally) *reversible* genidentical causal process such as the rolling of a ball on the floor of a room along a path connecting points *P* and *P′* of the floor. So long as we confine ourselves to processes which are both nomologically and de facto reversible, we forsake all reliance on the anisotropy of time, since the latter depends—as we shall see in Chapter Eight—on processes which are de facto (and statistically) *irreversible*. Hence in our present temporally *isotropic* context, which excludes every kind of irreversible process, there is no objective physical basis for singling out *one* of two causally connected events as "the" cause and the other as "the" effect. For the designation of one of these events as "the" cause depends on its being the *earlier* of the two, and, in the absence of irreversible processes, the latter characterization expresses not an objective physical relation sus-

[7] The order of points on an undirected circle which we have here called "separation-closure" is generally called "separation of point pairs" in the literature. It is the order of the points on a *closed*, undirected line with respect to a tetradic relation *ABCD* obtaining between points *A, B, C, D* in virtue of the separation of the pair *BD* by the pair *AC*. Axioms for separation of point pairs and for o-betweenness are given in E. V. Huntington: "Inter-Relations Among the Four Principal Types of Order," *Transactions of the American Mathematical Society*, Vol. XXXVIII (1935), pp. 1–9. Cf. also E. V. Huntington and K. E. Rosinger: "Postulates for Separation of Point-Pairs (Reversible Order on a Closed Line)," *Proceedings of the American Academy of Arts and Sciences* (Boston), Vol. LXVII (1932), pp. 61–145, and J. A. H. Shepperd: "Transitivities of Betweenness and Separation and the Definitions of Betweenness and Separation Groups," *Journal of the London Mathematical Society*, Vol. XXXI (1956), pp. 240–48.

tained by it but only the convention that a lower time-number was assigned to it. By contrast, in the context of irreversible processes, the following results obtain: first, the relation term "earlier than" names an *objective* physical relation between two states which is *different* from the converse objective relation named by "later than," and second, the designation of *one* of two causally connected events as "the cause" of the other names a physically different relational attribute from the one named by the converse designation "the effect." Let us return to our reversible genidentical causal process of the rolling ball from whose description we have eliminated all reliance on the anisotropy of time. It is now clear that this renunciation of reference to all attributes depending on the anisotropy of time renders the rudimentary *causal* relation uniting the events of our genidentical reversible causal chain *symmetric*.[8] And the properties of that *symmetric* causal relation would exhaust all the properties of causality in a world all of whose processes are de facto reversible.

Is it possible to provide an explicit definition of our symmetric causal relation without using any of the temporal concepts which that relation is first intended to define? Every attempt to do so known to or made by the writer has encountered insurmountable difficulties which are closely related to those familiar from the study of lawlike subjunctive conditionals by N. Goodman and others.[9] Suppose, for example, that we attempted to define the symmetric causal relation between two genidentically related events E and E′ by asserting that either of these two events is existentially a sufficient and a necessary condition for the other's occurrence in the following sense: if the set U of events constituting the universe contains an event of the kind E, then it also contains an event of the kind E′, and if U does not contain an

[8] In Chapter Eight, we shall give reasons for rejecting the objections to this conclusion set forth by Reichenbach in *DT, op. cit.*, p. 32.

[9] Cf. N. Goodman: *Fact, Fiction and Forecast* (Cambridge: Harvard University Press; 1955), esp. pp. 13–31.

It is noteworthy that even in a context which, unlike ours, does presuppose the temporal concept of "later," C. I. Lewis reaches the conclusion (cf. his *Analysis of Knowledge and Valuation* [LaSalle: Open Court Publishing Co.; 1946], pp. 226–27) that the "if...then" encountered in causal statements does not yield to analysis, and therefore he speaks of the undefined "if...then of real connection."

E-like event, then it also will not contain an E′-like occurrence, no assumptions being made at all as to one of these two events being *earlier* than the other in the sense of presupposing the anisotropy of time. But this attempt at definition won't do for several reasons as follows: (1) the statement of the existential sufficiency of E is either tautological or self-contradictory, depending upon whether it is understood in its antecedent that the set U does or does not contain E′, (2) the corresponding statement of existential necessity is either self-contradictory or tautological depending on whether in its antecedent, U is or is not construed as having E′ among its members, and (3) attempting to turn these tautological or self-contradictory statements into true synthetic ones by making these assertions not about U itself but about appropriate proper *subsets* of U founders on (a) the need to utilize *temporal* criteria to circumscribe the membership of these subsets and (b) our inability to specify *all* of the *relevant conditions* which must be included in the antecedent, if the statement of the existential sufficiency of E is to be true.

The numerous difficulties besetting the specification of the relevant conditions[1] are not removed but only baptized by giving them a name—to borrow a locution from Poincaré—by the physicist's reference to the total state of a *closed system* in the antecedents of his causal descriptions of physical processes. Instead of enlightening us concerning the content of the *ceteris paribus* assumption, the invocation of the concept of closed system merely shifts the problem over to ascertaining the *cetera* which must materialize throughout the vast Minkowski light cones *outside* the system or on the spatial enclosure of the system in order that the system be *closed.*

These considerations suggest that we introduce the symmetric causal relation under discussion as a *primitive* relation for the purpose of then defining temporal betweenness and simultaneity. The reader will ask at once why the reduction of these ordinal concepts of time to such a primitive is not to be rejected as demanding too high a price epistemologically. Several weighty replies can be given to this question as follows:

(1) Despite the difficulties besetting the various versions of

[1] Many of these difficulties are discussed searchingly by N. Goodman: *Fact, Fiction and Forecast, op. cit.*, pp. 17–24.

the causal theory of time beginning with Leibniz's, the Einstein-Minkowski formulation of special relativity has sought to ground the temporal order of the physical world on its causal order by its reliance on influence chains (signals). Although the theory of relativity, like any geometry or other scientific theory, can be axiomatized in several different ways (at least in principle) by the use of different sets of primitives, the axiomatization of the relativistic topology of time on the basis of causal (signal) chains gives telling testimony of the desirability of formulating a philosophically viable version of the causal theory of time.

(2) The very explanation of some of the temporal features of the physical world on the basis of its causal features and the embeddedness of man's organism in this causal physical world lead us to expect that the causal acts of intervention (even insofar as they are reversible) which enter into man's *testing* of the Einstein-Minkowski theory will be part of the temporal order and will require for their practical execution by us *conscious* organisms recourse to the deliverances of our *psychological* sense of temporal order.

(3) My motivation for advancing below a particular version of the causal theory of time in which the attempt is made to dispense entirely with these psychological deliverances in the axiomatic foundations derives from the following two premises: (a) the thesis of astrophysics (cosmogony) and of the biological theory of human evolution that temporality is a significant feature of the physical world independently of the presence of man's conscious organism and hence might well be explainable as a purely physical attribute of those preponderant regions of space-time which are not inhabited by conscious organisms, and (b) the view of philosophical naturalism that man is part of nature and that those features of his conscious awareness which are held to be isomorphic with or likewise ascribable to the inanimate physical world must therefore be explained by the laws and attributes possessed by that world independently of human consciousness. When coupled with certain results of statistical thermodynamics, the version of the causal theory of temporal order to be advanced below provides a unified account of certain basic features of physical and psychological time. Since man's body participates in those purely physical processes which con-

fer temporality on the inanimate sector of the world and which are elucidated in part by the causal theory of time, that theory contributes to our understanding of some of the traits of psychological time.

So much for the justification of our use of the rudimentary symmetric causal relation as a primitive and of eschewing reliance on the deliverances of psychological time.[2]

Now if we are confronted with a situation in which two actual events E and E′ are *genidentical* and hence causally connected in our rudimentary *symmetric* sense—or "k-connected" as we shall say for brevity—then we are able to use the properties of genidentical chains, which include causal continuity, to define both temporal betweenness and simultaneity.

For reasons which will become apparent, the causal definition of temporal betweenness to be offered will be given, pending the introduction of further requirements, so as to allow that time be topologically either open in the sense of being a system of o-betweenness like a Euclidean straight line, or closed in the sense of being a system of separation closure like a circle. In order to make the statement of the definition correspondingly general, we require the following preliminaries in which we use the abbreviation "iff" for "if and only if":

(1) We shall call the quadruplet of events E L E′ M (where

---

[2] The renunciation of essential recourse to *phenomenalistic* relations in the conceptual elaboration of the theory of temporal order was championed by Reichenbach in 1928, when he wrote (*Philosophie der Raum-Zeit-Lehre, op. cit.*, p. 161; cf. also pp. 327–28): "it is in principle impossible to use subjective feelings for the determination of the [temporal] order of external events. We must therefore establish a different criterion." But in his last publication (*DT, op. cit.*, pp. 33–35) he invokes *direct observation* of nearby quasi-coincidences as a basis for giving meaning to the local order of temporal betweenness of such coincidences. If this observational criterion is intended to involve man's subjective time sense, then one wonders what considerations persuaded Reichenbach to abandon his earlier opposition to it. And it is not clear how he would justify having recourse to it in the very context in which he claims to be providing a "causal definition" of the order of temporal betweenness "by means of reversible processes" (*DT, op. cit.*, p. 32). However, if he is making reference here not to the temporal deliverances of consciousness but rather to the directly observable indications of a material clock, then he has merely displaced the difficulty by posing but not solving the problem of showing how the causal features of a reversible clock furnish the definition of temporal betweenness.

$E \neq E'$) an "n-quadruplet," iff given the actual occurrence of E and E′, it is necessary that either L or M occur in order that E and E′ be k-connected, L and M being genidentical with E and E′, and "or" being used in the inclusive sense. It is essential to note that "it is necessary that either L or M occur" does not entail "either the occurrence of L is necessary or the occurrence of M is necessary." The sentence "E L E′ M are an n-quadruplet" will be abbreviated to "n(E L E′ M)."

(2) We utilize the property of causal continuity possessed by genidentical causal chains and assert: for any two genidentically related (and hence k-connected) distinct events E and E′, there exist sets $\hat{X}$ and $\hat{F}$ of events genidentical with them, each of which has the cardinality $\aleph$ of the continuum and such that for each X belonging to $\hat{X}$ and each F belonging to $\hat{F}$, we have n(E X E′ F). Thus, there are $\aleph$ n-quadruplets.

(3) Given E and E′, we shall call a set α an "n-chain" connecting E and E′, iff the members X of α are given by the following condition:

$$X \in \alpha \text{ iff } (\exists F)\,[n(E\,X\,E'\,F) \cdot \sim n(E\,F\,E'\,F)].$$

All the members of n-chains connecting a given pair of genidentical events E and E′ are thus genidentical with E and E′.

We are now able to define *temporal betweenness* as follows: any event belonging to an n-chain connecting a pair of genidentical events E and E′ is said to be temporally between E and E′.[3]

[3] It might be objected that there is a difficulty in applying this definition to our earlier paradigmatic example of the rolling ball, for it would be possible to have someone place the ball at point *P* at the appropriate time even though that ball never was and never will be at the point *P′* and to have a different ball placed appropriately at the latter point. Or a critic might say that even if the original ball were to be at one of the points *P* or *P′*, it could always be prevented from reaching the other point by being suitably intercepted while in transit. The aim of such objections would be to show that in order to rule out these alleged counter-examples, our construction would become circular by having to invoke temporal concepts which it is avowedly not entitled to presuppose. But the irrelevance of these objections becomes clear, when cognizance is taken of the fact that instead of exhibiting a circularity in our causal definition of temporal betweenness, they merely call attention to the existence of pairs of events which do not fulfill the conditions for the applicability of our definition, because they are not causally connected (or connectible) in the requisite genidentical way.

Once a definition of topological simultaneity becomes available, the definition just given admits of a generalization so as to allow that events which are not genidentical with E and E′ but which are simultaneous with any event both temporally between and genidentical with them will likewise be temporally between E and E′. These topological definitions can then be particularized by the metrical definitions of simultaneity used in particular reference frames.

A very important feature of our definition of temporal betweenness is that it leaves open the question as to which one of the following alternatives prevails:

(1) The n-quadruplets which E and E′ form with pairs of members of the n-chains connecting E and E′ have the formal properties of the tetradic relation of separation, thus yielding a system of separation closure as the temporal order, or

(2) The membership of the n-chains connecting E and E′ is such as to yield a system of o-betweenness as the temporal order.

In order to articulate this neutrality of the definition, we note, for purposes of comparison, the following *partial* sets of properties of the two alternative types of order in question. Letting "ABC" denote the triadic relation of betweenness, "ABCD" the tetradic relation of separation and "→" the relation of logical entailment, we have:

o-betweenness

1. ABC → CBA (symmetry in the end-points or "undirectedness")
2. ABC → ~ BCA (preclusion of closure)

separation-closure

If all four elements A, B, C, and D are distinct, then

1. ABCD → DCBA ("undirectedness")
2. ABCD → BCDA ("closure")

The corresponding *partial* properties of *cyclic* betweenness, as exemplified by the class of points on a *directed* closed line, are:

cyclic betweenness

1. ABC → ~CBA (preclusion of symmetry in the endpoints or "directedness")
2. ABC → BCA (closure)

But we are not concerned with the species of closed order represented by cyclic betweenness, since we shall argue that it is not relevant to temporal betweenness.

The neutrality which we have claimed for our definition of temporal betweenness in regard to both o-betweenness and separation closure will be clarified by dealing with an objection which Dr. Abner Shimony has suggested to me for consideration.

If the definition is to allow time to be a system of o-betweenness and the subsequent introduction of a serial temporal order in which the members E, X, Y, E′, F, and G of a genidentical causal chain are ordered as shown by the order of their names, then our definition ought not to entail that events like F, which are *outside* the time interval bounded by E and E′, are temporally between E and E′. Nor should it entail that *inside* events like X and Y turn out not to be temporally between E and E′. Now, our definition of the membership of n-chains α was designed to preclude precisely such entailments as well as to allow the order of separation closure, in which every event is temporally between every other pair of events. The question is whether it could not be objected that we have failed nonetheless on the following grounds: events such as F and G, which are outside the interval EE′, are each necessary for the respective occurrences of E and of E′, just as much as inside events like X and Y are thus necessary; and does it not follow then that every "outside" event like F does enter into an n-quadruplet n(E F E′ F), thereby qualifying, no less than do "inside" events in this case, as necessary for the k-connectedness of E and E′?

If this conclusion did in fact follow, then our definition would indeed preclude o-betweenness by making the n-chains α empty. For in that case, there would be no genidentical event whatever satisfying the requirement of not being necessary for the k-connectedness of E and E′, as demanded by our definition of α. But the reasoning of the objection breaks down at the point of inferring that outside events like F and G do form n-quadruplets with E and E′. For although all events genidentical with E and E′—be they inside ones or outside ones—are necessary for the respective occurrences of E and of E′, only the inside ones have the further property of being necessary for the k-connectedness of E and E′, given that the latter do occur. By noting the distinction between the properties of (1) being necessary for the respective occurrences of E and E′ and (2) being necessary for the k-connectedness of events E and E′ whose occurrence is otherwise granted, we see that the objection to our definition derived its plausibility but also its lack of cogency from inferring the second property from the first.

The fact that our definition of temporal betweenness does not itself discriminate between closed and open time becomes further evident upon considering universes to each of which it applies but which have topologically different kinds of time.

(A) CLOSED TIME

Let there be a universe consisting of a platform and one particle constantly moving in a circular path without friction. And be sure not to introduce surreptitiously into this universe either a conscious human observer or light enabling him to see the motion of the particle. Then the motion might be such that the temporal betweenness which it defines would exhibit closedness rather than openness, because there would be no physical difference whatever between a given passage of the particle through a fixed point *A* and its so-called "return" to the same state at *A*: instead of appearing periodically at the same place *A* at different instants in open time, the particle would be "returning"—in a highly Pickwickian sense of that term—to the selfsame event at the same instant in closed time. This conclusion rests on Leibniz's thesis that if two states of the world have precisely the same attributes, then we are not confronted by distinct states at different times but merely by two different names for the same state at one time. And it is this Leibnizian consideration which renders the following interpretation inadmissible as an alternative characterization of the time of our model universe: the same kind of set of events (circular motion) keeps on recurring eternally, and the time is topologically open and infinite in both directions. The latter interpretation is illegitimate since a difference in identity is assumed among events for which their attributes and relations provide no basis whatever. Hence that interpretation cannot qualify as a legitimate rival to our assertion that the events of our model world form an array which is topologically closed both spatially and temporally.

Ordinary temporal language is infested with the assumption that time is open, and a description of the closed time of our model world in that language would take the misleading form of saying that the same sequence of states keeps recurring all the time. This description can generate pseudo-contradictions or puzzles in this context, because it suggests the following structure.

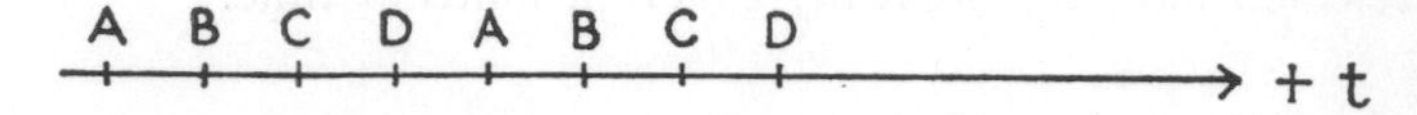

Here there are distinct sets of events ABCD which are merely the same in kind with respect to one or more of their properties. But this is not the structure of closed time. A closed time is very counterintuitive psychologically for reasons which will emerge later on. And hence the assumption of the closedness of time is a much stranger one intuitively than is a cyclic theory of history. In a cyclic theory of history, one envisions the periodic recurrence of the same kind of state at different times. And this conception of cyclic recurrence affirms the openness of time. Perhaps the lack of psychological imaginability as distinct from theoretical intelligibility of a closed time accounts for the fact that its logical possibility is usually overlooked in theological discussions of creation. This failure of imagination is unfortunate, however, since there could be no problem of a beginning or creation if time were to be cosmically closed.

Three kinds of objections might be raised to my claim that our model universe does provide a realization of a topologically closed kind of time: First, it might be argued that my earlier caveat concerning the need for a highly Pickwickian construal of the term "returning" actually begs the question in the following sense: the mere contemplation of the model universe under consideration compels the conclusion that the particle does indeed return in the proper sense of that term to the same point A at different instants of open time. And it might be charged that to think otherwise is to disregard a plain fact of our contemplative experience. Second, the complete circular motion could be subdivided into a finite number n of (equal) parts or submotions (episodes). And this has been claimed to show that instead of being topologically closed, the time of our model universe has the open topology of a finite segment of a straight line which is bounded by distinct endpoints, the first and n^{th} submotions allegedly being the ones which contain the two terminal events. And third, it has been said that it is of the essence of time to be open. Noting that my characterization of the model universe at issue as having a closed time depends on the invocation of Leibniz's principle of the identity of indiscernibles, a critic has maintained that instead of showing that a closed time is logically possible, this model shows that Leibniz's principle must be false!

As to the first of these objections, which rests on the deliverances of our contemplation of the model universe, my reply is that the objector has tacitly altered the conditions that I had postulated for that model universe, thereby tampering with the very features on which my affirmation of the closure of its time had been predicated. For the objector has not only introduced a conscious organism such as himself, whom he presumes to have distinct memories of two passages at *A*, but he has also surreptitiously brought in another physical agency needed to make these distinct memories possible: a light source such as a candle which enables him to see and which distinguishes an earlier and a later passage at *A* by being more dissipated or burnt-out at the time of the later passage. Thus, the objection is untenable, because the objector assumes a universe differing from my hypothetical one so as not to have a closed time.

The second objection, which adduces the n submotions, is vitiated by the following gratuitous projection of the ordinal properties of numerical names onto the events (or submotions) to which they are assigned in a counting procedure: the divisibility of the motion of the particle on the platform into n subintervals of events—where n is a cardinal number—does not make for the possession of any objective property of being *temporal termini* by the particular subintervals which were assigned the numbers 1 and n respectively. For no two of the n subintervals of events—whatever the particular numbers that happened to have been assigned to them in the counting—are ordinally distinguished qua termini from any of the others. If therefore we count them by arbitrarily assigning the number 1 to some one of them and by then assigning the remaining n-1 numbers, this cannot serve to establish that temporally the ordinal properties of the particular subintervals thus accidentally named 1 and n respectively are any different from those of the subintervals which are thereby assigned natural numbers between 1 and n.

The justification for this rebuttal can be thrown into still bolder relief by noting the objective differences between our model universe of the particle moving "perpetually" in a circular path and the following different model universe whose time does indeed have the open topology of a finite segment of a straight line bounded by two endpoints: a platform universe differing from the first one only by exchanging the particle moving in a closed (circular) path for a simple pendulum oscillating "perpetually" and frictionlessly over the platform as if under the action of terrestrial gravity. Let the oscillation be through a fixed small amplitude $\theta = 2\,\alpha$ between two fixed points whose angular separation from the vertical is $+\,\alpha$ and $-\,\alpha$ respectively. Then the finitude of the time of this pendulum universe is assured by the fact that the "perpetual periodic returns" of the pendulum bob to the same points over the platform are, in fact, identical events by Leibniz's principle. For this identity prevents this latter model universe from qualifying as an infinitely periodic universe whose time is open. But the two fixed points at angular distances $-\,\alpha$ and $+\,\alpha$ respectively from the vertical uniquely confine the motion spatially to the points between them. And hence the events constituted by the presence of the pendulum bob at $-\,\alpha$ and $+\,\alpha$ are objectively distinguished as termini from all other events belonging to the motion of the pendulum, although no one of these two events is distinguished from the other as the first rather than the last, since the motion is reversible. Accordingly, the time of the pendulum universe has the topology of an undirected finite segment of a straight line bounded by two termini. But there is no foundation for the objection that our first model universe of the particle in the circular path has the latter kind of time.

The third objection, which has the spirit of an argument based on Kant's presuppositional method, suffers from precisely the same well-known logical defects as does the claim that we know a priori that the universe must be spatially infinite (topologically open) rather than finite (closed but unbounded). For the defender of the a priori assertibility of the infinite Euclidean topology would offer the following corresponding argument regarding space: if it appeared that a spatial geodesic of our universe were traversable in a finite number of equal steps terminating at the same spatial point, the certification of the sameness of that point via Leibniz's principle would have to be rejected. And the a priori proponent of the Euclidean as opposed to the spherical topology of space would then adduce these circumstances to show that Leibniz's principle is false. Although the rise of the non-Euclidean geometries has issued in the displacement of the Kantian conception of the topology of physical *space* by an empiricist one in most quarters, a vestigial Kantianism persists in many quarters with respect to the topology of *time.* And this lingering topological apriorism seems to be nurtured by the following failure of imagination: the neglect to envision having to divest the topology of the time of a model universe—or the cosmic time of our actual universe—of some of the topological properties of the cosmically local time of our everyday experience. Such divesture would be necessary, if Einstein's field equations were to allow solutions yielding temporally closed geodesics, as has been claimed by Gödel.[4] If correct, Gödel's result would seem to indicate that temporally closed geodesics can also be possessed by a deterministic universe containing sentient beings like ourselves. But Chandrasekhar and

[4] In recent papers ("An Example of a New Type of Cosmological Solutions of Einstein's Field Equations of Gravitation," *Reviews of Modern Physics,* Vol. XXI [1949], p. 447, and "A Remark About the Relationship of Relativity Theory and Idealistic Philosophy," in P. A. Schilpp [ed.] *Albert Einstein: Philosopher-Scientist, op. cit.,* pp. 560–62), Gödel has maintained that there exist solutions of Einstein's field equations which assert the existence of *closed* timelike world-lines. Einstein says (*ibid.,* p. 688) that "It will be interesting to weigh whether these [solutions] are not to be excluded on physical grounds." Reichenbach himself points out (*PST, op. cit.,* 141, 272–73) that in a world of closed timelike world lines, his causal criterion of temporal order becomes self-contradictory in the large (*Axiomatik der relativistischen Raum-Zeit-Lehre, op. cit.,* p. 22). And, as Einstein remarks (*ibid.*), in such a world, irreversibility also fails to hold in the large.

Wright[5] have argued very recently that Gödel's claim is mathematically unfounded.

There is a rather simple way of seeing how manlike beings might discover that the cosmic time of their universe is closed, despite the seriality of the local segment of cosmic time accessible to their daily experience. Suppose that all the equations governing the temporal evolution of the states of physical and biological systems are deterministic with respect to the properties of events and that these equations are formulated in terms of a time variable t ranging over the real numbers, it thereby being assumed to begin with that time is topologically open. Now postulate further that the boundary conditions of this deterministic world are such that all of the variables of state pertaining to it (including those variables whose values characterize the thoughts of scientists living in it) assume precisely the same values at what are prima facie the time t and a very much later time $t + T$. Then upon discovering this result by calculations, these scientists would have to conclude that the two different values of the time variable for which this sameness of state obtains do not denote two objectively distinct states but are only two different numerical names for what is identically the same state. In this way, they would discover that their universe is temporally closed, much as a scientist who begins by assuming that the universe is spatially infinite may then find that, in the large, it is spatially closed. But there is an important difference between the psychological intuitability of a closed space and that of a closed time: a cosmically closed time could *not everywhere* be locally serial with respect to "earlier than."

As previously emphasized, it is of crucial importance in this context, if pseudo-puzzles and contradictions are to be avoided, that the term "returning" and all of the preempted temporal language which we tend to use in describing a world whose time (in the large) is closed be divested of all of its tacit reference to an external serial super-time. Awareness of the latter pitfall now enables us to see that as between the two kinds of closed order which we have mentioned, separation closure and not

[5] S. Chandrasekhar and J. P. Wright: "The Geodesics in Gödel's Universe," *Proceedings of the National Academy of Sciences*, Vol. XLVIII (1961), pp. 341–47, and esp. p. 347.

cyclicity must be held to be the order of a closed time. For in the context of physical states, cyclic betweenness depends for its directional anisotropy on an appeal to a serial and hence open time and derives its closure only from a spatial periodicity. Thus, a closed physical time must exhibit separation-closure and cannot meaningfully be cyclic. And if we are to give a concise characterization of a physically plausible closed time, it should read: every state of the world is temporally between every other pair of states of the universe in a sense of "between" given by our definition for the case of separation closure. It might be asked why we have been assuming that the structure of a closed time would have to be that of a knot-free circle rather than that given by the self-intersecting closed line in the numeral 8. The reply is that the framework of our models of a closed time is deterministic and that the course of the phase curve representing a finite (closed) mechanical or other deterministic system is uniquely determined by any one of its phase points.

Our characterization of a closed time is to apply not only to a thoroughly uninteresting model world like that of the solitary particle moving in a circular path on a platform but to a cosmos whose total states are not elementary events but large classes of coincidences of many genidentical objects. It is therefore essential that we specify the meaning of topological simultaneity, which is presupposed by the concept of a state of the world. Reserving comment until we discuss open time and, in particular, the species of open time affirmed by the special theory of relativity, on the serviceability of the following definition in that context, we define: two events are *topologically simultaneous*, iff it is not physically possible that they be connected by genidentical causal chains. It is to be noted in connection with this definition that a light ray directly connecting a pair of macroscopic events is held to qualify as an entity possessing genidentity, although this assumption no longer holds in contexts in which the Bose-Einstein statistics applies to photons.

(B) OPEN TIME

Since our definition of topological simultaneity is completely neutral as to the closedness or openness of time, it is apparent that we can also utilize the definition of topological simultaneity for the description of a universe whose time is characterized by o-betweenness, such as that of Newton or of special relativity. In the latter Einsteinian world, the limiting role played by electromagnetic causal chains makes for the fact that the topological definition of simultaneity leaves a good deal of latitude for the synchronization rule and thereby for the metrical definition of simultaneity. And hence in that case, our purely ordinal definition would have to be particularized in each Galilean frame, as we shall see in detail in Chapter Twelve, so as to render metrically simultaneous in any given frame only those pairs of causally non-connectible events which conform to the criterion of that particular frame.

To appreciate the role of boundary conditions in conferring openness on time, we first recall our pendulum world model of a finite open time and now proceed to provide models of an infinite open time as follows: first a very simple kind of world and then a world having far greater relevance to the actual world in which we live.

Let there be a universe consisting of a platform, material clocks and at least two simple pendulums *X* and *Y* which have incommensurable periods of oscillation so that after once being in the same phase they are permanently out-of-phase with each other. Then these motions would define an infinite open time in virtue of Leibniz's non-identity of discernibles: in this case, a given passage E_p of the bob of pendulum *X* through a fixed point *P* would be physically different from any other such passage E_q in virtue of E_p's being simultaneous with a different phase of pendulum *Y* from the one with which E_q is simultaneous, thereby giving rise to an order of o-betweenness for time. In the over-all light of the construction of physical time presented in this book, this assertion involves a philosophical commitment to a Leibnizian criterion for the individuality of the events belonging to the causal chains formed by genidentical classical material particles or macro-objects, along with a non-Leibnizian primitive concept of material genidentity for the entities whose relations generate these events. I do not see any inconsistency or circularity in this feature of the construction. In particular, it seems to me that only a confusion of the context of justification with the context of discovery (in Reichenbachian terminology) or of the factual reference with the evidential base (in Feigl's parlance) can inspire the charge that it is circular to use the concept of (material) genidentity as a primitive in a reconstruction of the temporal order of physical events, on the alleged grounds that the meaning of the temporal order is already presupposed in our *re*cognitions of objects as the same upon encountering them in different places at different times.

Now consider a finite universe or a large finite quasi-closed portion of our actual universe, if the latter be spatially infinite, to which the Maxwell-Boltzmann gas statistics is roughly applicable. The concept of what constitutes an individual micro-state ("arrangement" or "complexion") in the Maxwell-Boltzmann statistics depends crucially on the assumption that material genidentity can be ascribed to particles (molecules) and involves the very Leibnizian conception of the individuality of events which we set forth by reference to our illustrative examples of the simple model universes. An individual instant of time is thus defined for this universe by one of its particular micro-states. And, on this criterion, it will therefore be quite meaningful to speak, as I shall later on, of the occurrence of the same *macro*-state ("distribution") and hence of the same entropy at different times, provided that the respective underlying *micro*-states are different. But whether or not a universe constituted by a Maxwell-Boltzmann gas will exhibit a set of micro-states which define an open time rather than a time which, in the large, is closed depends not on the causal character of the motions of the constituent molecules but on the boundary conditions governing their motions! And whether the time thus defined will, if open, also be infinite, depends on the micro-states having the degree of specificity represented by points in phase space rather than by the mere cells used to compute the coarse-grained probabilities of various macro-states. For a finite closed mechanical system of constant energy is at least quasi-periodic and can possibly qualify as aperiodic only with respect to a punctual characterization of its micro-states.

Hence, the symmetric kind of causality affirmed by the equations of mechanics themselves, as distinct from prevailing nomologically contingent boundary conditions to which they apply, *allows but does not assure* that the temporal betweenness (and simultaneity) defined by genidentical causal chains is that of an open rather than a closed ordering.[6]

This analysis of the physical basis of open time requires the addition of a comment on the meaning of the "reversibility" of mechanical motions. We observed that the total states of our pendulums *X* and *Y* whose periods are incommensurable define an infinite open time by not exhibiting any "reversals." What is meant, therefore, when we attribute reversibility to the motions of either of the individual pendulums is that either elementary constituent of the total system can (under suitable boundary conditions of its own) give rise to the same *kind* of event at different times. We do not mean that the pendulum in question has "returned" to the selfsame event, since the total states of the complete physical system assure via the Leibnizian *non*-identity of discernibles that the events belonging to the individual pendulum form an infinite open order of time. Thus, the reversibility of the laws of mechanics has a clear meaning in the context of an infinite open time. And the reversibility of the elementary processes in our Maxwell-Boltzmann universe, which is affirmed by these laws, is therefore entirely compatible with the infinite openness of the time defined by the total micro-states of that universe. We shall soon see, however, that the mere *non-reversal* of the *total* micro-states, which assures the infinite openness of time, is a much weaker property of these states than the kind of *irreversibility* that is a sufficient condition for the anisotropy of time.

We have emphasized the neutrality of our definition of temporal betweenness on the basis of causal continuity in regard to

---

[6] Mehlberg ["Essai sur la théorie causale du temps," *op. cit.*, Vol. I, p. 240] correctly calls attention to the fact that the principle of causal continuity is independent of whether physical processes are reversible or not. But then, affirming the complete reversibility of the physical world and thereby the isotropy of physical time (*ibid.*, p. 184), he takes it for granted that the principle of causal continuity as such always defines an open betweenness (*ibid.*, pp. 239–40, and *op. cit.*, Vol. II, pp. 179, 156–57, 168–69). But, as we just saw, in a completely reversible world, it can happen that the betweenness defined by the principle of causal continuity for genidentical chains is that of the closed variety associated with separation closure. Thus, instead of being isotropic while being open, time would then be both isotropic and closed.

the rival possibilities of open and closed betweenness for time. The second feature of our definition of temporal betweenness which needs to be noted is that it utilizes the concept of physical possibility and thereby runs the logical risk mentioned earlier in this chapter, *if* we expect it to order B as temporally between A and C even in those cases where B would be between A and C if A and C *were* genidentically connected but are actually not. This risk also applies to our definition of (topological) simultaneity for noncoincident events.

In the theories of pair-production due to Wheeler, Feynman, and Stückelberg,[7] some of the phenomena under investigation may be described in the usual macro-language by saying that a "particle" can travel both "forward" and "backward" in macrotime. Thus, in that description a "particle" can violate the necessary condition for simultaneity given in our definition here by being at two different places at instants which are macroscopically simultaneous. But this fact does not disqualify our definition of simultaneity. For the topology of the time whose physical bases our analysis is designed to uncover is defined by (statistical) *macro*-properties for which these difficulties do not arise. The macro-character of our concept of simultaneity is evident from the fact that it depends on the concept of material or quasi-material genidentity. Precisely this concept and the associated classical concept of a particle trajectory are generally no longer applicable to micro-entities, so that the consequences of their inapplicability in the Wheeler-Feynman theory and in the Bose-Einstein statistics need not occasion any surprise. Furthermore, the macro-character of the anisotropy of time will emerge from the analysis to be given later in Chapter Eight.

We see that since our definitions of temporal betweenness and simultaneity employed a concept of causal connection joining two events which made no reference to *one* of the events being *the* cause of the other by being *earlier*, these definitions do not presuppose in any way the anisotropy of time. Nevertheless, they do exhaust the contribution which the causality of reversible processes can make to the elucidation of the structure of time. If physical time is to be *anisotropic*, then we must look to fea-

---

[7] Cf. H. Reichenbach: *DT, op. cit.*, pp. 262–69.

tures of the physical world other than the causality of reversible processes as the source. In particular, since it has been shown[8] that if the micro-statistical analogue of entropy fails to confer anisotropy on time, all other micro-statistical properties of *closed systems for which an entropy is defined* will fail as well, we must turn to an examination of entropy to see in what sense, if any, it can supply attributes of physical time not furnished by causality.

Before doing so, however, I wish to make a concluding comment on the causal theory of time in its bearing on the modern mathematical resolution of Zeno's paradoxes of motion.

Having no recourse to the anisotropy or even to the openness of time, our definitions of temporal betweenness and simultaneity established a *dense* temporal order. For our construction entails that between any two events, there is always another. Now, as I have explained elsewhere in 1950,[9] it was the ascription of this denseness property to the temporal order by the modern mathematical theory of motion which prompted the Zenonian charge by such philosophers as William James and A. N. Whitehead that this theory of motion is neither physically meaningful nor consistent. Resting their case on the immediate deliverances of consciousness, which include "becoming," these philosophers maintained that the temporal order is discrete rather than dense, the events of nature occurring *seriatim* or, as Paul Weiss put it, "pulsationally."[1] We see that the causal theory of time as here presented refutes their polemic on the issue of the denseness of the temporal order of the physical world and that this refutation is not dependent on the logical viability of Reichenbach's unsatisfactory mark method, which I employed in my paper of 1950, when giving a critique of their arguments on the basis of the causal theory of time.[2]

---

[8] Cf. A. S. Eddington: *The Nature of the Physical World* (New York: The Macmillan Company; 1928), pp. 79–80. For details on the "principle of detailed balancing" relevant here, cf. R. C. Tolman: *The Principles of Statistical Mechanics* (Oxford: Oxford University Press; 1938), pp. 165, 521.

[9] A. Grünbaum: "Relativity and the Atomicity of Becoming," *The Review of Metaphysics*, Vol. IV (1950), pp. 143–86.

[1] Cf. Paul Weiss: *Reality* (New York: Peter Smith; 1949), p. 228.

[2] The analysis on the basis of the mark method is given on pp. 160–86 of my previously cited "Relativity and the Atomicity of Becoming." For a treatment not encumbered by the weaknesses of the mark method, cf. A. Grünbaum: "Modern Science and the Refutation of the Paradoxes of Zeno," *The Scientific Monthly*, Vol. LXXXI (1955), pp. 234–39.

Chapter 8

THE ANISOTROPY OF TIME

(A) IS THERE A THERMODYNAMIC BASIS FOR THE ANISOTROPY OF TIME?

Just as we can coordinatize one of the dimensions of space by means of real numbers *without* being committed to the anisotropy of that spatial dimension, so also we can coordinatize a topologically open time-continuum without prejudice to whether there exist irreversible processes which render that continuum anisotropic. For so long as the states of the world (as defined by some one simultaneity criterion) are ordered by a relation of temporal betweenness having the same formal properties as the spatial betweenness on a Euclidean straight line, there will be two senses which are opposite to each other. And we can then assign increasing real number coordinates in one of these senses and decreasing ones in the other by convention *without* assuming that these two senses are *further distinguished* by the *structural* property that some kinds of sequences of states encountered along one of them are never encountered along the other.

If the latter situation does indeed obtain because of the existence of irreversible kinds of processes, then the time continuum is anisotropic. By the same token, if the temporal inverses of all kinds of processes actually materialized, then time would be isotropic.

We shall have to determine what specific properties of the physical world, if any, confer anisotropy on the time of nature in the sense of structurally distinguishing the opposite directions of "earlier" and "later" on the time axis in a specifiable way. We shall then be ready to see in Chapter Ten whether any features of the universe such as the hypothetical one of indeterminism can give a physical meaning to the following further purported property of time: "passage," "coming into being," "becoming," "flux," or "flow," conceived as the *shifting* or transiency of the "Now" along the structurally distinguished direction of "later than" on the time axis.

It is clear that the anisotropy of time resulting from the existence of irreversible processes consists in the mere structural differences between the two opposite senses of time but provides no basis at all for singling out *one* of the two opposite senses as "*the* direction" of time. Hence the assertion that irreversible processes render time anisotropic is *not at all* equivalent to such statements as "time flows one way." And the metaphor of time's "arrow," which Eddington intended to refer to the anisotropy of time, can be misleading: attention to the head of the arrow to the exclusion of the tail may suggest that there is a "flow" in *one* of the two anisotropically-related senses.

It will be essential to begin by giving an analysis of the concept of a temporally irreversible process in its bearing on the anisotropy of time with a view to then determining the following: to what extent, if any, do the kinds of physical processes obtaining in our universe confer anisotropy on time?

In what sense do the failure of the dead to rise from their graves and the failure of cigarettes to reconstitute themselves from their ashes render the processes of dying and cigarette burning "irreversible"? There is both a weak sense and a strong sense in which a process might be claimed to be "irreversible." The weak sense is that the temporal inverse of the process in fact never (or hardly ever) occurs with increasing time for the following reason: certain particular de facto conditions ("initial" or "boundary" conditions) obtaining in the universe independently of any law (or laws) combine with a relevant law (or laws) to render the temporal inverse de facto non-existent, although no law or combination of laws itself disallows that inverse process. The strong sense of "irreversible" is that the temporal inverse is impossible in virtue of being ruled out by a law alone or by a combination of laws. In contexts calling for the distinction between these two senses of "irreversible," we shall use essentially the terminology of H. Mehlberg,[1] and shall speak of the stronger, law-based kind of irreversibility as "nomological" while referring to the weaker kind of irreversibility as being "de facto" or "nomologically contingent." In the absence of these qualifications, the ascription of irreversibility to a process commits us to no more than the non-occurrence or virtual non-occurrence of its temporal inverse and leaves it open whether the irreversibility is de facto or nomological in origin. The neutrality of our use of the term "irreversible" as between the nomological and de facto senses is an asset in our concern with the anisotropy of time. For what is decisive for the obtaining of that anisotropy is not whether the non-existence of the temporal inverses of certain processes is merely de facto rather than nomological; instead, what matters here is whether the temporal inverses of these processes always (or nearly always) fail to occur, whatever the reason for that failure! Thus, the processes of masticating food and mixing cream with coffee are irreversible in this *neutral* sense. Hence if a silent film of a dinner party were to show a whole beef steak being reassembled from "desalivated" chewed pieces or the mixture of coffee and cream unmixing, we would know that it has been played backward.

The distinction between the weak and strong kinds of irreversibility has a clear relevance to those physical theories which allow a sharp distinction between laws and boundary conditions in virtue of the repeatability of specified kinds of events at different places and times. But it is highly doubtful that this distinction can be maintained throughout *cosmology*. For what criterion is there for presuming a spatially ubiquitous and temporally permanent feature of the universe to have the character of a boundary condition rather than that of a law?

Let us consider graphically the significance of the presumed fact that there are kinds of sequences of states ABCD, occurring with increasing time, such that the opposite sequence DCBA would not also occur with increasing time. Suppose, for example, ABCD in the diagram are successive kinds of states of a house that burns down completely with increasing or later time. Then

[1] H. Mehlberg: "Physical Laws and Time's Arrow," in: H. Feigl and G. Maxwell (eds.), *Current Issues in the Philosophy of Science, op. cit.*, pp. 105–38.

there will be no case of the inverse kind of sequence DCBA with increasing time, since the latter would constitute the resurrection of a house from debris. Thus, this opposite kind of sequence DCBA would exist only in the direction of decreasing or earlier time, while the first kind of sequence ABCD would not obtain in the latter direction.

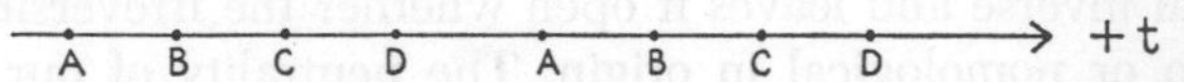

Accordingly, comparison of the structures of the opposite directions of time shows that, at least for the segment of cosmic time constituting the current epoch in our spatial region of the world, the kinds of sequences of states exhibited by the one direction are different from those found in the other. Hence we say that, at least locally, time is anisotropic rather than isotropic. It will be noted that the anisotropy of physical time consists in the mere structural differences between the opposite directions of physical time and constitutes no basis at all for singling out one of the two opposite directions as "the" direction of time.

The dependence of such anisotropy as is exhibited by time on the irreversible character of the processes obtaining in the universe can be thrown into still bolder relief by noting what kind of time there would be, if there were no irreversible processes at all but only reversible processes whose reversibility is not merely nomological but also de facto. That is to say the temporal inverses would not only be allowed by the relevant laws but would actually exist in virtue of the obtaining of the required initial (boundary) conditions. To forestall misunderstandings of such a hypothetical eventuality, it must be pointed out at once that our very existence as human beings having memories would then be impossible, as will become clear later on. Hence, it would be entirely misconceived to engage in the inherently doomed attempt to imagine the posited eventuality within the framework of our actual memory-charged experiences, and then to be dismayed by the failure of such an attempt. As well try to imagine the visual color of radiation in the infra-red or ultra-violet parts of the spectrum.

To see that if all kinds of natural processes were actually de facto reversible, time would indeed be *isotropic,* we now consider an example of such a reversible physical process: the frictionless rolling of a ball over a path AD in accord with Newton's laws, say from A to D.[2] This motion is both nomologically and de facto reversible, because (1) Newton's laws likewise allow another motion over the same path but from D to A, which is the temporal inverse of the motion from A to D, and (2) there are actual occurrences of this inverse motion, since the initial conditions requisite to the occurrence of the inverse motion do obtain.

Let us plot on a time axis the special case in which a particular ball rolls from A to D and is reflected so as to roll back to A, the zero of time being chosen for the event of the ball's being at the point D. The letters "A," "B," "C," and "D" on the time axis in our diagram denote the respective *events* of being at the point A in space, etc., thereby representing the sequence of states (events) ABCD of the "outgoing" motion and then the states DCBA of the "return" motion.

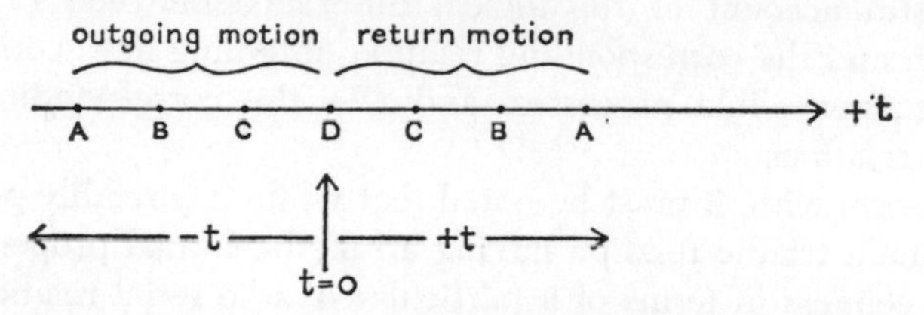

Mathematically, the nomological reversibility of the processes allowed by Newton's laws expresses itself by the fact that the form of the Newtonian equations of motion remains unaltered or invariant upon substituting −t for +t in them. We say, there-

[2] In the present context of exclusively reversible processes, the relation of "earlier than" implicitly invoked in the assertion that a ball moves "*from* A *to* D" (or, in the opposite direction, "*from* D *to* A") must be *divested* of its customary anchorage in an anisotropic time. For in the world of exclusively reversible processes now under discussion, the assertion that a given motion of a ball was "*from* A *to* D" rather than "*from* D *to* A" expresses not an objective physical relation between the two terminal events of the motion, but only the convention that we have assigned a lower time-number to the event of the ball's being at A than to its being at D. And this absence here of an objective physical basis for saying that the motion was "from" one of two points "to" the other makes for the fact, noted in Chapter Seven above, that, if all the processes of nature are de facto reversible, there is no objective physical basis for singling out one of two motion states of a ball as "the" cause of the other.

fore, that Newton's laws for frictionless motions are "time-symmetric." And hence our diagram shows that for every state of the ball allowed by Newton's laws at a time +t, these laws allow precisely the same kind of state at the corresponding time −t. In other words, in the case of reversible processes, the sequences of (allowed) states along the opposite directions of the time-axis are, as it were, mirror-images of each other. Hence, if all of the processes of nature were de facto reversible, time would be isotropic.

Thus, it is further apparent that the structure of time is not something which is apart from the particular kinds of processes obtaining in the universe. Instead, the nature of time is rooted in the very character of these processes.

Our remark in the last footnote on the logical status of the relation "earlier than" in a world of exclusively reversible processes must now be amplified by reference to Reichenbach's unsuccessful account of the logical difference between (1) that relation and the corresponding relation obtaining in a world containing irreversible processes, and (2) the corresponding two causal relations.

To begin with, it must be noted that while it is readily possible to define a triadic relation having all of the formal properties of o-betweenness in terms of a particular dyadic serial relation, the converse deduction is not possible, since in a given system of serial order, we can distinguish one "direction" from its opposite, whereas the system of o-betweenness does not, by itself, enable us to make such a differentiation. The case of the Euclidean straight line will illustrate this fact. The points of the straight line form a system of o-betweenness. This order is intrinsic to the straight line in the sense that its specification involves no essential reference to an external viewer and his particular perspective. The serial ordering of the points with respect to a concrete relation "to the left of" is extrinsic in the sense of requiring reference to an external viewer, at least for the establishment of an asymmetric dyadic relation "to the left of" between two given arbitrarily selected reference points U and V. Once we thus introduce an asymmetric dyadic relation between two such points, then, to be sure, we can indeed use the intrinsic system of o-betweenness on the line to define a serial order throughout the line.[3] To say that a given serial order on the line with respect to the relation "to the left of" is conventional is another way of saying that it is extrinsic in our sense. For a particular external perspective, it is of course not arbitrary whether a given point x is to the left of another point y or conversely. In contrast to the "extrinsic" character of the serial ordering of the points on the line with respect to the relation "to the left of," the serial ordering of the real numbers with respect to "smaller than" is intrinsic in our sense, since for any two real numbers, the ordering with respect to magnitude requires no reference to entities outside the domain.

It is essential not to overlook, as Reichenbach did and as the writer did in an earlier publication,[4] that a serial ordering always establishes a difference in direction independently of whether it is intrinsic or extrinsic! Confusing extrinsicality of a serial relation with undirectedness, Reichenbach maintains that the relation "to the left of" on the line, though serial, is not "unidirectional," thereby allegedly failing to distinguish two opposite directions from one another, whereas he regards the relation "smaller than" among real numbers as both serial and "unidirectional."[5] But he found himself driven to this contention only because he failed to note that, by being asymmetric, a serial relation is automatically a directed one even when the seriality has an extrinsic basis. And this oversight led him to distinguish relations which are serial while allegedly being undirected from directed serial relations. The latter error, in turn, issued in his

[3] The relevant details on these formal matters can be found in Lewis and Langford: *Symbolic Logic* (New York: The Century Company; 1932), pp. 381–87, and in E. V. Huntington: "Inter-Relations Among the Four Principal Types of Order," *Transactions of the American Mathematical Society*, Vol. XXXVIII (1935), Sec. 3.1, p. 7.

[4] H. Reichenbach: "The Philosophical Significance of the Theory of Relativity," in P. A. Schilpp (ed.) *Albert Einstein: Philosopher-Scientist, op. cit.*, pp. 304–05, and *DT, op. cit.*, pp. 26–27; A. Grünbaum: "Time and Entropy," *American Scientist*, Vol. XLIII (1955), p. 551.

[5] For a telling critique of Reichenbach's account of the alleged general logical features of "unidirectional" as opposed to "merely" serial relations, cf. H. Mehlberg: "Physical Laws and Time's Arrow," in H. Feigl and G. Maxwell (eds.) *Current Issues in the Philosophy of Science, op. cit.*, pp. 109–11.

false distinction between the supposed mere seriality of time, which he called "order," and its "direction," a distinction which he attempted to buttress by pointing to the seriality of the time of Newton's mechanics and of special relativity in the face of the complete time-symmetry of the fundamental laws of these theories. But Reichenbach's distinction should be replaced by the distinction between intrinsically isotropic and anisotropic kinds of time, which we shall now explain.

De facto reversible processes intrinsically define a temporal order of mere o-betweenness under suitable boundary conditions, but the symmetric causal relation associated with these processes provides no physical basis for an intrinsic serial order of time. Just as it was possible, however, in the case of the Euclidean line, to introduce a serial ordering in its system of o-betweenness by means of an extrinsically grounded asymmetric dyadic relation between two chosen reference points U and V, so also it is possible to choose two reference states in a time that is intrinsically merely a system of o-betweenness, and extrinsically render this time serial by making one of these two states later than the other through the assignment of suitable real numbers as temporal names. It is in this sense that a world containing no kinds of irreversible processes can nonetheless be legitimately and significantly described by a serial time.

We see that if a universe contains no nomologically or de facto irreversible kinds of processes but is nonetheless claimed to have a serial time, this seriality must have an *extrinsic* component. For, in the presence of suitable boundary conditions, this hypothetically reversible universe will intrinsically define only a temporal order of o-betweenness. And this latter order is isotropic in the following twofold sense: first, all elementary processes are de facto reversible, and second, there is no one property of physical systems such as the entropy whose values intrinsically define a dyadic relation between pairs of states of these systems such that the class of states forms a serial order with respect to that relation.

But in a non-equilibrium world to which the non-statistical second law of classical thermodynamics is actually applicable, precisely the latter kind of property does exist in the form of the entropy. And hence such a world is temporally anisotropic: its time exhibits a special kind of *difference in direction* arising from the directed, intrinsically grounded serial relation of "later than." It is apparent that if we say that the processes in such a world are "irreversible," our assertion differs logically from affirming the "non-return" of any of the total states of a universe whose time is therefore open and perhaps also infinite (in one or both directions). For the *latter* kind of time can be *intrinsically isotropic,* and indeed will be isotropic if the universe possessing it contains only de facto reversible kinds of processes. By contrast, in the former, irreversible kind of universe, whose time is anisotropic, the following two features are present: first, the classical entropy law precludes the occurrence of the same (*non*-equilibrium) *macro*-state at different times rather than merely asserting that the *micro*-states define an *open* order of time in virtue of obtaining boundary conditions, and second, that law makes a specific assertion about the way in which macro-states occurring at different times do differ with respect to a single property.

Although the serial relation "later than" itself does have *a* "direction" in the obvious sense of being asymmetric, the set of states ordered by it does not have *a* direction but rather exhibits a special *difference* or anisotropy as between the *two opposite directions.* Thus, when we speak of the anisotropy of time, this must not be construed as equivalent to making assertions about "the" direction of time. J. J. C. Smart and Max Black have correctly pointed out[6] that reference to "the" direction of time is inspired by the notion that time "flows." In particular, as we shall see in Chapter Ten, Reichenbach's assertions about "the" direction of time rest on his incorrect supposition that there is a physical basis for becoming in the sense of the shifting of a physically defined "now" along one of the two physically distinguished directions of time. By contrast, our characterization of physical time as anisotropic involves no reference whatever to a transient division of time into the past and the future by a "now" whose purported unidirectional "advance" would define "the" direction of time. In fact, we shall argue in Chapter Ten

[6] J. J. C. Smart: "The Temporal Asymmetry of the World," *Analysis,* Vol. XIV (1954), p. 81, and M. Black: "The 'Direction' of Time," *Analysis,* Vol. XIX (1959), p. 54.

that the concept of becoming has no significant application to physical time despite its relevance to psychological and common sense time, because the "now" with respect to which the distinction between the past and the future acquires meaning depends crucially on the egocentric perspectives of a conscious organism for its very existence. Nevertheless, having entered this explicit caveat, we shall achieve brevity by using the locution "the direction of time" as a synonym not only for "the future direction" in psychological time but also for "the one of two physically distinguished directions of time which our theory calls 'positive.' "

Our analysis of the logical relations between symmetric causality, open time, extrinsic *vs.* intrinsic seriality of time, and anisotropy of time requires us to reject the following statement by Reichenbach:

> In the usual discussions of problems of time it has become customary to argue that only irreversible processes supply an asymmetrical relation of causality while reversible processes allegedly lead to a symmetrical causal relation. This conception is incorrect. Irreversible processes alone can define a direction of time; but reversible processes define at least an [serial] order of time, and thereby supply an asymmetrical relation of causality. The reader is referred to the discussion of the relation *to the left of* (. . .). The correct formulation is that only irreversible processes define a *unidirectional* causality.[7]

Reichenbach notes that while the causal processes of classical mechanics and special relativity are reversible, these "reversible" theories nonetheless employ a temporal order which is *serial.* He then infers that (a) even in a reversible world the causal relation must be asymmetric, and (b) in an irreversible world, we require a temporal relation which is not "merely" serial but also "unidirectional," as well as a causal relation which is both asymmetric and unidirectional. But Reichenbach seems to have overlooked completely that if a physical theory affirms the seriality of the time of a completely reversible world, then that seriality is extrinsic and the assignment of the lower of two real numbers as the temporal name to one of two causally connected events therefore does not express any objective asymmetry on the part of the causal relation itself.

[7] H. Reichenbach: *DT, op. cit.*, p. 32.

We are now ready to consider in detail whether thermodynamic processes furnish a physical basis for an anisotropy of time. More specifically, our problem is whether entropy, whose values are given by real numbers, succeeds, unlike the causality of reversible processes, in conferring anisotropy on open time by *intrinsically* defining a serial ordering in the class of states of a closed system. At first, we shall deal with this problem in the light of classical phenomenological thermodynamics. A separate treatment based on the entropy of statistical mechanics will then follow. The second part of this chapter will then consider the existence of *non*-entropic physical bases for the anisotropy of time. It will turn out that both the thermodynamic and the non-thermodynamic species of irreversibility which we shall find to obtain are *de facto* or nomologically-contingent in character.

I. *The Entropy Law of Classical Thermodynamics.*

Suppose that a physical system is created such that one end is hot and the other cold and is then essentially closed off from the rest of the world. In ordinary experience, we do not find that the hot end becomes hotter at the expense of the increased coolness of the cool end. Instead, the system tends towards an equilibrium state of intermediate temperature. And, at least as far as ordinary physical experience is concerned, this entire process of temperature equalization is irreversible. It is possible to characterize this irreversibility more precisely by associating with each momentary state of the closed system a certain quantity, called the "entropy." For the entropy provides the following relative measure of the degree of temperature-equalization attained by the system in the given state: the irreversible temperature-equalization associated with the transition from the initial to the final state corresponds to an increase in the entropy. Accordingly, for a closed system not already in equilibrium, the original *non*-statistical form of the second law of thermodynamics affirms an increase of the entropy with time.[8]

---

[8] Although we shall have occasion to see that Clausius's phenomenological principle of entropy increase does indeed require emendation in the light of the discoveries of statistical mechanics, I must dissent from the following judgment by H. Mehlberg ["Physical Laws and Time's Arrow," in H. Feigl and G. Maxwell (eds.) *Current Issues in the Philosophy of Science*, *op. cit.*, p. 115]: "the only rigorous axiomatization of phenomenological thermodynamics (due to Carathéodory) [C. Carathéodory: "Untersuchungen

We shall find the *non*-statistical form of this second law to be untenable in the light of statistical mechanics. It will nonetheless prove useful initially to take Clausius's original form of the law

---

über die Grundlagen der Thermodynamik," *Mathematische Annalen,* 1909] has also stripped the second phenomenological principle of thermodynamics of its irreversible and anisotropic implications. This important result has been clarified by Professor A. Landé in his illuminating presentation of Carathéodory's contribution [A. Landé: "Axiomatische Begründung der Thermodynamik durch Carathéodory," *Handbuch der Physik,* Vol. IX (1926), pp. 281–300]." But upon turning to the latter article, we find Professor Landé writing as follows: "The theory of quasi-static changes of state (existence of the entropy, etc.) is independent, however, of assertions that might be made about the behavior in the case of non-static processes; thus, for example, all theorems . . . for quasi-static processes would remain valid without change, even if in the case of non-static processes . . . irreversible processes were to run their course in a direction opposite to the actual one." [A. Landé, *ibid.,* p. 299] . . . "Hence our first conclusion is that it is irrelevant for the existence of the entropy whether the quasi-static processes themselves are reversible or not." [A. Landé, *ibid.,* p. 300, quoting from T. Ehrenfest]. "The second conclusion is that the existence of the entropy (or the quasi-static-adiabatic unattainability of neighboring points) is likewise independent of whether the non-static processes are reversible or not. . . . Thus the second law for *quasi-static* changes of state would not even be placed in jeopardy in the event that one were to succeed in making non-static processes time-reversible. To be sure, in that case the principle of Thomson and Clausius for *non-static* processes would become invalid. . . . This case would occur, if one were to include within the scope of one's consideration the temporally inverse processes which, according to the findings of kinetic theory, cannot be excluded." [A. Landé, *ibid.,* p. 300]. It would seem that the Carathéodory-Landé account of *phenomenological* thermodynamics does not justify the claim that Clausius's law for *non*-static processes allows these to be time-reversible. Landé's analysis allows only the following far weaker conclusion, which cannot be adduced as support for Mehlberg's contention: if one does go outside phenomenological thermodynamics and invokes the findings of statistical mechanics (kinetic theory) to affirm the reversibility of *non*-static processes, then one can still uphold the second law of thermodynamics for *quasi*-static processes, although, as Landé points out explicitly, "in that case the principle of . . . Clausius for *non-static* processes would become invalid."

Contrary to Mehlberg, it would appear, therefore, that such emendations as must be made in the second phenomenological law of thermodynamics because of its ascription of irreversibility to *non*-static processes do *not* derive at all from the rigor of Carathéodory's axiomatic account of that law within the framework of phenomenological thermodynamics; instead, the required emendations derive wholly from a domain of physical occurrences whose theoretical mastery requires recourse to *statistical* considerations falling outside the purview of phenomenological thermodynamics. Hence, I see no justification for Mehlberg's claim that Carathéodory "has also stripped the second phenomenological principle of thermodynamics of its irreversible and anisotropic implications."

at face value, since A. S. Eddington invoked that version of the law as the basis for attributing anisotropy to time. And although Eddington's account of the anisotropy of time will turn out to lack an adequate physical foundation, a statement of that account and a critique of P. W. Bridgman's misunderstanding of it will be instructive for our subsequent purposes.

The customary statement of the second law of thermodynamics given above has factual content in an obvious sense, *if* the direction of increasing time is defined independently of the entropy-increase. It has been suggested that the independent criterion of time increase be provided either by the continuous matter-energy accretion (as distinct from energy dispersion) postulated by the "new cosmology"[9] or—in the spatially limited and cosmically brief career of man—by reliance on the subjective sense of temporal order in human consciousness. But I reject both of these proposed criteria. For I am unwilling to base the factual content of so earthy a macroscopic law as that of Clausius on a highly speculative cosmology in which the matter-energy *accretion* manifests itself macro-*empirically* to us merely by the existence of a *steady state!* And we shall see that some important features of man's subjective sense of temporal order can be explained on the basis of the participation of his organism in processes governed by the *entropy statistics* of space ensembles of temporarily closed systems. Would one then have to regard the second law as a tautology, if one were to follow Eddington[1] in resting the anisotropy of physical time on the supposed fact that in one of the two opposite directions of time—which is called the direction of "later"—the entropy of a closed system *increases*, whereas in the opposite direction, the entropy *decreases?* No, this conclusion could not be sustained even if one were to ignore the existence of the viable non-entropic criteria of "later than" to be discussed in Part B of the present chapter. To be sure, if one were to restrict oneself to a *single* closed system and were to say then that of two of its entropy states, the state of greater

[9] For details on the "new cosmology," see H. Bondi: *Cosmology* (2nd Edition, Cambridge: Cambridge University Press; 1961). A brief digest is given in A. Grünbaum: "Some Highlights of Modern Cosmology and Cosmogony," *The Review of Metaphysics*, Vol. V (1952), pp. 493–98.

[1] Cf. A. S. Eddington: *The Nature of the Physical World, op. cit.*, pp. 69ff.

entropy will be called the "later" of the two, then indeed one would have failed to render the factual content of the second, phenomenological law of thermodynamics. But just as in the case of other specifications of empirical indicators – specifications which constitute "definitions" only in the *weak* sense set forth in Chapter One à propos of the "definition" of congruence– Eddington's *non*-statistical entropic "definition" of "later than" was prompted by the presumed *empirical* fact that it does not give rise to contradictions or ambiguities when *different* closed systems are used. For, apart from statistical modifications which we are ignoring for the present, there is concordance in the behavior of *all* closed systems: given any two such systems A and B, neither of which is in thermodynamic equilibrium, if an entropy state S_{Aj} of A is simultaneous with a state S_{Bj} of B, then there is no case of a state S_{Ak} being simultaneous with a state S_{Bi}, such that $S_{Ak} > S_{Aj}$ while $S_{Bi} < S_{Bj}$.[2] It was this presumed entropic concordance of thermodynamic systems which prompted Eddington's attempt[3] to use the second law of thermodynamics in order to account for the anisotropy of time. But by his very unfortunate choice of the misleading name "time's arrow" for the latter feature of physical time, he ironically invited the very misunderstanding which he had been at pains to prevent,[4] viz., that he was intending to offer a thermodynamic basis for the "unidirectional flow" of psychological time. Eddington maintained that the entropic behavior of closed physical systems distinguishes the two *opposite* directions of time *structurally* in regard to *earlier* and *later* respectively as follows: of two states of the world, the *later* state is the one coinciding with the *higher*

---

[2] Examples of other "definitions" which have a corresponding factual foundation in the concordant behavior of different bodies are: first, the "definition" of the metric of time on the basis of the empirical law of inertia, and second, the "definition" of congruence for spatially separated bodies on the basis of the fact that two bodies which are congruent at a given place will be so everywhere, independently of the respective paths along which they are transported individually. Cf. M. Schlick: "Are Natural Laws Conventions," in: H. Feigl and M. Brodbeck (eds.) *Readings in the Philosophy of Science* (New York: Appleton-Century-Crofts, Inc.; 1953), p. 184; H. Reichenbach: "Ziele und Wege der physikalischen Erkenntnis," *Handbuch der Physik*, Vol. IV (1929), pp. 52–53, and *PST*, *op. cit.*, pp. 16–17.

[3] A. S. Eddington: *The Nature of the Physical World*, *op. cit.*, pp. 69ff.

[4] *Ibid.*, pp. 68, 87–110.

entropy of any one closed non-equilibrium system, whereas the *earlier* state corresponds to the state of *lower* entropy. Thus, according to Eddington, the anisotropy of physical time would derive from the supposed fact that in one of the two opposite directions of time—which we call the direction of "later"—the entropy of any closed system *increases*, whereas in the opposite direction, the entropy *decreases*.

Since the relation "larger than" for real numbers is serial, Eddington's entropic definition of "later than" intrinsically renders the seriality of time, once its over-all openness is assured by suitable boundary conditions, openness and seriality being attributes concerning which the causal theory of time had to be noncommittal. But, as Eddington seems to have neglected to point out, that theory of time can play an essential role in his entropic "definition" of "later than." For the latter presupposes coordinative "definitions" for the concepts of "temporally between" and "simultaneous" in the very statement of the second law of thermodynamics: this law uses the concept of "closed system," thereby presupposing the concept of "temporal betweenness," as we saw in footnote 9 on page 186 in Chapter Seven, and furthermore the second law requires the concept of simultaneity by making reference to the entropy of an *extended* system *at a certain time*, and implicitly, to the *simultaneous* entropy states of several systems.

P. W. Bridgman has offered a critique of Eddington's entropic "definition" of "later than" which takes as its point of departure the conceptual commitments of the pre-scientific temporal discourse of common sense in which the "Now" of conscious experience is enshrined.[5] And we shall therefore now find that among the things which render P. W. Bridgman's critique of A. S. Eddington's account of the anisotropy of physical time nugatory, there is Bridgman's erroneous identification of physical time with the "unidirectional flow" of common sense and psychological time.

Bridgman contends that the entropy increase cannot be regarded

[5] A detailed articulation of the logical commitments of pre-scientific temporal discourse is included in W. Sellars's penetrating "Time and the World Order," *Minnesota Studies in the Philosophy of Science* (Minneapolis: University of Minnesota Press; 1962), Vol. III, pp. 527–616.

with Eddington as a fundamental indicator of the relation "later than." Speaking of the significance which Bridgman believes Eddington to have ascribed to the invariance of the laws of mechanics under time-reversal, Bridgman states his objection to Eddington as follows:

> The significance that Eddington ascribes to it is that the equation is unaffected by a reversal of the direction of flow of time, which would mean that the corresponding physical occurrence is the same whether time flows forward or backward, and his thesis is that in general there is nothing in ordinary mechanical occurrences to indicate whether time is flowing forward or backward. In thermodynamic systems on the other hand, in which entropy increases with time, time enters the differential equation as the first derivative, so that the direction of flow of time cannot be changed without changing the equation. This is taken to indicate that in a thermodynamic system time must flow forward, while it might flow backward in a mechanical system.[6]
>
> . . . how would one go to work in any concrete case to decide whether time were flowing forward or backward? If it were found that the entropy of the universe were decreasing, would one say that time was flowing backward, or would one say that it was a law of nature that entropy decreases with time?[7]

We see that Bridgman takes Eddington to have offered an entropic basis of the "forward flow" of psychological time rather than of the anisotropy of physical time, because Bridgman falsely identifies these two different concepts. And we shall now show that his purported *reductio ad absurdum* argument against Eddington's entropic account of the anisotropy of physical time

[6] P. W. Bridgman: *Reflections of a Physicist* (New York: Philosophical Library; 1950), p. 163.

[7] *Ibid.*, p. 165. Lest Bridgman's point here be misunderstood, I must point out that in the same essay (pp. 169, 175–77, 181–82) he has explicitly rejected as unfounded the conclusion of statistical mechanics that the entropy of a closed system will *decrease* markedly after being in equilibrium for a very long time. For he rejects as gratuitous the assumption that the micro-constituents of thermodynamic systems can be held to behave reversibly in accord with time-symmetric laws whose observational foundation is only macroscopic. Hence Bridgman's hypothetical finding of an entropy *decrease* is predicated on an actual human observation of an overall entropy decrease with increasing psychological time.

derives its plausibility but also its lack of cogency from the conjunction of precisely this illegitimate identification with a contrary-to-fact assumption. Thus, we ask Bridgman: under what circumstances would it be found that the entropy of the universe "were decreasing"? This situation would arise in the contrary-to-fact eventuality that physical systems or the universe would exhibit *lower* entropy states at times which are *psychologically later,* and *higher* entropy states at times which are *psychologically earlier,* such as in the hypothetical case of finding that lukewarm water separates out into hot and cold portions as time goes on psychologically. To appreciate the import of this contrary-to-fact assumption, we note that an experience *B* is *psychologically later* than an experience *A* under one of the following two conditions: First, the awareness-and-memory content constituting experience *A* is a *proper part* of the memory-content of experience *B,* or second, experience *B* contains the memory of *the fact* of the occurrence of another experience *A* (e.g., the fact of having dreamt), but the memories ingredient in *B* do not contain the *content* of experience *A* (e.g., the details of the dream having been forgotten).[8] Thus, psychologically later times are either times at which we do, in fact, have more memories or information than at the correspondingly earlier ones, or they are times at which it would be possible to have a richer store of memories even if the latter did not, in fact, materialize because of partial forgetting. Accordingly, Bridgman's posit of our finding that the entropy "were decreasing" would require the entropy increase among physical systems and the future direction of psychological time to be *temporally counter-directed* as follows: temporally, the direction of increasing entropy among *physical* systems would *not also* be the direction of actual or

[8] I am indebted to my colleague Professor A. Janis for pointing out to me (by reference to the example of having dreamt) that the first of these conditions is only a sufficient and not also a necessary condition for the obtaining of the relation of being psychologically later. This caution must likewise be applied to the following two assertions by William James, if they are to hold: "our perception of time's flight . . . is due . . . to our *memory* of a content which it [i.e., time] had a moment previous, and which we feel to agree or disagree with its content now," (W. James: *The Principles of Psychology* [New York: Dover Publications, Inc.; 1950], p. 619), and "what is past, to be known as past, must be known *with* what is present, and *during* the 'present' spot of time." (*Ibid.,* p. 629)

possible *memory* (information) increase among *biological* organisms, since (actually or possibly) "richer" memory states would be coinciding temporally with *lower* entropy states of physical systems.

What is the logical force of Bridgman's contrary-to-fact assumption as a basis for invalidating Eddington's account of the anisotropy of physical time? Bridgman's objection is seen to be devoid of cogency in the light of the following reasons. In the first place, quite apart from the fact that Eddington was not concerned to account for the "forward flow" of psychological time, *in actual fact* the very production of memories in biological organisms depends, as we shall see, on entropy *increases* in certain portions of the external environment. And since Eddington was offering his criterion as an account of what does, in fact, obtain, the adequacy of this account cannot be impugned by the contrary-to-fact logical possibility of counter-directedness envisioned by Bridgman. But even if the situation posited by Bridgman were to materialize in actuality, it would certainly not refute Eddington's claim that (1) the purported entropic behavior of physical systems renders the opposite directions of physical time anisotropic, because the entropy of each of these systems decreases in the one direction and increases in the other, and (2) the direction of increasing entropy can be called the direction of "later than" or time increase. Although Eddington had left himself open to being misunderstood by using the misleading term "time's arrow," he had also sought to spike the misunderstanding that entropically characterized physical time "is flowing forward" in the sense of there being a physical becoming. For he makes a special point of emphasizing[9] that this becoming, so familiar from *psychological* time, eludes conceptual rendition as an attribute of physical processes, because it involves the concept "now." Contrary to Bridgman, Eddington saw no problem of *physical* time flowing backward rather than forward, since, as Chapter Ten will show in detail, the metaphor "flow" has no relevance to physical time. *Physically,* certain states are later than others by certain amounts of time. But there is no "flow" of *physical* time, because physically there is no egocentric (psychological) transient *now.* Moreover, as applied

[9] A. S. Eddington: *The Nature of the Physical World, op. cit.*, pp. 68–69 and Chapter iv.

to *psychological* time, the locution "flow backward" is self-contradictory, since the assertion that the now shifts forward (in the future direction) is a tautology, as Chapter Ten will demonstrate. A fluid can flow spatially up or down, because the meaning of spatial "flow" is independent of the meaning of "spatially up" or "down." But as applied to psychological time, the meaning of the action-verb metaphor "flow" *involves* the meaning of the metaphor "forward," i.e., of "from earlier to later." For the *flowing* here denotes metaphorically the shifting of the "now" from earlier to later or "forward." Hence if Bridgman's hypothetical situation of counter-directedness could actually materialize, we would say that the entropy is decreasing with increasing psychological time without damage to Eddington's account and not that time is "flowing backward."

Furthermore, if the situation envisioned by Bridgman did arise, we might well not survive long enough to be troubled by it. Poincaré and de Beauregard[1] have explained, in a qualitative way, why prediction and action would probably become impossible under the circumstances posited by Bridgman: two bodies initially at the same temperature would then acquire different temperatures at psychologically later times, while we would be unable to anticipate which of these bodies will become the warmer one, and thus we might be burnt severely if we happen to be in touch with the one that turns out to be the hot one. Or imagine taking a bath in lukewarm water and then not being able to predict which portion of the bathtub will turn out to be the boiling hot end. By the same token, whereas in actuality friction is a *retarding* force, because its dissipation of mechanical energy issues in an increase of the entropy with increasing psychological time, on Bridgman's contrary-to-fact assumption friction would be an *accelerating* force that sets stationary bodies into motion in unpredictable directions. Thus, with increasing psychological time, heat energy would convert itself into mechanical energy of a previously stationary body such as a heavy rock, and prediction of the direction in which the rock would

---

[1] H. Poincaré: *The Foundations of Science, op. cit.*, pp. 399–400; O. C. de Beauregard: "L'Irréversibilité Quantique, Phénomène Macroscopique," in A. George (ed.) *Louis de Broglie, Physicien et Penseur* (Paris: Albin Michel; 1953), p. 403, and *Théorie Synthétique de la Relativité Restreinte et des Quanta* (Paris: Gauthier-Villars; 1957), Chapter xiii, esp. pp. 167–71.

start moving would then be well-nigh impossible. And even if we escaped destruction most of the time by not being in the paths of these unpredictable motions, we might well succumb to the anxiety induced by our inability to anticipate and control daily developments in our environment which would constantly threaten our survival.

Finally, suppose that we supplement the hypothetical conditions posited by Bridgman by assuming that in addition to his hypothetical human species *A* whose members are supposed to experience *higher* entropy states of physical systems as psychologically *earlier* than lower ones, there is another human species *B* possessing our actual property of experiencing these same higher entropy states as psychologically later. Then, as Norbert Wiener has noted, a very serious difficulty would arise for communication between our two species *A* and *B* whose psychological time senses are counter-directed. Wiener writes:

> It is a very interesting intellectual experiment to make the fantasy of an intelligent being whose time should run the other way to our own. To such a being all communication with us would be impossible. Any signal he might send would reach us with a logical stream of consequents from his point of view, antecedents from ours. These antecedents would already be in our experience, and would have served to us as the natural explanation of his signal, without presupposing an intelligent being to have sent it. . . . Our counterpart would have exactly similar ideas concerning us. *Within any world with which we can communicate, the direction of time is uniform.*[2]

In amplification of Wiener's statement, consider a situation in which our species *A* and *B* have distinct habitats, which are represented respectively by the regions *A* and *B* of our diagram.

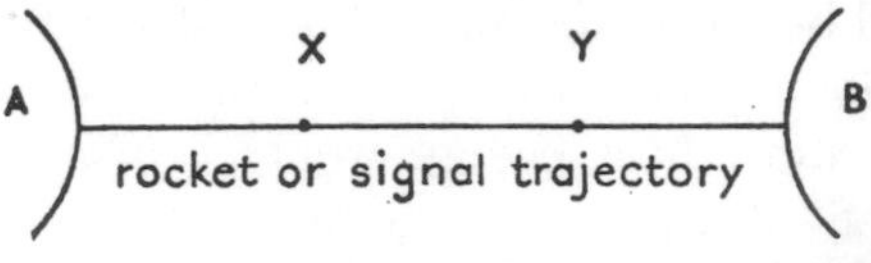

A X Y B
rocket or signal trajectory

We can then show that any particle or signal which would be regarded as *outgoing* by one of the two species would likewise be held to be *departing* by the other, and any object or message which is *incoming* in the judgment of either species will also be

[2] N. Wiener: *Cybernetics* (New York: John Wiley & Sons, Inc.; 1948), p. 45. A second edition was published in 1961.

held to be *arriving* by the other. For—to take the case of the *outgoing* influence—suppose that as judged by the members of A, the particle reaches the point Y of its trajectory (see diagram) *later than* the point X and is therefore held to be departing by the men in A. Then the members of B will conclude that the particle is *leaving* them as well, since they will judge that it reaches point X *after* reaching point Y. And thus, if as judged by the A-men, they hurl a rock toward the B region such that the rock comes to rest and remains at rest in B indefinitely, then the B-men, in turn, will judge that a rock, having been at rest in their region all along, suddenly left their habitat and traveled to A, where it was then received by the A-men with ready, open arms. And if the B-men were struck by the discrepancy between the dynamical behavior of that rock and the behavior of other rocks in their habitat—assuming that the *latter* obey the familiar dynamical principles—they might conceivably conclude after a number of such experiences that the dynamically aberrant rocks are linked to the presence elsewhere of temporally counter-directed beings.[3] In order to draw this conclusion, however, the

[3] Reichenbach (*DT*, *op. cit.*, pp. 139–40) discusses a situation envisaged by Boltzmann in which there are two entropically counter-directed galactic systems each of which contains intelligent beings whose positive sense of psychological time is geared to the direction of entropy increase in its own galactic environment. Says he (*DT*, *op. cit.*, p. 139):

> Let us assume that among the many galaxies there is one within which time goes in a direction opposite to that of our galaxy. . . . In this situation, some distant part of the universe is on a section of its entropy curve which for us is a downgrade; if, however, there were living beings in that part of the universe, then their environment would for them have all the properties of being on an upgrade of the entropy curve.

And Reichenbach then offers the following quite incomplete hints as to how there might be physical interaction between the two sets of intelligent beings such that either set of beings would be able to secure information indicating the temporal counter-directedness of the other (*DT*, *op. cit.*, pp. 139–40):

> That such a system is developing in the opposite time direction might be discovered by us from some radiation traveling from the system to us and perhaps exhibiting a shift in spectral lines upon arrival . . . the radiation traveling from the system to us would, for the system . . . not leave that system but arrive at it. Perhaps the signal could be interpreted by inhabitants of that system as a message from our system telling them that our system develops in the reverse time direction. We have here a connecting light ray which, for each system, is an arriving light ray annihilated in some absorption process.

B-men would have to assume that the entropy decrease involved in the rock's spontaneous acquisition of kinetic energy from the sand is less probable than the presence elsewhere of temporally counter-directed beings.

Our earlier recognition of the dependence of Eddington's "definition" of "later" on the prior concepts of "temporally between" and "simultaneous" enables us to reply to a further objection raised by Bridgman.[4]

Bridgman claims, on operational grounds, that Eddington's definition is circular and that reliance on the *psychological* sense of what times are "later" is logically indispensable. Says he:

> in any operational view of the meaning of natural concepts the notion of time must be used as a primitive concept, which cannot be analyzed, and which can only be accepted, . . . I see no way of formulating the underlying operations without assuming as understood the notion of earlier or later in time.[5]

In an endeavor to show that the specification of the entropy of a closed system at a given instant presupposes the use of the psychological sense of "later," Bridgman says:

> [Consider what] is involved in specifying a thermodynamic system. One of the variables is the temperature; it is not sufficient merely to read at a given instant of time an instrument called a thermometer, but there are various precautions to be observed in the use of a thermometer, the most important of

[4] P. W. Bridgman: *Reflections of a Physicist, op. cit.*, pp. 162–67. The rebuttal about to be offered to Bridgman's critique of Eddington's definition also applies to L. Susan Stebbing's arguments against it, as set forth in her *Philosophy and the Physicists* (London: Methuen & Company, Ltd.; 1937), Chapter xi, esp. pp. 262–63.

I should emphasize, however, that my defense of Eddington's "definition" against Bridgman's criticisms must not be construed as agreement with either his general philosophy of science or with his view (cf. *The Nature of the Physical World, op. cit.*, pp. 84–85) that the entire universe's supposed past state of *minimum* entropy constitutes a conundrum to which even theological ideas are relevant. For I not only deem theological considerations wholly unilluminating physically in any case (cf. my "Some Highlights of Modern Cosmology and Cosmogony," *The Review of Metaphysics, op. cit.*, esp. pp. 497–98, and my "Science and Ideology," *The Scientific Monthly*, Vol. LXXIX [1954], p. 13), but I also maintain that the *statistical* conception of entropy to be discussed below cuts the ground from under Eddington's assumption that the entire universe must have been *primordially* in a state of *minimum* entropy, an assumption essential to Eddington's puzzlement as to the origin of this presumed state.

[5] P. W. Bridgman: *Reflections of a Physicist, op. cit.*, p. 165.

> which is that one must be sure that the thermometer has come to equilibrium with its surroundings and so records the true temperature. *In order to establish this, one has to observe how the readings of the thermometer change as time increases.*[6]

Bridgman claims more than is warranted on precisely the point at issue. For to certify the existence of equilibrium at a certain instant t, we must assure ourselves of the absence of a change in the thermometer's reading during a time interval containing the instant t other than as an endpoint. But does the procurement of the assurance require a knowledge as to which of the two termini of such an interval is the earlier or later of the two with respect to the usual positive time direction? Is it not sufficient to ascertain the constancy of the reading *between* the terminal instants of the time-interval in question? Indeed, what is presupposed is merely temporal betweenness, which, as we saw, is defined by causal processes, independently of the concept of "later than." But this is hardly damaging to Eddington's "definition" of this concept. And it is irrelevant that, in practice, the experimenter may note also which one of the termini of the time-interval containing the instant t is the earlier of the two. For what is at issue is the *semantical* as distinct from the *pragmatic* anchorage of concepts, our inquiry being one in the context of justification and not in the context of discovery.[7] The complete dispensability of the experimenter's subjective sense of

---

[6] *Ibid.*, p. 167, *my italics.*

[7] For a discussion of Bridgman's unwarranted absorption of semantics within pragmatics, which is a new version of the Sophist doctrine that man is the measure of all things, see A. Grünbaum: "Operationism and Relativity," *The Scientific Monthly*, Vol. LXXIX (1954), pp. 228–31 [reprinted in P. Frank (ed.) *The Validation of Scientific Theories* (Boston: Beacon Press; 1957)]. The same absorption of semantics within pragmatics is found in the following statement by Bridgman ["Reflections on Thermodynamics," *American Scientist*, Vol. XLI (1953), p. 554]: "In general, the meaning of our concepts on the microscopic level is ultimately to be sought in operations on the macroscopic level. The reason is simply that we, for whom the meanings exist, operate on the macroscopic level. The reduction of the meanings of quantum mechanics to the macroscopic level has, I believe, not yet been successfully accomplished and is one of the major tasks ahead of quantum theory."

For a critique of the use of Bridgman's homocentrism in the interpretation of quantum mechanics, see H. Reichenbach: *DT, op. cit.*, p. 224 and A. Grünbaum: "Complementarity in Quantum Physics and Its Philosophical Generalization," *The Journal of Philosophy*, Vol. LIV (1957), p. 719.

earlier and later is further apparent from the fact that the experimenter could certify equilibrium at the instant t, if he were given a film strip showing the constancy of the reading during a time interval between t_1 and t_2 which contains t, *without* being told which end of the film strip corresponds to the earlier moment t_1. A similar reply can be given to Bridgman's argument[8] that the *physical* meaning of "velocity" presupposes the psychological sense of earlier and later. For we shall see that purely physical processes in nature define a difference in time-direction quite independently of human consciousness. And thus, for any given choice of the positive space-direction, physical processes themselves define the meanings of both the signs (directions) and magnitudes of velocities independently of man's psychological sense of earlier and later. Here again, Bridgman falsely equates and confuses two *different* meaning components of terms in physics: the physical or semantical with the psychological or pragmatic. The semantical component concerns the properties and relations of purely physical entities which are denoted (named) by terms like "velocity." By contrast, the pragmatic component concerns the activities, both manual and mental, of scientists in *discovering* or coming to *know* the existence of physical entities exhibiting the properties and relations involved in having a certain velocity. That statements about the velocities of masses do not derive their physical meaning from our psychological sense of earlier and later is shown by the fact that cosmogonic hypotheses make reference to the velocities of masses during a stage in the formation of our solar system which preceded the evolution of man and of his psychological time sense. In fact, even in a completely reversible world devoid of beings possessing a time sense, velocity would be a significant attribute of a body despite that hypothetical world's temporal isotropy. But this isotropy would have the consequence that the velocities in such a hypothetical world would not, of course, be anchored in an anisotropic time any more than the positive and negative directions in the presumably isotropic space of our actual world involve the anisotropy of space.

Let us recall Poincaré's and de Beauregard's explanation of why certain sorts of *prediction* and action would become well-

---

[8] P. W. Bridgman: *Reflections of a Physicist, op. cit.*, p. 167.

nigh impossible under the hypothetical conditions of Bridgman's contrary-to-fact assumption. While still remaining within the framework of the phenomenological second law of thermodynamics, it then becomes apparent that our *actual* world presents us with precisely the inverse temporal asymmetry of inferrability under analogous initial conditions. There are physical conditions in our actual world under which we cannot infer the past but can predict the future. The existence of this particular temporal asymmetry has been obscured by a preoccupation with both reversible processes whose past is as readily determinable as their future, and with conditions in intermittently *open non*-equilibrium systems under which the past can often be inferred from the present, as we shall see presently, whereas the future generally cannot.

Let us clarify the conditions which allow the prediction of the future while precluding the retrodiction of the past by reference to the equation describing a diffusion process, a process in which the entropy increases. This equation is of the form

$$\frac{\partial^2 \Psi}{\partial x^2} + \frac{\partial^2 \Psi}{\partial y^2} + \frac{\partial^2 \Psi}{\partial z^2} = a^2 \frac{\partial \Psi}{\partial t}$$

where a^2 is a real constant. This diffusion equation differs from the wave equation for a reversible process by having a first time derivative instead of a second. In the one-dimensional case of, say, heat-flow, the general solution of the equation governing the temperature Ψ is given by

$$\Psi\ (x, t) = \sum_{n=1}^{\infty} b_n \sin nx\ e^{\frac{-n^2}{a^2}t},$$

where the b_n are constants. The behavior of this equation is temporally asymmetric in the following twofold sense: First, if the physical system is in equilibrium at time $t = 0$, then we *cannot* infer what particular sequence of non-equilibrium states issued in the present equilibrium state, since no such sequence is unique,[9] and second, if the physical system is found in a non-

[9] This is not to say that there are not other initial conditions under which at least a finite portion of the system's past can be inferred. For a discussion of this case, see J. C. Maxwell: *Theory of Heat* (6th edition; New York: Longmans, Green, and Company; 1880), p. 264, and F. John: "Numerical Solution of the Equation of Heat Conduction for Preceding Times," *Annali di Matematica Pura ed Applicata*, Vol. XL (1955), p. 129.

equilibrium temperature state at t = 0, then it could not have been undergoing diffusion for all past values of t, although it can theoretically do so for all future values. Specifically, if external agencies impinge on the system and produce a non-equilibrium state of low entropy at time t = 0, then there is no basis for supposing that the system has been undergoing diffusion before t = 0, and then the diffusion equation cannot be invoked to infer the "prenatal" past of the system on the basis of its state at t = 0, although that equation can be used to predict its future as a closed system undergoing diffusion. Another illustration of the same temporal asymmetry of inferrability in equilibrating processes is given by the case of a ball rolling down the wall on the inside of a round bowl subject to friction: if the ball is found to be at rest at the bottom of the bowl, we *cannot retrodict* its particular motion prior to coming to rest, but if the ball is released at the inside wall near the top, we *can predict* its subsequent coming to rest at the bottom.

This possibility of prophesying the future states of an irreversible process in a closed system under the stipulated conditions in the face of the enigmatic darkness shrouding the non-equilibrium states of the past is so important that E. Hille, following J. Hadamard's analysis of Huyghens's principle in optics, has formulated the fundamental principle of scientific determinism as follows: "From the state of a [closed] physical system at the time t_0 we may deduce its state at a later [but not at an earlier] instant t."[1]

If this be the case, then it is natural to ask why it is that in so many cases involving irreversible processes, we seem to be far more reliably informed concerning the past than concerning the future. This question is raised by Schlick, who points out that human footprints on a beach enable us to infer that a person *was* there in the past but *not* that someone *will* walk there in

[1] E. Hille: *Functional Analysis and Semi-Groups* (New York: American Mathematical Society Publications; 1948) Vol. XXXI, p. 388. Mathematically, the difference between the temporal symmetry of determination in the case of reversible processes and the corresponding asymmetry for irreversible processes expresses itself in the fact that the equations of the former give rise to associated groups of linear transformations while the latter lead to *semi*-groups instead. See also M. S. Watanabe, "Symmetry of Physical Laws. Part III. Prediction and Retrodiction," *Reviews of Modern Physics*, Vol. XXVII (1955), pp. 179–86.

the future. His answer is that "the structure of the past is inferred not from the extent to which energy has been dispersed [i.e., not from the extent of the entropy increase] but from the spatial arrangement of objects."[2] And he adds that the spatial traces, broadly conceived, are always produced in accord with the entropy principle. Thus, in the case of the beach, the kinetic energy of the person's feet became dispersed in the process of arranging the grains of sand into the form of an imprint, which owes its (relative) persistence in part to the fact that the pedal kinetic energy lost its organization in the course of being imparted to the sand. To be sure, Schlick's claim that the process of leaving a trace occurs in accord with the entropy principle is quite true. But Schlick fails to articulate the logic of the invocation of entropic considerations in the retrodictive inference.

To exhibit the logic of our retrodictive inference that a person did walk on the beach, I briefly anticipate and utilize here results which will emerge from our discussion below of the *statistical* entropy of *temporarily* closed systems. The justification for the inference of the past incursion of the beach by a stroller derives from the following:

(1) most systems which we now encounter in an isolated state of relatively low entropy, behaving as if they might remain isolated, neither were in fact permanently closed in the past nor will remain isolated indefinitely in the future,

(2) in the case of such temporarily isolated or "branch" systems, we can reliably infer a past interaction of the system with an outside agency from a present ordered or low entropy state, an inference which is not feasible, as we shall see in detail, on the basis of the statistical version of the second law of thermodynamics as applied to a single, permanently closed system, and

(3) the retrodictive inference is based on the assumption that a transition from an earlier high entropy state to a presently given low one is overwhelmingly *im*probable in a system while

[2] M. Schlick: *Grundzüge der Naturphilosophie* (Vienna: Gerold & Company; 1948), pp. 106–07. J. J. C. Smart ["The Temporal Asymmetry of the World," *Analysis*, Vol. XIV (1954), p. 80] also discusses the significance of traces but reaches the following unwarrantedly agnostic conclusion: "So the asymmetry of the concept of trace has something to do with the idea of formlessness or chaos. But it is not easy to see what." See also Smart's paper in *Australasian Journal of Philosophy*, Vol. XXXIII (1955), p. 124.

it is *isolated,* and this assumption of improbability refers to the frequency of such transitions within a space-ensemble of *branch* systems, each of which is considered at two different times; this improbability does *not* refer to the time-sequence of entropy states of a single, *permanently* closed system.

As applied to the case of the stroller on the beach, these considerations take the following form. We assume that the beach itself was a quasi-closed system not unduly far from equilibrium (smoothness of its sandy surface) for a time in the recent past prior to our encountering it in a foot-shape-bearing state. And we are informed by the discovery of the latter state that the degree of order possessed by the grains of sand is higher and hence the entropy is lower than it should be, in all probability, if the beach had actually remained a quasi-closed system until our present encounter with it. For it is highly improbable that the beach, which is not a permanently closed system, evolved isolatedly from an earlier state of randomness (smoothness) to its present state of greater organization, although the statistical entropy principle for a single permanently closed system does call for precisely such behavior. Hence we conclude that, in all probability, after its initial state of relative smoothness, the beach was an open, interacting system whose increase in order was acquired at the expense of an at least equivalent decrease of organization in the external system with which it interacted (the stroller, who is metabolically depleted). It is clear, therefore, that our retrodictive inference of the stroller's incursion is not made to rest on the premise that in a *permanently* closed system, the entropy never decreases with time, a premise which turns out to be untenable in statistical mechanics, as we shall see presently.

II. *The Statistical Analogue of the Entropy Law.*

Our analysis of entropy so far has been in the macroscopic context of thermodynamics and has taken no adequate account of the important questions which arise concerning the serviceability of the entropy criterion as a basis for the anisotropy of time, when the entropy law is seen in the statistical light of both classical and quantum mechanics. These questions, which we must now face, derive from the attempt to uphold the phenomenological irreversibility of classical thermodynamics in the face of principles of statistical mechanics asserting that the motions of the microscopic constituents of thermodynamic systems are completely reversible.

If we compare a gas in a highly unequalized state of temperature with a near-equilibrium state from the standpoint of the kinetic theory of gases, we note that the molecular speeds will be much more equalized in the near-equilibrium state of high entropy than in the disequilibrium state of relatively low entropy.

Hence, a *high* entropy corresponds to

(1) a high degree of molecular equalization
(2) great homogeneity
(3) a well-shuffled state
(4) *low macro*-separation
(5) low order, where "order" means *not* smoothness or homogeneity, but rather inhomogeneity.

The application of the principles of Newtonian particle mechanics to the constituent molecules of idealized gases takes the following form: each of the n molecules of the gas in the closed system has a position and a velocity, or more accurately, three position coordinates x, y, and z, and three components of velocity. Hence the *micro-state* of the gas can be characterized at any given time by specifying the six position and velocity attributes corresponding to each of the n molecules, each value being given to within a certain small range. The micro-state of the gas at any given time may then be thought of as represented by points in the cells of a six-dimensional position-velocity space or "phase space." And each of the n molecules will then be in some one of the finite number m of cells compatible with the given volume and total energy of the gas.

A particular arrangement of the n individual molecules among the m cells constitutes a micro-state of the gas. Thus, if two individual gas molecules A and B were to exchange positions and velocities, a different arrangement would result. However, the *macroscopic* state of the gas, i.e., its being in a state of nearly uniform temperature or very uneven temperature, does not depend on whether it is molecule A or B that occupies a particular point in the container and has a given velocity. What matters *macroscopically* is whether more fast molecules are at one end of the container than at the other end or not, thereby making the one end hotter, or as hot as the other. In other words, the macro-state depends on *how many* molecules are at certain places in the container, as compared to the number in other places, and also depends on their respective velocities. Thus, the macro-state depends on the numerical spatial and velocity distribution of the molecules, not on the particular *identity* of the molecules having certain positional or velocity attributes. It follows that the *same* macro-state can be constituted by a *number* of *different* micro-states, as in the case of the mere interchange of the microscopic roles played by our two molecules A and B.

It is a basic postulate of statistical mechanics that each one of the m^n possible arrangements or micro-states occur with the same frequency in time or have *the same probability* $1/m^n$. This equi-probability postulate is called the quasi-ergodic hypothesis and gives the so-called probability metric of the Maxwell-Boltzmann statistics, since it asserts what occurrences are equally probable or frequent in time.[3]

It is of basic importance to see now that the number of micro-states W corresponding to a macro-state of near-equilibrium (uniform temperature) or high entropy is overwhelmingly *greater* than the number corresponding to a disequilibrium state of *non*-uniform temperature or quite low entropy. A drastically oversimplified example will make this fact evident.

Consider a position-velocity or phase-space of only four cells, and let there be just two different distributions (macro-states) of four particles among these cells as follows:

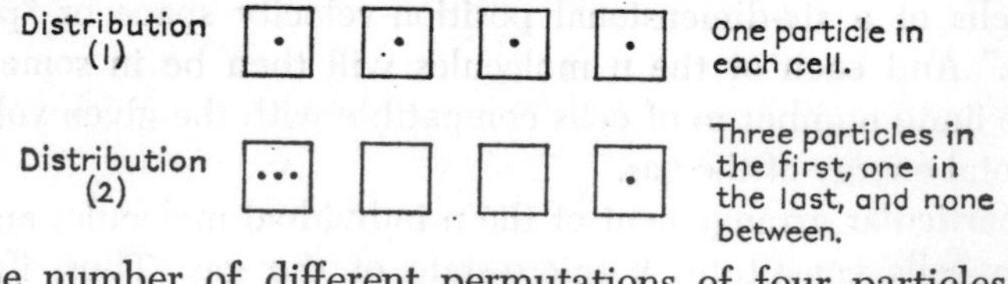

Distribution (1)	•	•	•	•	One particle in each cell.
Distribution (2)	•••			•	Three particles in the first, one in the last, and none between.

The number of different permutations of four particles in a row is $4! = 4{\cdot}3{\cdot}2{\cdot}1 = 24$. And thus the number W of different

[3] The so-called quasi-ergodic hypothesis is not an assertion based on our lack of knowledge as to the actual relative frequency of the different micro-states: instead it has the logical status of a theoretical claim concerning a presumed fact. What is a matter of our lack of knowledge in this context, however, is which one of the many micro-states that can underlie any given macro-state does, in fact, obtain when the system exhibits the specified macro-state at any one time.

arrangements or micro-states corresponding to the *homogeneous, equalized* macro-state given by distribution (1) is 24. But for the case of the second *inhomogeneous, unequalized* distribution, W is not 24, since the permutations of the three particles *within* the first cell do *not* issue in different arrangements. Hence, for the second case, W has the much smaller numerical value $\frac{4!}{3!} = 4$.[4] And since the entropy S is given by $S = k \log W$, where k is a constant, the entropy will be lower in the second case than in the first.

It now becomes evident that in the course of time, high entropy states of the gas are enormously more probable or frequent than low ones. For (1) all arrangements are assumed to be equiprobable, i.e., to occur with the same frequency, and (2) many more arrangements correspond to macro-states of high entropy than to states of low entropy. Saying that high entropy states are very probable means that the gas spends the overwhelming portion of its indefinitely long career in the closed system in states of high entropy or equilibrium. This then is the statistical analogue of the law of entropy increase, and it affirms that if a closed system is in a non-equilibrium state of relatively low entropy, then the increase of entropy with time is overwhelmingly probable by virtue of the approach of the particles to their equilibrium distribution. This statistical entropy law is also known as Boltzmann's "H-theorem," the quantity H being related to the entropy S by the relation $S = -kH$, so that an entropy increase corresponds to a decrease in H.

To appreciate the import of this statistical entropy law for the anisotropy of time, we recall that according to Newton's laws, the motions of particles are completely reversible. And all other known laws governing the behavior of the elementary constituents of physical processes likewise affirm the reversibility of

---

[4] More generally, Bernoulli's formula for W is $W = \frac{n!}{n_1!\, n_2!\, n_3! \ldots n_m!}$, where $\sum_i^m n_i = n$. If we wished to *normalize* the thermodynamic probability (which is a large number) so as to be less than 1, then we would have to divide it by the total number of arrangements for all distributions. Hence the (normalized) probability W_p of a *particular* distribution is given by $W_p = \frac{W}{m^n}$.

that behavior. Thus, Maxwell's equations for electromagnetic phenomena, and the fundamental probabilities of state transitions of quantum mechanical systems are time-symmetrical.

Accordingly, we can discuss the case of a gas constituted by Newtonian particles behaving reversibly as the paradigm case for answering the following question: can the statistical form of the entropy law for a permanently closed system form a basis for the anisotropy of time?

Our answer will now turn out to be in the negative. For soon after Boltzmann's enunciation of his theorem, it was felt that there is a logical hiatus in a deduction which derives the overwhelming probability of macroscopic irreversibility from premises attributing complete reversibility to micro-processes. For according to the principle of dynamical reversibility, which is integral to these premises, there is, corresponding to any possible motion of a system, an equally possible reverse motion in which the same values of the coordinates would be reached in the reverse order with reversed values for the velocities.[5] Thus, since the probability that a molecule has a given velocity is independent of the sign of that velocity, a molecule will have the velocity +v as frequently as the velocity −v in the course of time. And separation processes will occur just as frequently in the course of time as mixing processes, because micro-states issuing in the *unmixing* of hot and cold gases will occur as often as micro-states resulting in their mixing and achieving temperature equalization. J. Loschmidt therefore raised the *reversibility objection* to the effect that for any behavior of a system issuing in an increase of the entropy S with time, it would be equally possible to have an entropy decrease.[6] Accordingly, the fact that the gas spends most of its career in a state of high entropy does not at all preclude that, in the course of increasing time, the entropy will decrease as often as it increases. A criticism similar to that of Loschmidt was presented in the *periodicity objection*,

[5] Cf. R. C. Tolman: *The Principles of Statistical Mechanics, op. cit.*, pp. 102–04.

[6] J. Loschmidt: "Über das Wärmegleichgewicht eines Systems von Körpern mit Rücksicht auf die Schwere," *Sitzungsberichte der Akademie der Wissenschaften*, Vienna, Vol. LXXIII (1876), p. 139, and Vol. LXXV (1877), p. 67.

based on a theorem by Poincaré[7] and formulated by Zermelo.[8] Poincaré's theorem had led to the conclusion that the long-range behavior of an isolated system consists of a succession of fluctuations in which the value of S will decrease as often as it increases. And Zermelo asked how this result is to be reconciled with Boltzmann's contention that if an isolated system is in a state of low entropy, there is an overwhelming probability that the system is actually in a microscopic state from which changes in the direction of higher values of S will ensue.[9]

These logical difficulties were resolved by the Ehrenfests.[1] They explained that there is no incompatibility between (i) the assertion that *if* the system is in a low entropy state, then, *relative* to that state, it is highly probable that the system will soon be in a higher entropy state, and (ii) the contention that the system plunges down from a state of high entropy to one of lower entropy as frequently as it ascends entropically in the opposite direction, thereby making the *absolute* probability for these two opposite kinds of transition equal. The compatibility of the equality of these two absolute probabilities with a *high* relative probability for a future transition to a higher entropy

[7] H. Poincaré: "Sur le problème des trois corps et les equations de la dynamique," *Acta Mathematica*, Vol. XIII (1890), p. 67.

[8] E. Zermelo: "Über einen Satz der Dynamik und der mechanischen Wärmetheorie," *Wiedmannsche Annalen*, [*Annalen der Physik und Chemie*], Vol. LVII (1896), p. 485.

[9] For additional details on these objections and references to Boltzmann's replies, see P. Epstein: "Critical Appreciation of Gibbs's Statistical Mechanics," in: A. Haas (ed.) *A Commentary on the Scientific Writings of J. Willard Gibbs* (New Haven: Yale University Press; 1936), Vol. II, pp. 515–19. Cf. also C. Truesdell: "Ergodic Theory in Classical Statistical Mechanics," in: P. Caldivole (ed.) *Ergodic Theories* (New York: Academic Press; 1961), pp. 21–56.

[1] P. and T. Ehrenfest: "Begriffliche Grundlagen der statistischen Auffassung in der Mechanik," *Encyklopädie der mathematischen Wissenschaften*, IV, 2, II, pp. 41–51. See also R. C. Tolman: *The Principles of Statistical Mechanics, op. cit.*, pp. 152–58, esp. p. 156; R. Fürth: "Prinzipien der Statistik," in: H. Geiger and K. Scheel (eds.) *Handbuch der Physik* (Berlin: J. Springer; 1929), Vol. IV, pp. 270–72, and H. Reichenbach: "Ziele und Wege der physikalischen Erkenntnis," *op. cit.*, pp. 62–63.

The classical investigations of the Ehrenfests have recently been refined and extended to include quantum theory in D. Ter Haar's important paper "Foundations of Statistical Mechanics," *Reviews of Modern Physics*, Vol. XXVII (1955), pp. 289–338.

becomes quite plausible, when it is remembered that (i) the low entropy states to which the high relative probabilities of subsequent increase are referred are usually at the low point of a trajectory at which changes back to higher values are initiated, and (ii) the Boltzmann H-theorem therefore does not preclude such a system's exhibiting decreases and increases of S with equal frequency. Thus, if we consider a large number of low entropy states of the gas, then we will find that the vast majority of these will soon be *followed* by high entropy states. And in that sense, we can say that it is highly probable that a *low* entropy state will soon be followed by a high one. But it is no less true that a low entropy state was *preceded* by a state of high entropy with equally great probability! The time variation of the entropy, embodying these two claims compatibly, can be visualized as an entropy staircase curve.[2]

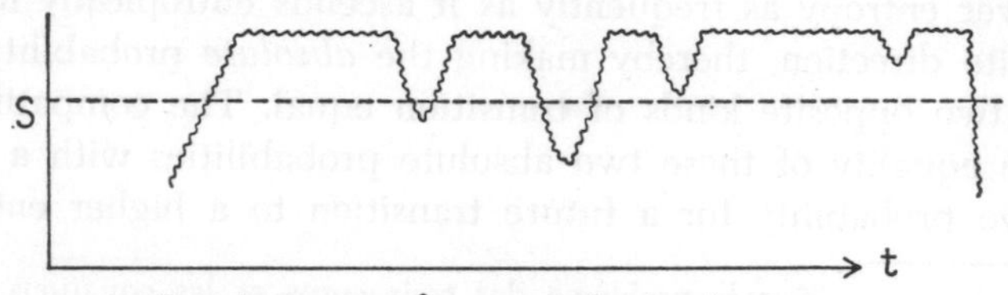

The entropy curve of a permanently closed system.

Boltzmann's H-theorem can thus be upheld in the face of the reversibility and periodicity objections, but only if coupled with a very important proviso: the affirmation of a high probability of a future entropy increase must *not* be construed to assert a *high* probability that low entropy values were *preceded by still lower* entropies *in the past*. For, as we saw, the relative probability that a *low* entropy state was *preceded* by a state of *higher* entropy is just as great as the relative probability that a low state will be *followed* by a higher state. The fulfillment of the proviso demanded by these results has *two* consequences of fundamental importance whose deduction depends on the statistical entropy law for a *permanently* closed system and which we shall now consider in turn:

1. It destroys the thermodynamic basis for supposing that

[2] Cf. R. Fürth: "Prinzipien der Statistik," in: H. Geiger and K. Scheel (eds.) *Handbuch der Physik, op. cit.*, p. 272.

merely because a present state in a system is one of *low* entropy, this fact is itself sufficient ground for believing that ordered state to be a *veridical trace* of earlier states of still lower entropy which were initiated by a specifiable kind of *interaction* of the system with an outside agency. Von Weizsäcker has rightly suggested[3] that, in the absence of other grounds to the contrary, the statistical entropy law itself provides every reason for regarding a present ordered state in a system as a *randomly achieved* low entropy state rather than as a veridical trace of an actual past interaction: it is statistically far more probable in the time sequence of states of a permanently closed system that present low entropy states are mere chance fluctuations rather than the *continuous* successors of actual earlier states of still lower entropy. But present low entropy states in a system cannot serve as articulate documents of the contiguous past of that system, unless we may assume that they directly evolved from specifiable antecedents and hence render veridical testimony of the occurrence of these antecedents in the contiguous past. Hence von Weizsäcker's considerations would suggest that our belief in the *thermodynamic* inferrability of the past is rendered untenable by the verdict of the H-theorem that the entropic behavior of a single, permanently closed system is time-symmetric! Yet it would be a serious error to abandon our ordinary practice of inferring the contiguous past from present low entropy states and to no longer interpret these as traces of interactions with outside agencies. For our supposition that a low entropy state in which we encounter a system is due to the system's past openness or interaction with the outside is based on grounds *other than* the statistically *untenable* assumptions that (i) a system which has been closed for a *very long* time could not now be in a low entropy state or that (ii) the entropy of a permanently closed system never decreases with time. In fact, we shall see that the statistics of the space ensembles of *branch* systems, which we mentioned earlier merely in passing, provide a sound empirical basis for our thermodynamic inferences concerning the past. The existence of this empirical basis, as well

[3] C. F. von Weizsäcker: "Der zweite Hauptsatz und der Unterschied von Vergangenheit und Zukunft," *Annalen der Physik*, Vol. XXXVI (1939), p. 281.

as difficulties of its own, undercut the subjectivistic a priori justification of our inferences concerning the past offered by von Weizsäcker on the basis of the alleged transcendental conditions of all possible experience, disclosed in this context by the application of Kant's presuppositional method.

2. Of crucial importance to whether there is an *entropic* basis for the anisotropy of time is the fact that the time-symmetric behavior of the entropy of a permanently closed system would seem to suggest a negative answer to this question. For if we confine ourselves to a *permanently* closed system, statistical mechanics yields the following fundamental results: first, it is impossible to say with Eddington that the *lower* of two given entropy states is the *earlier* of the two, a low entropy state being *preceded* by a high state no less frequently than it is *followed* by a high state,[4] and second, there is no contradiction between the high relative probabilities of Boltzmann's H-theorem and the equality among the absolute probabilities of the reversibility and periodicity objections, but the aforementioned time-symmetry of the statistical results on which these objections are based shows decisively that *no pervasive anisotropy could possibly be conferred on time by the entropic evolution of a single, permanently closed system.* Hence even if the entire universe qualified as a system whose entropy is defined—which a spatially infinite universe does not, as we shall see later in this chapter—its entropic behavior could not confer any pervasive anisotropy on time. An attempt to find a thermodynamic basis for the anisotropy of time has been made by Max Born by his refusal to affirm the reversibility of elementary processes.[5] Noting that Boltzmann's averag-

---

[4] Cf. H. Reichenbach: *DT, op. cit.,* pp. 116–17. It will be recalled that a system can be held to be in the *same macro*-state at different times t and t′ and hence to have the same entropy at t and t′ while the underlying *micro*-states at these times are *different.*

[5] M. Born: *Natural Philosophy of Cause and Chance* (Oxford: Oxford University Press; 1949), pp. 59, 71–73, and 109–14. In G. Bergmann's review of this work (*Philosophy of Science,* Vol. XVII [1950]), the latter remarks (p. 198) that Born's point can be rendered more clearly by the statement that "in a very relevant sense of the terms statistical mechanics is not mechanics. If, in applying it to, say, the gas, one predicts from a distribution the probability of other distributions, one abandons the idea of orbits and, therefore, deals with 'particles' only in the attenuated sense of using a theory whose fundamental entities have the formal properties of position-momentum coordinates."

In a different vein, L. L. Whyte has since suggested that the reversibility

ing is the expression of our ignorance of the actual microscopic situation, he maintains that the reversibility of mechanics is supplanted by the irreversibility of thermodynamics as a result of "a deliberate renunciation of the demand that in principle the fate of every single particle be determined. You must violate mechanics in order to obtain a result in obvious contradiction to it." He therefore finds that "the statistical foundation of thermodynamics is quite satisfactory even on the basis of classical mechanics."[6] But it is precisely in the domain of elementary processes that classical mechanics must be superseded by quantum theory. Born therefore attempts to solve the problem by asserting that the new theory "has accepted partial ignorance already on a lower level and need not doctor the final laws" and then offers a derivation of Boltzmann's H-theorem from quantum mechanical principles.[7]

Schrödinger has commented on Born's attempt and has offered an alternative to it. I shall therefore defer considering the status of irreversibility in quantum mechanics until after giving attention to Schrödinger's views.[8] The assessment of the bearing of quantum mechanics on irreversibility will then be followed by an account of what I consider to be a viable thermodynamic basis for a statistical anisotropy of time.

Referring to Born's account of irreversibility, Schrödinger says:

---

of elementary processes may have to be abandoned in future physical theory. Says he: "We should give up the long struggle with the question: 'How does irreversibility arise if the basic laws are reversible?' and ask instead: 'If the laws are of a one-way character, under what . . . conditions can reversible expressions provide a useful approximation?' " ("One-way Processes in Physics and Biophysics," *British Journal for the Philosophy of Science,* Vol. VI [1955], p. 110.) The successful implementation of Whyte's proposal would readily provide a previously unsuspected physical basis for the anisotropy of time. But it must not be overlooked that there is such confirmation for fundamental reversibility as the fact that the experimentally substantiated reciprocity law is deducible, as Onsager showed, from the reversibility of elementary collisions. Cf. J. M. Blatt: "Time Reversal," *Scientific American,* August, 1956, pp. 107–14 and R. G. Sachs: "Can the Direction of Flow of Time Be Determined?" *Science,* Vol. CXL (1963), pp. 1284–90.

[6] M. Born: *Natural Philosophy of Cause and Chance, op. cit.,* pp. 72–73.

[7] *Ibid.,* pp. 110, 113–14.

[8] E. Schrödinger: "Irreversibility," *Proceedings of the Royal Irish Academy,* Vol. LIII (1950), Sec. A, p. 189, and "The Spirit of Science," in: J. Campbell (ed.) *Spirit and Nature* (New York: Pantheon Books; 1954), pp. 337–41.

"to my mind, in this case, as in a few others, the 'new doctrine' which sprang up in 1925/26 has obscured minds more than it has enlightened them."[9] His proposal to deal with the issue without a "philosophical loan from quantum mechanics" does not take the form of deriving the increase of entropy with time from some kind of general reversible model. He rejects that approach on the grounds that he is unable to devise a model sufficiently general to cover all physical situations and also suitable for incorporation in all future theories. Neither does he wish to confine himself to a refutation of the arguments directed against Boltzmann's particular reversible model of a gas whose macro-behavior is irreversible.[1] Instead of *deriving* irreversibility, Schrödinger offers to "reformulate the laws of phenomenological irreversibility, thus certain statements of thermodynamics, in such a way, that the logical contradiction *any* derivation of these laws from reversible models seems to involve is removed once and for ever."[2]

To implement this program, he makes use of the fact that if, during a period of over-all entropy increase or decrease, a system has separated into two subsystems that are isolated from one another, then the respective entropies of the latter will either increase monotonically in both of them (apart from small fluctuations) or will so decrease in both of them. And, instead of considering merely a single isolated system, he envisages at least two systems, called "1" and "2," *temporarily isolated from the remaining universe for a period not greatly exceeding the age of our present galactic system.* Specifically, using a time variable t whose relation to phenomenological time will become clear below, he assumes that systems 1 and 2 are isolated from one another between the moments t_A and t_B, where $t_B > t_A$, but in contact for $t < t_A$ and $t > t_B$. Denoting the entropy of system 1 at time t_A by "S_{1A}" and similarly for the other entropy states, Schrödinger then formulates the entropy law as

$$(S_{1B} - S_{1A})(S_{2B} - S_{2A}) \geqslant 0.$$

[9] E. Schrödinger: "Irreversibility," *op. cit.*, p. 189.

[1] Born (*Natural Philosophy of Cause and Chance, op. cit.*, p. 59) points out that it was not until recently that the H-theorem was proven for cases other than Boltzmann's model of a gas.

[2] E. Schrödinger: "Irreversibility," *op. cit.*, p. 191.

Since the law is always applied to those *pairs* of systems having a common branching origin, the product of the entropy differences in it will yield the arithmetical "arrow" of the inequality even in the case of negative entropy differences.

Can Born's quantum mechanical approach or Schrödinger's alternative to it furnish a basis for the anisotropy of time? We saw that Born is guided by the view that since probability enters in quantum mechanics in a fundamental way *ab initio*, the derivation of the probabilistic macroscopic irreversibility affirmed by the H-theorem is feasible in that discipline and not liable to the charge, leveled against Boltzmann's classical derivation, that the deduction depended upon the addition of extraneous probability assumptions to the reversible dynamical equations. But Born's argument is open to important criticisms. To state these, we note first the requirements constituting the quantum mechanical analogue of the classical conditions for the *reversal* of the motion of a closed system: a system N can be said to behave in a manner reverse to that of a system M, if at time t it exhibits the same probability for specified values of the coordinates, the same probability for specified values of the momenta taken with reversed sign, and the same expectation value for any function of the coordinates and reversed momenta as would be exhibited by system M at time $-t$. Now, it has been shown by reference to the Schrödinger equation governing the change of isolated (conservative) quantum mechanical systems with time that all three of these conditions are satisfied by such systems.[3] The Schrödinger equation for a single free particle relevant here is of the form

$$\frac{\partial^2 \Psi}{\partial x^2} + \frac{\partial^2 \Psi}{\partial y^2} + \frac{\partial^2 \Psi}{\partial z^2} = \frac{4\pi m}{i\,h}\,\frac{\partial \Psi}{\partial t}, \text{ where } i \equiv \sqrt{-1}$$

and thus would *seem* to belong formally to the same class as the diffusion equation, which we considered earlier. Due to the presence, however, of an imaginary constant in the Schrödinger equation in place of the real constant in the diffusion equation, it turns out[4] that the Schrödinger equation describes a *reversible*

[3] Cf. R. C. Tolman: *The Principles of Statistical Mechanics, op. cit.*, pp. 396–99, and H. Reichenbach: *DT, op. cit.*, pp. 207–11.

[4] See A. Sommerfeld: *Partial Differential Equations in Physics,* trans. E. G. Straus (New York: Academic Press, Inc.; 1949), pp. 34–35.

oscillation while the diffusion equation describes an *irreversible* equalization.

What is the physical meaning of the purely formal reversibility of the Schrödinger equation? Instead of being confronted with the classical reversibility of the elementary processes themselves, we now have a *two-wayness* of the transitions between two sets of *probability* distributions of measurable quantities as follows: if nature permits a system which is characterized by the state function Ψ' and the associated set s' of probability distributions at time t_1 to evolve so as to acquire the state function Ψ'' and the associated set s'' of probability distributions at time t_2, then it also permits the inverse transition from s'' at time t_1 to s' at time t_2.[5] S. Watanabe was therefore able to demonstrate that Born's deduction of a monotonic entropy increase with time from the basic principles of quantum mechanics is just as vulnerable to Loschmidt's reversibility objection as the corresponding classical derivation.[6] And the resulting irrelevance of Born's invocation of the non-deterministic character of the fundamental principles of quantum mechanics is now apparent from the following lucid statement by L. Rosenfeld, who writes:

> The introduction of the quantal description of the elementary constituents as a basic assumption instead of the classical picture does not make the least difference to the fundamental structure of statistical thermodynamics; for the quantal laws, just as the classical ones, are reversible with respect to time, and the problem of establishing the macroscopic irreversibility by taking account of the statistical element involved in the concept of macroscopic observation remains unchanged and is again solved by ergodic theorems. The issue has been obscured by the fact that quantum theory itself, in contrast to classical theory, introduces a statistical element at the microscopic level; and it has

---

[5] Cf. O. Costa de Beauregard: "Complémentarité et Relativité," *Revue Philosophique*, Vol. CXLV (1955), pp. 397–400.

[6] S. Watanabe: "Réversibilité contre Irréversibilité en Physique Quantique," *Louis de Broglie, Physicien et Penseur* (Paris: Albin Michel; 1953), p. 393. Cf. also that author's earlier *Le Deuxième Théorème de la Thermodynamique et la Mécanique Ondulatoire* (Paris: Hermann & Cie.; 1935), esp. Chapter iv, Sec. 3, where he shows that, like Newtonian mechanics, quantum mechanics can furnish an irreversible thermodynamics only by adding a distinctly statistical supplementary postulate to its fundamental dynamical principles.

> sometimes been confusedly argued that it is this elementary quantal statistics which provides the basis of macroscopic irreversibility. In reality, we have here two completely distinct statistical features, which are not only logically independent of each other, but also without physical influence upon each other. The question whether the elementary law of change is deterministic (as in classical physics) or statistical (as in quantum theory) is entirely irrelevant for the validity of the ergodic theorems.[7]

It will be noted that in articulating the physical meaning of the formal reversibility of the Schrödinger time-equation, we spoke only of two-way transitions *from present to future* states and made no statement concerning inferences from a present state regarding the values we would have obtained in hypothetical *past* measurements, if we had carried them out earlier. There is a very important reason for this deliberate omission, and this reason is the source of a lack of isomorphism between classical reversibility and its quantum mechanical analogue: according to the orthodox version of quantum mechanics, the interaction between the system under observation and the measuring device discontinuously and irreversibly *changes* the Ψ-function characterizing the system before the measurement by imposing a random phase factor on that earlier Ψ-function, and this discontinuous change in Ψ is not governed by Schrödinger's equation. Thus, when the quantum mechanical system is subjected to observation by being coupled indivisibly to a classically describable macroscopic system, a present state function obtained by a measurement of one of the eigenvalues of an observable may be utilized in Schrödinger's equation to determine future but not past values of Ψ. Hence the irreversible alteration of the Ψ-function prevailing before the measurement by the act of measurement, i.e., the *irreversible* changes which take place both in the observed physical system and in the macroscopic measuring apparatus while the latter secures observational information, enter integrally into the (orthodox version of) quantum theory

[7] L. Rosenfeld: "On the Foundations of Statistical Thermodynamics," *Acta Physica Polonica,* Vol. XIV (1955), p. 9, and "Questions of Irreversibility and Ergodicity," in: P. Caldivola (ed.) *Ergodic Theories* (New York: Academic Press; 1961), pp. 1–20. Cf. also G. Ludwig: "Zum Ergodensatz und zum Begriff der makroskopischen Observablen. I," *Zeitschrift für Physik,* Vol. CL (1958), p. 346.

in marked contrast to classical mechanics and electrodynamics.[8]

We can now understand A. Landé's argument that if we construe reversibility to mean that there are temporal "mirror-images" of physical processes such that the original process and its inverse each comprise an initial state, intermediate states, and a final state, then it is *incorrect* to suppose that the Schrödinger time-equation warrants the ascription of reversibility to elementary quantum mechanical processes for the following reasons: First, actual states are ascertained by particular tests (e.g., states of energy or position), whereas Ψ is not a state but a *statistical link* between two states, and second, the Schrödinger time equation "does not describe processes from an initial to a final state via intermediate [measured] states actually passed through."[9] But Landé notes elsewhere[1] that the *metrogenic* entropy increase of quantum mechanics is only statistical in the sense that "In reality, the entropy values yielded by successive tests will oscil-

---

[8] Details on metrogenic irreversibility in quantum mechanics are given in J. von Neumann: *Mathematische Grundlagen der Quantenmechanik* (Berlin: J. Springer; 1932), pp. 191, 202–12; English trans. R. T. Beyer (Princeton: Princeton University Press; 1955). Cf. also S. Watanabe: "Prediction and Retrodiction," *op. cit.*, p. 179; Watanabe's essay for the de Broglie *Festschrift*: "Réversibilité contre Irréversibilité en Physique Quantique," *op cit.*, p. 389; D. Bohm: *Quantum Theory* (New York: Prentice-Hall; 1951), Chapter xxii, and S. Watanabe: "Le Concept de Temps en Physique Moderne et la Durée Pure de Bergson," *Revue de Métaphysique et de Morale*, Vol. LVI (1951), pp. 134–35. Reichenbach does not take cognizance of quantum mechanical metrogenic irreversibility in his theory of the direction of time (cf. *DT*, *op. cit.*, Chapter xxiv, and "Les Fondements Logiques de la Mécanique des Quanta," *Annales de l'Institut Henri Poincaré*, Vol. XIII [1953], pp. 148–54).

In an article "Philosophical Problems Concerning the Meaning of Measurement in Physics" (*Philosophy of Science*, Vol. XXV [1958]), H. Margenau has contested the "orthodox" conception of the process of measurement as the reduction of a wave packet. And thus he rejects the necessity of associating with the process of measurement a discontinuous change in the ψ-function that is not governed by the Schrödinger equation. On this "unorthodox" view, the claims made above on the basis of the "orthodox" version would have to be revised accordingly. Cf. also H. Margenau: "Measurements and Quantum States," *Philosophy of Science*, Vol. XXX (1963), pp. 1–16.

[9] A. Landé: "The Logic of Quanta," *British Journal for the Philosophy of Science*, Vol. VI (1956), p. 300, esp. pp. 305–07 and 311.

[1] A. Landé: "Wellenmechanik und Irreversibilität," *Physikalische Blätter*, Vol. XIII (1957), pp. 312–14. For a general discussion of these issues, see M. M. Yanase: "Reversibilität und Irreversibilität in der Physik," *Annals of the Japan Association for Philosophy of Science*, Vol. I (1957), pp. 131–49.

late up and down just as the classical entropy values of the Ehrenfest curve."

A brief comment concerning the epistemological status of metrogenic irreversibility in quantum mechanics must precede our assessment of its capabilities to account for the anisotropy of our macro-time.

Guided by the precepts of philosophical idealism, Watanabe erroneously equates the observer qua recorder of physically registered observational data with the observer qua *conscious* organism. He then infers that the metrogenic irreversibility of quantum mechanics shows "decisively" that "there is no privileged direction in the time of physics, and that, if one finds a unique direction in the evolution of physical phenomena, this is merely the projection of the flow of our psychic time . . . the increase in entropy is not a property of the external world left to itself, but is the result of the union of the subject and the object."[2] Treating the seriality inherent in psychological time as autonomous and *sui generis,* he nevertheless admits that the uniformity of psychological time-directions as between different living organisms requires explanation, being too remarkable to be contingent. But he seeks the explanation along the lines of Bergson's very questionable conception that living processes obey autonomous principles.[3] In his mentalistic interpretation of metrogenic irreversibility in quantum mechanics, Watanabe requires, as a crucial tacit premise in his argument, the traditional idealist characterization of the status of such material common sense objects as the classically describable pieces of apparatus used, in one way or another, in all quantum mechanical measure-

---

[2] S. Watanabe: "Le Concept de Temps en Physique Moderne et la Durée Pure de Bergson," *op. cit.,* pp. 134–36. Cf. also that author's contribution to the de Broglie *Festschrift*: "Réversibilité contre Irréversibilité en Physique Quantique," *op. cit.,* pp. 385, 392, 394.

[3] For a detailed discussion of the role of *physical* irreversibility in biological processes, see H. F. Blum: *Time's Arrow and Evolution* (2nd edition; Princeton: Princeton University Press; 1955); E. Schrödinger: *What is Life?* (New York: The Macmillan Company; 1945), Chapter vi; and R. O. Davies: "Irreversible Changes: New Thermodynamics from Old," *Science News* (May, 1953), No. 28. Attempts to prove the autonomy of living processes can draw no support from instances of entropy decrease in the human body. For, being an open system, that body's entropy can decrease or increase in complete conformity with even the *non*-statistical second law of thermodynamics.

ments. But this idealist premise is altogether unconvincing, and without it, there is every reason to regard the interaction between physical systems and the observational devices used in quantum mechanics as an entirely physical matter devoid of psychological ingredients of any kind. For, as has been explained by von Neumann[4] and more recently, by Ludwig,[5] the demand that cognizance be taken of the disturbances produced by measurements and observation can be adequately met in quantum mechanics without including the human observer's retina or body in the analysis, let alone his stream of consciousness. In regard to the macroscopic system which undergoes irreversible changes in the course of registering the results of microphysical measurements, Ludwig points out that, in principle, the perception of its readings by a conscious subject is irrelevant. Says he: "in principle, it is not necessary that it was a physicist [i.e., human observer] who built the apparatus for the purpose of measurement. It can also be a system on which the microscopic object impinges, entirely in the natural course of events."[6] Thus, as far as the role of the human observer *qua conscious organism* is concerned, there is no epistemological difference between quantum mechanics and classical physics.

Although quantum irreversibility is an entirely physical matter and a quantum world precludes our ascribing certain kinds of physical properties to a physical system apart from the interaction of that system with specifiable measuring devices, the irreversibility of our ordinary environment cannot be held to be attributable to metrogenic quantum irreversibility alone. Bohr's principle of complementarity must be taken in conjunction with his own emphasis in the correspondence principle that the meas-

---

[4] J. von Neumann: *Mathematische Grundlagen der Quantenmechanik, op. cit.*, pp. 187, 223–37, esp. pp. 223–24. Cf. also D. Bohm: *Quantum Theory, op. cit.*, pp. 584–85, 587–90, 600–09.

[5] G. Ludwig: "Der Messprozess," *Zeitschrift für Physik*, Vol. CXXXV (1953), p. 483, esp. p. 486; see also his *Die Grundlagen der Quantenmechanik* (Berlin: J. Springer; 1954), pp. 142–59, 178–82, and his "Die Stellung des Subjekts in der Quantentheorie," in *Veritas, Justitia, Libertas*, Festschrift zur 200–Jahr Feier der Columbia University (Berlin: Colloquium Verlag; 1954), pp. 261–71. See also H. Reichenbach: *DT, op. cit.*, pp. 223–24, and *Philosophic Foundations of Quantum Mechanics* (Berkeley: University of California Press; 1948), pp. 15ff.

[6] G. Ludwig, "Der Messprozess," *Zeitschrift für Physik, op. cit.*, p. 486.

uring devices which constitute the epistemological basis of quantum mechanics are themselves describable by the principles of classical physics. The actual irreversibility of our macro-environment is set in a context in which Planck's constant h may be considered negligibly small and in which the classical view that the physical system can be said to have definite physical properties independently of any coupling to a measuring instrument is legitimately applicable.[7]

We can therefore endorse Schrödinger's *rejection* of the use of quantum mechanical metrogenic irreversibility as a basis for explaining the phenomenological (macro-) irreversibility of our environment. Says he: "Surely the system continues to exist and to behave, to undergo irreversible changes and to increase its entropy in the interval between two observations. The observations we might have made in between cannot be essential in determining its course."[8]

Granted then that quantum mechanics does not furnish the required account of the temporal anisotropy of our macroscopic environment in its "current" non-equilibrium state, does Schrödinger's own non-quantal account succeed in doing so? He avowedly made no attempt to deduce irreversibility. But he does explain that if at least one of the entropy differences in his formulation of Clausius's principle is positive, then it is the parametric time t which corresponds to phenomenological time and that, alternatively, if at least one such difference is negative, it is −t that corresponds to phenomenological time. Schrödinger's perceptive guiding idea that the attempt to characterize the phenomenological anisotropy of time entropically without run-

---

[7] For an interesting discussion of the conditions governing such applicability, see L. Brillouin: *Science and Information Theory* (New York: Academic Press; 1956), pp. 229–32.

[8] E. Schrödinger: "Irreversibility," *op. cit.*, p. 190. Even less helpful than quantum irreversibility as a basis for the anisotropy of time is the Wheeler-Feynman-Stückelberg analysis of pair production in quantum electrodynamics (cf. H. Margenau: "Can Time Flow Backwards?" *Philosophy of Science*, Vol. XXI [1954], p. 79), since it involves indeterminacies even in regard to those order properties of time which are defined by reversible macro-processes (cf. H. Reichenbach: *DT*, *op. cit.*, pp. 262–69, and "Les Fondements Logiques de la Mécanique des Quanta," *op. cit.*, pp. 150–53). See also C. W. Berenda: "Determination of Past by Future Events," *Philosophy of Science*, Vol. XIV (1947), p. 13.

ning afoul of the reversibility and periodicity objections can succeed only if we regard the entropy law as an assertion about *at least two temporarily* closed systems was developed independently by Reichenbach. And the valid core—but only the valid core—of Reichenbach's version of this idea seems to me to provide a foundation for an entropic basis of a statistical anisotropy of physical time. Believing that Reichenbach's account requires significant modification in order to be satisfactory, I shall now set forth what I consider to be a corrected elaboration of his principal conception.

Actual physical experience presents us with entropy increases in quasi-isolated systems overwhelmingly more frequently than with corresponding decreases: if 10,000 people sat down together to dinner and each poured some cream into a cup of black coffee, it is an incontestably safe bet that the cream will mix with the coffee in all cases and that no one will report a subsequent *unmixing* of them *for ordinary intervals of time,* i.e., prior to the consumption of the creamed coffee. This kind of phenomenon of temporally asymmetric entropy increase is, of course, not incompatible with the statistical form of the entropy law for a permanently closed system, since we restricted ourselves to ordinary intervals of time. Let us inquire, therefore, whether this kind of phenomenon of entropy increase confers anisotropy at least on the time of our galactic system during the current epoch. We shall now see in detail that it is indeed correct to conclude that such phenomena do furnish a viable physical basis for a statistical anisotropy of time. To do so, we must first describe certain features of the physical world *having the character of initial or boundary conditions* within the framework of the theory of statistical mechanics. The sought-after entropic basis of a statistical anisotropy of time will then emerge from principles of statistical mechanics relevant to these de facto conditions.

The universe around us exhibits striking disequilibria of temperature and other inhomogeneities. In fact, we live in virtue of the nuclear conversion of the sun's reserves of hydrogen into helium, which issues in our reception of solar radiation. As the sun dissipates its reserves of hydrogen via the emission of solar radiation, it may heat a terrestrial rock embedded in snow during the daytime. At night, the rock is no longer exposed to the sun but is left with a considerably higher temperature than the snow surrounding it. Hence, at night, the warm rock and the cold snow form a quasi-isolated subsystem of either our galactic or solar system. And the relatively low entropy of that subsystem was purchased at the expense of the dissipation of the sun's reserves of hydrogen. Hence, *if* there is some quasi-closed system comprising the sun and the earth, the branching off of our subsystem from this wider system in a state of low entropy at sunset involved an entropy increase in the wider system. During the night, the heat of the rock melts the snow, and thus the entropy of the rock-snow system increases. The next morning at sunrise, the rock-snow subsystem merges again with the wider solar system. Thus, there are subsystems which branch off from the wider solar or galactic system in a state of relatively low entropy, remain quasi-closed for a *limited* period of time, and then merge again with the wider system from which they had been separated. Following Reichenbach,[9] we have been using the term "branch system" to designate this kind of subsystem.

Branch systems are formed not only in the natural course of things, but also through human intervention: when an ice cube is placed in a glass of warm gingerale by a waiter and then covered for hygienic purposes, a subsystem has been formed. The prior freezing of the ice cube had involved an entropy increase through the dissipation of electrical energy in some large quasi-closed system of which the electrically run refrigerator is a part. While the ice cube melts in the covered glass subsystem, that quasi-closed system increases its entropy. But it merges again with another system when the then chilled gingerale is consumed by a person. Similarly for a closed room that is closed off and then heated by burning logs.

Thus, our environment abounds in branch systems whose initial relatively low entropies are the products of their earlier coupling or interaction with outside agencies of one kind or another. This rather constant and ubiquitous formation of a branch system in a relatively low entropy state resulting from interaction often proceeds at the expense of an entropy increase in some wider quasi-closed system from which it originated. And the de facto, nomologically contingent occurrence of these branch systems has the following *fundamental consequence,* at

---

[9] Cf. H. Reichenbach: *DT, op. cit.,* pp. 118–43.

least for our region of the universe and during the current epoch: among the quasi-closed systems whose entropy is relatively low and which behave as if they might remain isolated, the vast majority have not been and will not remain permanently closed systems, being branch systems instead.

Hence, upon encountering a quasi-closed system in a state of fairly low entropy, we know the following to be overwhelmingly probable: the system has not been isolated for millions and millions of years and does not just *happen* to be in one of the infrequent but ever-recurring low entropy states exhibited by a permanently isolated system. Instead, our system was formed not too long ago by branching off after an interaction with an outside agency. For example, suppose that an American geologist is wandering in an isolated portion of the Sahara desert in search of an oasis and encounters a portion of the sand in the shape of "Coca Cola." He would then infer that, with overwhelming probability, a kindred person had interacted with the sand in the recent past by tracing "Coca Cola" in it. The geologist would not suppose that he was in the presence of one of those relatively low entropy configurations which are assumed by the sand particles spontaneously but very rarely, if beaten about by winds for millions upon millions of years in a state of effective isolation from the remainder of the world.

There is a further de facto property of branch systems that concerns us. For it will turn out to enter into the temporally asymmetrical statistical regularities which we shall find to be exhibited in the entropic behavior of these systems. This property consists in the following *randomness* obtaining *as a matter of nomologically contingent fact* in the distribution of the W_1 micro-states belonging to the initial macro-states of a space-ensemble of branch systems each of which has the same initial entropy $S_1 = k \log W_1$: For each class of *like* branch systems having the *same* initial entropy value S_1, the micro-states constituting the identical initial macro-states of entropy S_1 are *random samples* of the set of all W_1 micro-states yielding a macro-state of entropy S_1.[1] This attribute of randomness of micro-states on the part of the initial states of the members of the *space-*

[1] Cf. R. C. Tolman: *The Principles of Statistical Mechanics, op. cit.*, p. 149.

ensemble will be recognized as the counterpart of the following attribute of the micro-states of one single, permanently closed system: there is equi-probability of occurrence among the W_1 micro-states belonging to the *time*-ensemble of states of equal entropy $S_1 = k \log W_1$ exhibited by one single, permanently closed system.

We can now state the statistical regularities which obtain as a consequence of the de facto properties of branch systems just set forth, when coupled with the principles of statistical mechanics. These regularities, which will be seen to yield a *temporally asymmetric* behavior of the entropy of *branch* systems, fall into two main groups as follows.[2]

Group 1. In most space-ensembles of quasi-closed branch systems each of which is initially in a state of non-equilibrium or relatively *low* entropy, the majority of branch systems in the ensemble will have *higher* entropies *after* a given time t. But these branch systems simply did not exist as quasi-closed, distinct systems at a time t *prior to* the occurrence of their initial, branching off states. Hence, not existing then as such, the branch systems did in fact *not* also exhibit the same higher entropy states at the *earlier* times t, which they would indeed have done then had they existed as closed systems all along. In this way, the space-ensembles of branch systems do *not* reproduce the entropic time-symmetry of the single, permanently closed system. And whatever the behavior of the components of the branch systems prior to the latter's "birth," that behavior is irrelevant to the entropic properties of branch systems as such.

The increase after a time t in the entropy of the overwhelming majority of branch systems of initially low entropy—as confirmed abundantly by observation—can be made fully intelligible. To do so, we note the following property of the *time*-ensemble of entropy values belonging to a single, permanently closed system and then affirm that property of the space-ensembles of branch systems: since *large* entropic downgrades or decreases are *far less* probable (frequent) than moderate ones, the *vast majority* of *non*-equilibrium entropy states of a permanently closed system

[2] Cf. R. Fürth: "Prinzipien der Statistik," *op. cit.*, pp. 270 and 192–93. The next-to-the-last sentence on p. 270 is to be discounted, however, since it is self-contradictory as it stands and incompatible with the remainder of the page.

are located either at or in the immediate temporal vicinity of the *bottom* of a *dip* of the one-system entropy curve. In short, the vast majority of the *sub*-maximum entropy states are on or temporally very near the *upgrades* of the one-system curve. The application of this result to the space-ensemble of branch systems whose initial states exhibit the aforementioned de facto property of *randomness* then yields the following: among the initial low entropy states of these systems, the vast majority lie at or in the immediate temporal vicinity of the bottoms of the one-system entropy curve at which an upgrade begins.

Group 2. A decisive *temporal asymmetry* in the statistics of the temporal evolution of branch systems arises from the further result that in most space ensembles of branch systems each of whose members is initially in a state of *equilibrium* or very *high* entropy, the vast majority of these systems in the ensemble will *not* have *lower* entropies *after* a finite time t, but will still be in equilibrium. For the aforementioned randomness property assures that the vast majority of those branch systems whose initial states are equilibrium states have maximum entropy values lying somewhere *well within* the plateau of the one-system entropy curve, rather than at the extremity of the plateau at which an entropy *decrease* is initiated.

Although the decisive asymmetry just noted was admitted by H. Mehlberg,[3] he dismisses it as expressing "merely the factual difference between the two relevant values of probability." But an asymmetry is no less an asymmetry for depending on de facto, nomologically contingent boundary conditions rather than being assured by a *law* alone. Since our verification of laws generally has the same partial and indirect character as that of our confirmation of the existence of certain complicated de facto boundary conditions, the assertion of an asymmetry depending on de facto conditions is generally no less reliable than one wholly grounded on a law. Hence when Mehlberg[4] urges against Schrödinger's claim of asymmetry that for every pair of branch systems which change their entropy in one direction, "there is nothing to prevent" another pair of closed subsystems from changing their entropy in the opposite direction, the reply is:

[3] H. Mehlberg: "Physical Laws and Time's Arrow," *op. cit.*, p. 129.
[4] *Ibid.*, p. 117, n. 30.

Mehlberg's criticism can be upheld only by gratuitously neglecting the statistical asymmetry admitted but then dismissed by him as "merely" factual. For it is the existence of the specified boundary conditions which statistically *prevents* the existence of entropic time-symmetry in this context.

We see therefore that in the vast majority of branch systems, either one end of their finite entropy curves is a point of low entropy and the other a point of high entropy, or they are in equilibrium states at both ends as well as during the intervening interval. And it is likewise apparent that the statistical distribution of these entropy values on the time axis is such that the vast majority of branch systems have the *same direction of entropy increase* and hence also the same opposite direction of entropy decrease. Thus, the statistics of entropy increase among branch systems assure that *in most space ensembles the vast majority* of branch systems will increase their entropy in *one* of the two opposite time directions and decrease it in the other: in contradistinction to the entropic time-symmetry of a single, permanently closed system, the probability within the space-ensemble that a low entropy state s at some given instant be *followed* by a higher entropy state S at some given later instant is much *greater* than the probability that s be *preceded* by S. In this way *the entropic behavior of branch systems confers the same statistical anisotropy on the vast majority of all those cosmic epochs of time during which the universe exhibits the requisite disequilibrium and contains branch systems satisfying initial conditions of "randomness."*

Let us now call the direction of entropy increase of a *typical representative* of the aforementioned kind of cosmic epoch the direction of "later," as indeed we have done from the outset by the mere assignment of higher time numbers in that direction but *without prejudice* to our findings concerning the issue of the anisotropy of time. Then our results pertaining to the entropic behavior of branch systems show that the directions of "earlier than" and "later than" are not merely opposite directions bearing decreasing and increasing time coordinates respectively but are statistically *anisotropic* in an objective physical sense. For we saw earlier in this chapter that increasing real numbers can be assigned as time coordinates in a physically meaningful way *without* any commitment to the existence of (de facto or nomologically) irreversible kinds of processes. In fact, the use of the real number continuum as a basis for coordinatizing time no more entails the anisotropy of time than the corresponding coordinatization of one of the three dimensions of space commits us to the anisotropy of that spatial dimension.

It should be noted that I have characterized the positive direction of time as the direction of entropy increase in branch systems for a typical representative of all those epochs of time during which the universe exhibits the requisite disequilibrium and contains branch systems satisfying initial conditions of "randomness." Accordingly, it is entirely possible to base the customary temporal description of fluctuation phenomena, i.e., the assertion that the entropy *decreases* with *positive time* in some systems or other, on entropic counter-directedness of the latter systems with respect to the *majority* of branch systems, a description which is not liable to the *reductio ad absurdum* offered by Bridgman in his unsuccessful attempt to show that there can be no entropic basis for the anisotropy of time. The very statistics of branch systems which invalidate Bridgman's attempted *reductio* likewise serve to discredit the following denial by K. R. Popper of the relevance of entropy statistics to the anisotropy of time, despite the soundness of Popper's objection to Boltzmann's original formulation:

> The suggestion has been made (first by Boltzmann himself) that the arrow of time is, either by its very nature, or by definition, connected with the increase in entropy; so that entropy cannot decrease in time because a decrease would mean a reversal of its arrow, and therefore an increase relative to the reversed arrow. Much as I admire the boldness of this idea, I think that it is absurd, especially in view of the undeniable fact that thermodynamic fluctuations do exist. One would have to assert that, within the spatial region of the fluctuation, all clocks run backwards if seen from outside that region. But this assertion would destroy that very system of dynamics on which the statistical theory is founded. (Moreover, most clocks are non-entropic systems, in the sense that their heat production, far from being essential to their function, is inimical to it.)
>
> I do not believe that Boltzmann would have made his suggestion after 1905, when fluctuations, previously considered no

> more than mathematically calculable near-impossibilities, suddenly became the strongest evidence in favour of the physical reality of molecules. (I am alluding to Einstein's theory of Brownian motion.) As it is, a statistical theory of the arrow of time seems to me unacceptable.[5]

I have contended against Mehlberg and Popper that the entropic behavior of branch systems confers the *same* statistical anisotropy on the vast majority of all those cosmic epochs of time during which the universe exhibits the requisite disequilibrium and contains branch systems satisfying the specified initial conditions of "randomness." My claim of statistical anisotropy departs significantly from Reichenbach's "hypothesis of the branch structure"[6] in the following ways: (1) I do not assume that the entropy is defined for the entire universe such that the universe as a whole can be presumed to exhibit the entropic evolution of the statistical entropy curve for a permanently closed, *finite* system, an assumption which leads Reichenbach to affirm the parallelism of the direction of entropy increase of the universe and of the branch systems at any time, and therefore, (2) I do not conclude, as Reichenbach does, that cosmically the statistical anisotropy of time is only local by "fluctuating" in the following sense: the supposed alternations of epochs of entropy increase and decrease of the universe go hand-in-hand with the alternations of the direction of entropy increase in the ensembles of branch systems associated with these respective epochs, successive disequilibrium epochs allegedly being entropically *counter*-directed with respect to each other.

In view of the reservations which Reichenbach himself expressed[7] concerning the reliability of assumptions regarding the universe as a whole in the present state of cosmology, one wonders why he invoked the entropy of the universe at all instead of confining himself, as I have done, to the much weaker assumption of the existence of states of disequilibrium in the universe. More fundamentally, it is unclear how Reichenbach thought he could reconcile the assumption that the branch systems satisfy initial conditions of randomness during whatever cosmic epoch

---

[5] K. R. Popper: *Nature*, Vol. CLXXXI (1958), p. 402.
[6] H. Reichenbach: *DT*, *op. cit.*, p. 136.
[7] *Ibid.*, pp. 132–33.

they may form—an assumption which, as we saw, makes for the *same* statistical anisotropy on the part of *most* disequilibrium epochs of the universe—with the following claim of alternation made by him: "When we come to the downgrade [of the entropy curve of the entire universe], always proceeding in the same direction [along the time-axis], the branches begin at states of high entropy . . . and they end at points of low entropy."[8] Contrary to Reichenbach, we saw in our statement of the consequences of the postulate of randomness under *Group 2* above that in the vast majority of cases, branch systems beginning in a state of equilibrium (high entropy) will *remain* in equilibrium for the duration of their finite careers instead of decreasing their entropies!

An inherent limitation on the applicability of the Maxwell-Boltzmann entropy concept to the entire universe lies in the fact that it has no applicability at all to a *spatially infinite* universe for the following reasons. If the infinite universe contains a denumerable *infinity* of atoms, molecules or stars, the number of complexions W becomes infinite, so that the entropy is not defined, and a fortiori no increase or decrease thereof.[9] And if the number of particles in the infinite universe is only finite, then (a) the equilibrium state of maximum entropy cannot be realized by a *finite* number of particles in a phase-space of *infinitely* many cells, since these particles would have to be *uniformly* distributed among these cells, and (b) the quasi-ergodic hypothesis, which provides the essential basis for the probability metric ingredient in the Maxwell-Boltzmann entropy concept, is presumably false for an infinite phase space.[1] If the universe *were finite and* such that an entropy is defined for it as

[8] *Ibid.*, p. 126.

[9] Cf. K. P. Stanyukovic: "On the Increase of Entropy in an Infinite Universe," *Doklady, Akademiia Nauk* SSSR, N.S., Vol. LXIX (1949), p. 793, in Russian, as summarized by L. Tisza in *Mathematical Reviews,* Vol. XII (1951), p. 787.

[1] For additional doubts concerning the cosmological relevance of the entropy concept, cf. E. A. Milne: *Sir James Jeans* (Cambridge: Cambridge University Press; 1952), pp. 164–65, and *Modern Cosmology and the Christian Idea of God* (Oxford: Clarendon Press; 1952), pp. 146–50; also L. Landau and E. Lifschitz: *Statistical Physics* (2nd ed.; Reading: Addison-Wesley; 1958), pp. 22–27.

a whole which conforms to the one-system entropy curve of statistical mechanics, then my contention of a *cosmically pervasive statistical anisotropy* of time could no longer be upheld. For I am assuming that the vast majority of branch systems in most epochs increase their entropy in the same direction and that space ensembles of branch systems do form during most periods of disequilibrium. And if one may further assume that the entropy of a finite, spatially-closed universe depends *additively* on the entropies of its component subsystems, then the assumed temporal asymmetry of the entropy behavior of the branch systems would appear to contradict the complete *time-symmetry* of the one-system entropy behavior of the finite universe. This conclusion, if correct, therefore poses the question —which I merely wish to ask here—whether in a closed universe, the postulate of the randomness of the initial conditions would not hold. For in that case, the cosmically *pervasive* statistical anisotropy of time which is assured by the randomness postulate would not need to obtain; instead, one could then assume initial conditions in branch systems that issue in Reichenbach's cosmically local kind of anisotropy of time, successive overall disequilibrium epochs having *opposite* directions of entropy increase both in the universe and in the branch systems associated with these epochs.

We shall show in Chapter Nine that our account of the entropic basis for the anisotropy of time has the following further important ramifications: First, it provides an empirical justification for interpreting present ordered states as veridical traces of actual past interaction events, a justification which the entropic behavior of a single, permanently closed system was incompetent to furnish, as we saw, and, as will likewise be shown in Chapter Nine, second, it explains why the subjective (psychological) and objective (physical) directions of positive time are parallel to one another by noting that man's own body participates in the entropic lawfulness of space ensembles of physical branch systems in the following sense: man's *memory,* just as much as all purely physical recording devices, accumulates "traces," records or information in a direction dictated by the statistics of physical branch systems. Contrary to Watanabe's conception of man's psychological time sense as *sui generis,* it will turn out in Chapter Nine that the future direction of psychological time is parallel to that of the accumulation of traces (increasing information) in interacting systems, and hence parallel to the direction defined by the positive entropy increase in the branch systems. Thus the examination of the anisotropy of psychological time will also have shown that Spinoza was in error when he wrote Oldenburg that "*tempus non est affectio rerum sed merus modus cogitandi.*"

We have now completed our discussion of the anisotropy of time insofar as it depends on systems for which an entropy is defined and whose entropy changes in a temporally asymmetric way. It remains to consider whether there are not also *non*-entropic kinds of physical processes which contribute to the anisotropy of time. We shall find that the answer is indeed affirmative. Moreover, it will turn out that just as the *entropically* grounded statistical anisotropy of time was *not* assured by the laws alone but also depended on the role played by specified kinds of boundary conditions, so also the *non*-entropic species of irreversibility is only de facto rather than nomological.

(B) ARE THERE NON-THERMODYNAMIC FOUNDATIONS FOR THE ANISOTROPY OF TIME?

In a series of notes, published in *Nature* during the years 1956–1958, K. R. Popper[2] has expounded his thesis of the "untenability of the widespread, though surely not universal, belief that the 'arrow of time' is closely connected with, or dependent upon, the law that disorder (entropy) tends to increase" (*II*). Specifically, he argues in the first three of his four notes that there exist *irreversible* processes in nature whose irreversibility does not depend on their involvement of an entropy increase. Instead, their irreversibility is *nomologically contingent* in the following sense: the *laws* of nature governing elementary processes do indeed allow the temporal inverses of these irreversible

[2] K. R. Popper: *Nature,* Vol. CLXXVII (1956), p. 538; Vol. CLXXVIII (1956), p. 382; Vol. CLXXIX (1957), p. 1297; Vol. CLXXXI (1958), p. 402. These four publications will be cited hereafter as "*I,*" "*II,*" "*III,*" and "*IV*" respectively.

processes, but the latter processes are de facto irreversible, because the *spontaneous* concatenation of the initial conditions requisite for the occurrence of their temporal inverses is well-nigh physically impossible. Noting that "Although the arrow of time is not implied by the fundamental equations [laws governing elementary processes], it nevertheless characterizes most solutions" (*I*), Popper therefore rejects the claim that "every non-statistical or 'classical' mechanical process is reversible" (*IV*). In the fourth of his communications, he maintains that the statistical behavior of the entropy of physical systems not only fails to be the *sole* physical basis for the anisotropy of time, as Boltzmann had supposed, but does not qualify *at all* as such a basis.[3] For, as we saw earlier, Popper argues that, if it did, the temporal description of fluctuation phenomena would entail absurdities of several kinds.

In response to the first two of Popper's four notes, E. L. Hill and I published a communication[4] in which we endorsed Popper's contention of the existence of non-entropic, nomologically contingent irreversibility in the form of an existential claim constituting a generalization of Popper's contention.

In view of Popper's criticism (*III*) of the latter generalization, my aim is to deal with non-entropic irreversibility as follows:

(1) To appraise Popper's criticism.

(2) To show that the generalization put forward in the paper by Hill and myself has the important merit of dispensing with the *restriction* on which Popper predicates his affirmation of nomologically contingent irreversibility. This restriction is constituted by the requirement of the *spontaneity* of the concatenation of the initial conditions requisite to the occurrence of the temporal inverses of the thus *conditionally* irreversible processes.

(3) To assess the import of my appraisal of Popper's claims for Mehlberg's denial of the anisotropy of time.[5]

---

[3] In view of the misleading potentialities of Eddington's metaphor "the arrow of time," which is also employed by Popper, I prefer to substitute the non-metaphorical expression "the anisotropy of time" in my account of Popper's views.

[4] E. L. Hill and A. Grünbaum: "Irreversible Processes in Physical Theory," *Nature*, Vol. CLXXIX (1957), p. 1296.

[5] Cf. H. Mehlberg: "Physical Laws and Time's Arrow," *op. cit.*

Independently of O. Costa de Beauregard, who had used the same illustration before him,[6] Popper (*I*) considers a large surface of water initially at rest into which a stone is dropped, thereby producing an outgoing wave of decreasing amplitude spreading concentrically about the point of the stone's impact. And Popper argues that this process is irreversible in the sense that the "spontaneous" (*IV*) concatenation on all points of a circle of the initial conditions requisite to the occurrence of a corresponding *contracting* wave is physically impossible, a "spontaneous" concatenation being understood to be one which is not brought about by coordinated influences emanating from a common center. Being predicated on the latter spontaneity, this nomologically contingent irreversibility is of a *conditional* kind.

Now, one might object that the attribution of the irreversibility of the outgoing wave motion to non-thermodynamic causal factors is unsound. The grounds would be that the statistical entropy law is *not irrelevant* to this irreversibility, because the diminution in the amplitude of the outgoing wave is due to the superposition of *two* independent effects as follows: First, the requirements of the law of conservation of energy (*first* law of thermodynamics), and second, an entropy increase in an essentially closed system through dissipative viscosity. To be sure, the entropy increase through dissipative viscosity is a *sufficient* condition (in the *statistical* sense of part A. of this chapter!) for the irreversibility of the outgoing wave motion, i.e., for the absence of a corresponding (spontaneously initiated) contracting wave motion. But this fact cannot detract from the soundness of Popper's claim that another, independent *sufficient condition* for the conditional kind of de facto irreversibility affirmed by him is as follows: the nomologically contingent non-existence of the spontaneous occurrence of the coordinated initial conditions requisite for a contracting wave motion. We see that Popper rightly adduces the need for the *coherence* of these initial conditions as his basis for denying the possibility of their *spontaneous* concatenation, i.e., their concatenation without first having been coordinated by an influence emanating from a central

[6] O. C. de Beauregard: "L'Irréversibilité Quantique, Phénomène Macroscopique," *op. cit.*, p. 402.

source. Says he (*III*): "Only such conditions can be causally realized as can be organized from one centre . . . causes which are not centrally correlated are causally unrelated, and can co-operate [i.e., produce coherence in the form of *isotropic* contraction of waves to a precise point] only by accident. . . . The probability of such an accident will be zero."

In view of the aforementioned *conditional* character of Popper's nomologically contingent irreversibility, E. L. Hill and I deemed it useful to point out the following:[7] there does indeed exist an important class of processes in *infinite* space whose *irreversibility* is (1) non-entropic and nomologically contingent, hence being of the kind correctly envisioned by Popper, yet (2) *not conditional* by not being predicated on Popper's proviso of spontaneity. Without presuming to speak for Professor Hill, I can say, for my part, that in making that existential claim I was guided by the following considerations:

1. Popper (*II*) briefly remarks correctly that the eternal expansion of a very thin gas from a center into a spatially *infinite* universe does not involve an entropy increase, and the de facto irreversibility of this process is therefore *non*-entropic. For the statistical Maxwell-Boltzmann entropy is not even defined for a spatially infinite universe: the quasi-ergodic hypothesis, which provides the essential basis for the probability-metric ingredient in the Maxwell-Boltzmann entropy concept, is presumably *false* for an *infinite* phase-space, since *walls* are required to produce the collisions which are essential to its validity. In the absence of some kind of wall, whose very existence would assure the *finitude* of the system, the rapidly moving particles will soon overtake the slowly moving ones, leaving them ever further behind for all future eternity instead of *mixing* with them in a space-filling manner. Moreover, as we noted earlier, if the number of particles in the infinite universe is only finite, the equilibrium state of maximum entropy cannot be realized, since a finite number of particles cannot be *uniformly* distributed in a phase space of infinitely many cells. Furthermore, if the number of particles is denumerably infinite, the number W of micro-

[7] Cf. E. L. Hill and A. Grünbaum: "Irreversible Processes in Physical Theory," *op. cit.*

scopic complexions in S = k log W becomes infinite, and no entropy increase or decrease is defined. Corresponding remarks apply to the case of the Bose-Einstein entropy of a gas of photons (bosons) of various frequencies (energies) which is not confined by an enclosure but has free play in an infinite space.[8]

2. Though *allowed* by the *laws* of mechanics, there seem to exist no "implosions" at all which would qualify as the temporal inverses of eternally progressing "explosions" of very thin gases from a center into infinite space. In the light of this fact, one can assert the de facto irreversibility of an *eternal* "explosion" *unconditionally*, i.e., *without* Popper's restrictive proviso of *spontaneity* in regard to the production of the *coherent* initial conditions requisite for its inverse. For in an infinite space, there is even no possibility at all of a *non*-spontaneous production of the coherent "initial" conditions for an implosion having the following properties: the gas particles converge to a point after having been moving through infinite space for all past eternity in a manner constituting the temporal inverse of the expansion of a very thin gas from a point for all future eternity. There can be no question of a *non*-spontaneous realization of the "initial" conditions required for the latter kind of implosion, since such a realization would involve a self-contradictory condition akin to that in Kant's fallacious First Antinomy: the requirement that a process which has been going on for all infinite past time must have had a finite beginning (production by past *initial* conditions) after all.

By contrast, in a spatially *finite* system it is indeed possible to produce *non*-spontaneously the initial conditions for contracting waves and for implosions of gas particles which converge to a point. Thus, assuming negligible viscosity, there are expanding water waves in *finite* systems of which the temporal inverses could be produced *non*-spontaneously by dropping a very large circular object onto the water surface so that all parts of the circular object strike the water surface simultaneously. And hence there are conditions under which contracting waves do exist in finite systems. But there is no need whatever for Popper's

---

[8] For a treatment of classical entropic aspects of light propagation within a *finite* system, cf. A. Landé: "Optik und Thermodynamik," *Handbuch der Physik* (Berlin: J. Springer; 1928), Vol. XX, esp. pp. 471–79.

spontaneity proviso to assert the de facto irreversibility of the eternal expansion of a *spherical light wave* from a center through infinite space! If space is infinite, the existence of the latter process of expansion is assured by the facts of observation in conjunction with electromagnetic theory; but despite the fact that the *laws* for a homogeneous and isotropic medium *allow* the *inverse* process no less than the actual one,[9] we never encounter the inverse process of spherical waves closing in isotropically to a sharp point of extinction.

In view of the decisive role of the infinitude or "openness" of a physical system (the universe)—as opposed to the finitude of closed systems—in rendering Popper's spontaneity proviso dispensable, Hill and I made the following existential claim concerning processes in "open" (infinite) systems whose irreversibility is *non*-entropic and de facto:

> In classical mechanics the closed systems have quasi-periodic orbits, whereas the open systems have at least some aperiodic orbits which extend to infinity. . . . there exists a fundamental distinction between the two kinds of system in the following sense. In open systems there always exists a class of allowed elementary processes the inverses of which are unacceptable on physical grounds by requiring a *deus ex machina* for their production. For example, in an open universe, matter or radiation can travel away indefinitely from the "finite" region of space, and so be permanently lost. The inverse process would require matter or radiant energy coming from "infinity," and so would involve a process which is not realizable by physical sources. Einstein's example of an outgoing light wave and Popper's analogous case of a water wave are special finite illustrations of this principle.[1]

It will be noted that Hill and I spoke of there being "at least some aperiodic orbits which extend to infinity" in the classical mechanics of open systems and that we were careful not to assert that every such allowed process extending to infinity is a

[9] Cf. G. J. Whitrow: *The Natural Philosophy of Time* (London: Thomas Nelson & Sons Ltd.; 1961), pp. 8–10 and 269; also E. Zilsel: "Über die Asymmetrie der Kausalität und die Einsinnigkeit der Zeit," *Naturwissenschaften*, Vol. XV (1927), p. 283.

[1] E. L. Hill and A. Grünbaum: "Irreversible Processes in Physical Theory," *op. cit.*

de facto irreversible one. Instead, we affirmed the existence of a de facto irreversibility which is not predicated on Popper's spontaneity proviso by saying: "there always exists a class of allowed elementary processes" that are thus de facto irreversible. And, for my part, I conceived of this claim as constituting an extension of Popper's recognition of the essential role of coherence in de facto irreversibility to processes of the following kind: processes whose de facto irreversibility is not conditional on Popper's finitist requirement of spontaneity, because these processes extend to "infinity" in open systems and would hence have inverses in which matter or energy would have to come from "infinity" *coherently* so as to converge upon a point.

I was therefore quite puzzled to find that the communication by Hill and myself prompted the following dissent by Popper (*III*):

> In this connexion, I must express some doubt as to whether the principle proposed by Profs. Hill and Grünbaum is adequate. In formulating their principle, they operate with two ideas: that of the "openness" of a system, and that of a *deus ex machina.* Both seem to be insufficient. For a system consisting of a sun, and a comet coming from infinity and describing a hyperbolic path around the sun, seems to me to satisfy all the criteria stated by them. The system is open; and the reversion of the comet on its track would require a *deus ex machina* for its realization: it would "require matter . . . coming from 'infinity.'" Nevertheless, this is an example of just that kind of process which, I take it, we all wish to describe as completely reversible.

Popper's proposed counter-example of the comet coming from "infinity" into the solar system seems to me to fail for the following reasons: First, neither the actual motion of the comet nor its inverse involve any *coherence,* a feature which I, for my part, had conceived to be essential to the obtaining of non-entropic de facto irreversibility in open systems. In my own view, the fact that a particle or photon *came from* "infinity" in the course of an infinite past does not per se require a *deus ex machina,* any more than does their *going to* "infinity" in the course of an infinite future: in this context, I regard as innocuous the asymmetry that a particle which has already come from infinity can be said to have traversed an infinite space *by now,* whereas a particle now embarking on an infinite future journey will only have traversed a *finite* distance at *any one* time in the future. It is a *coherent* "implosion" from infinity that I believe to require a *deus ex machina,* i.e., to be de facto non-existent, while coherent "explosions" actually do exist. Second, even ignoring that the motion of Popper's comet does not involve coherence, the issue is not, as he seems to think, whether it would require a *deus ex machina* to realize the reversal of any given actual comet in its track; rather the issue is whether no *deus ex machina* would be needed to realize the actual comet motion while a *deus ex machina* would have been needed to have another comet execute *instead* a motion inverse to the first one. The answer to this question is an emphatic "no": unlike the case of outgoing and contracting waves (explosions and implosions), the two comet motions, which are temporal inverses of each other, are on a par with respect to the role of a *deus ex machina* in their realization. And even the reversal of the motion of an actual comet at a suitable point in its orbit might in fact be effected by an elastic collision with an oppositely moving other comet of equal mass and hence would not involve, as Popper would have it (*III*), "a *deus ex machina* who is something like a gigantic tennis player."

It seems to me, therefore, that far from being vulnerable to Popper's proposed counter-example, the existential claim by Hill and myself is fully as viable as Popper's, while having the further merit of achieving generality through freedom from Popper's spontaneity proviso. I therefore cannot see any justification at all for the following two assertions by H. Mehlberg: First, Mehlberg states incorrectly that Hill and I have claimed de facto irreversibility for "the class of all conceivable physical processes provided that the latter meet the mild requirement of happening in an 'open' physical system," and second, Mehlberg asserts that "Popper has shown the untenability of the Hill-Grünbaum criterion by constructing an effective counter example which illustrates the impossibility of their sweeping generalization of his original criterion."[2]

---

[2] H. Mehlberg: "Physical Laws and Time's Arrow," *op. cit.*, p. 128.

Mehlberg's critical estimate of Popper's own affirmation of non-entropic de facto irreversibility likewise seems to me to be unconvincing in important respects. After asking whether the irreversibility asserted by Popper is "lawlike or factlike"—a question to which the answer is: "avowedly factlike"—Mehlberg[3] concludes that Popper's temporal asymmetry "seems to be rather interpretable as a local, factlike particularity of the terrestrial surface than as a universal, lawlike feature . . . which may be expected to materialize always and everywhere." There are two points in Mehlberg's conclusion which invite comment: First, the significance which he attaches to the circumstance that the irreversibility of certain classes of processes is de facto or factlike rather than nomological or lawlike, when he assesses the bearing of that irreversibility on the issue of anisotropy *vs.* isotropy of time, and second, the striking contrast between the *epistemological parsimony* of his characterization of Popper's irreversibility as a "local . . . particularity of the terrestrial surface" and the *inductive boldness* of Mehlberg's willingness to do the following: affirm a cosmically pervasive nomological *isotropy* of time on the basis of attributing *cosmic* relevance, both spatially and temporally, to the fundamental *time-symmetric* laws which have been confirmed in modern man's limited sample of the universe.

As to the first of these two points in Mehlberg's denial of the anisotropy of time, I note preliminarily that human hopes for an eternal biological life are no less surely frustrated if all men are indeed de facto mortal, i.e., mortal on the strength of "boundary conditions" which do obtain permanently, than if man's mortality were assured by some law. Moreover, we saw in Chapter Seven that properties of time other than anisotropy depend on boundary conditions and not on the laws alone: the topological openness as opposed to closedness of time is a matter of the boundary conditions, if the laws are deterministic, as is the particular kind of open time that may obtain (finite, half infinite or infinite in both directions). By the same token, I see no escape from the conclusion that *if* de facto irreversibility does actually obtain everywhere and forever, such irreversibility confers anisotropy on time. And this anisotropy prevails not one iota less than

[3] *Ibid.*, p. 126.

it would, if its existence were guaranteed by temporally *asymmetrical* fundamental *laws* of cosmic scope.

For what is decisive for the anisotropy of time is not whether the non-existence of the temporal inverses of certain processes is factlike or lawlike; instead, what is relevant for temporal anisotropy is whether the required inverses do actually ever occur or not, whatever the reason. It is of considerable independent interest, of course, that such irreversibility as obtains in nature is de facto rather than nomological, *if* the distinction between a law (nomic regularity) and a non-nomic regularity arising from boundary conditions can always be drawn in a conceptually clear way. But, in my view, when evaluating the evidence for the anisotropy of time, Mehlberg commits the following error of misplaced emphasis: he wrongly discounts de facto irreversibility vis-à-vis nomological irreversibility by failing to show that our warrant for a cosmic extrapolation of time-symmetric laws is actually greater than for a corresponding extrapolation of the factlike conditions making for observed de facto irreversibility. For on what grounds can it be maintained that the ubiquitous and permanent existence of the de facto probabilities of "boundary conditions" on which Popper rests his affirmation of temporal anisotropy is less well confirmed than those laws on whose time-symmetry Mehlberg is willing to base his *denial* of the anisotropy of time? In particular, one wonders how Mehlberg could inductively justify his contention that we are only confirming a "particularity of the terrestrial surface," when we find with Popper (III) that:

> Only such conditions can be causally realized as can be organized from one centre. . . . causes which are not centrally correlated are causally unrelated, and can co-operate [i.e., produce coherence in the form of *isotropic* contraction of waves to a precise point] only by accident . . . The probability of such an accident will be zero.

If this finding cannot be presumed to hold on all planet-like bodies in the universe, for example, then why are we entitled to assume with Mehlberg that time-symmetric laws of mechanics, for example, are exemplified by the motions of binary stars throughout the universe? Since I see no valid grounds for Mehlberg's double standard of inductive credibility of pervasiveness as between laws and factlike regularities, I consider his negative estimate of Popper's non-entropic de facto anisotropy of time as quite unfounded.

Mehlberg's misplaced emphasis on the significance of lawlike vis-à-vis de facto irreversibility likewise seems to me to vitiate the following account which he gives of the import of de facto irreversibility in optics, a species of irreversibility which he admits to be of cosmic scope. He writes:

> A less speculative example of cosmological irreversibility is provided by the propagation of light *in vacuo*, which several authors have discussed from this point of view. . . . In accordance with Maxwell's theory of light conceived as an electromagnetic phenomenon, they point out that light emitted by a pointlike source, or converging towards a point, can spread on concentric spherical surfaces which either expand or contract monotonically. Yet, independently of Maxwell's theory, the incidence of expanding optical spheres is known to exceed by far the incidence of shrinking spheres. The reason for this statistical superiority of expanding optical spheres is simply the fact that pointlike light-emitting atoms are much more numerous than perfectly spherical, opaque surfaces capable of generating shrinking optical spheres, mainly by the process of reflection. If true, this ratio of the incidences of both types of light waves would provide a cosmological clue to a pervasive irreversibility of a particular class of optical processes.
>
> The bearing of this optical irreversibility upon time's arrow was often discussed. A long time before the asymmetry of expanding and contracting light waves was promoted to the rank of time's arrow, Einstein[4] pointed out that the asymmetry of these two types of optical propagation holds only on the undulatory theory of light. Once light is identified instead with a swarm of photons, the asymmetry vanishes. This conclusion holds at least for a spatially finite universe or for optical phenomena confined to a finite spatial region.
>
> Once more, however, the decisive point seems to be that the asymmetry between the two types of light waves depends on factual, initial conditions which prevail in a given momentary

[4] "A. Einstein: Über die Entwicklung unserer Anschauungen über die Konstitution und das Wesen der Strahlung,' *Physikalische Zeitschrift*, Vol. X (1910), pp. 817–28."

cross section of cosmic history or at the "boundaries" of a finite or infinite universe rather than on nomological considerations concerning this history: any other ratio of the incidences of expanding and shrinking light waves would also be in keeping with the relevant laws of nature contained in Maxwell's theory of electromagnetic phenomena. Of course, the aforementioned non-nomological conditions, responsible for the factual ratio of these incidences, are not "local" either, since the whole world is involved—they belong to cosmology. These conditions are nevertheless factlike rather than lawlike, as a comparison with the pertinent laws which can be derived from Maxwell's theory clearly shows.[5]

Contrary to Mehlberg, the decisive point appears not to be that "the asymmetry between the two types of light waves depends on factual, initial conditions . . . rather than on nomological considerations." He also asserts that "at least for a spatially finite universe or for optical phenomena confined to a finite spatial region" the corpuscularity of the photon, as conceived by Einstein, invalidates the optical asymmetry which obtains on the undulatory theory of light. I believe, however, that this claim should be amended as follows: "the optical asymmetry vanishes, *if at all, only* in a *finite* space." For suppose that one assumes with Einstein that the elementary radiation process is one in which a single emitting particle transfers its energy to only a single absorbing particle. In that case, the fantastically complicated coherence needed for the formation of a continuous contracting undulatory spherical shell of light is no longer required. Instead, there is then a need for the less complicated coherence among emitting particles located at the walls of a finite system and emitting converging photons. But, as Hill and I pointed out,[6] the de facto irreversibility of the spatially symmetrical eternal propagation of a pulse of light from a point source into *infinite* space does not depend on whether the light pulse is undulatory instead of being constituted by a swarm of photons.

Neither does *this* irreversibility depend on the acceptance of

---

[5] H. Mehlberg: "Physical Laws and Time's Arrow," *op. cit.*, pp. 123–24.

[6] E. L. Hill and A. Grünbaum: "Irreversible Processes in Physical Theory," *op. cit.*

the steady state theory of cosmology which, in the words of T. Gold, offers the following explanation for the fact that the universe is a non-reflecting sink for radiation:

> It is this facility of the universe to soak up any amount of radiation that makes it different from any closed box, and it is just this that enables it to define the arrow of time in any system that is in contact with this sink. But why is it that the universe is a non-reflecting sink for radiation? Different explanations are offered for this in the various cosmological theories and in some schemes, indeed, this would only be a temporary property.[7] In the steady state universe it is entirely attributed to the state of expansion. The red shift operates to diminish the contribution to the radiation field of distant matter; even though the density does not diminish at great distances, the sky is dark because in most directions the material on a line of sight is receding very fast. . . .[8]

What Gold appears to have in mind here is that due to a very substantial Doppler shift of the radiation emitted by the receding galaxies, the frequency ν becomes very low or goes to zero, and since the energy of that radiation is given by $E = h\nu$, very little if any is received by us converging from these sources. And he goes on to say:

> This photon expansion going on around most material is the most striking type of asymmetry, and it appears to give rise to all other time asymmetries that are in evidence. The preferential divergence, rather than convergence, of the world lines of a system ceases when that system has been isolated in a box which prevents the expansion of the photons out into space. Time's arrow is then lost.

We see that Gold's account includes an appreciation of the decisive role played by the infinitude of the space in rendering irreversible the radiation spreading from a point. To be sure, he does emphasize that the Doppler shift due to the expansion makes for the darkness of the sky at night, which would other-

[7] Presumably Gold is referring here to models of spatially closed or finite universes.

[8] T. Gold: "The Arrow of Time," in *La Structure et L'Évolution de l'Univers,* Proceedings of the 11th Solvay Congress (Brussels: R. Stoops; 1958), pp. 86–87.

wise be lit up by strong radiation. But the crucial point is the following: even if the energy of radiation from receding galaxies were not drastically attenuated by the Doppler shift, such radiation would still not be the inverse of a process in which a pulse of photons from a point source forever spreads symmetrically into infinite space from that point source. The non-existent inverse of the latter process of outgoing radiation would be a contracting configuration of photons that has been coming from "infinity," i.e., from no sources at all, and has been converging on a point for all infinite past time.

We see again that the complete time-symmetry of the basic laws like those of dynamics or electromagnetism is entirely compatible with the existence of contingent irreversibility. In the concise and apt words of Penrose and Percival, the reason for this compatibility is that "Dynamics relates the states of a system at two different times, but it puts no restriction whatever on the state at any one time, nor on the probability distribution at any one time."[9]

The anisotropy arising from the non-entropic de facto irreversibility which we have been considering is (1) more pervasive temporally than the merely statistical anisotropy guaranteed by the thermodynamics of branch systems, and (2) more ubiquitous spatially than any exclusively large-scale anisotropy of time, such as would be guaranteed, for example, by the universe's being a monotonically expanding spherical 3-space which had a singular state without temporal predecessor in the finite past. For the non-entropic de facto irreversibility which we have been considering assures uniform temporal anisotropy for local intervals in the time continuum no less than for that continuum in the large. And, unlike such anisotropy as would arise from the monotone expansion of a spherical 3-space type of universe, the anisotropy guaranteed by Popper's non-entropic de facto irreversibility is manifest within the small spatial regions accessible to our daily experience.

The scope of the preceding discussion could be enlarged to include more detailed allowance for the results of further physical theories such as the steady-state cosmology. With respect to

[9] O. Penrose and I. C. Percival: "The Direction of Time," *Proceedings of the Physical Society*, Vol. LXXIX (1962), p. 606.

that wider class of physical theories, the connection between the thermodynamic and non-thermodynamic kinds of irreversibility may well turn out to be deeper than I have claimed. Specifically, the connection may go beyond the fact that both kinds of irreversibility are due to boundary conditions rather than to laws and fail to obtain in the case of systems which are both spatially finite and permanently closed. Thus, Gold inseparably links all temporal asymmetries which make for the anisotropy of time with the time-asymmetry between the divergence and convergence of the geodesics associated with fundamental observers (expansion of the universe). For he relates the temporal asymmetry in all statistical processes to the tendency for radiation to diverge with positive time, a tendency which he relates, in turn, to the expansion of the universe. In this way, Gold links the thermodynamic and radiational time-asymmetries to a cosmological one.

Since *non*-entropic de facto irreversibility suffices to confer anisotropy on time, the statistically time-asymmetric entropic behavior of branch systems is not a necessary condition for the anisotropy of time. Accordingly, if one says that the latter entropic behavior statistically (and intrinsically) "defines" the relation "later than" for physical, as distinct from psychological (common sense) time, the term "defines" must be construed in the weak sense of "constitutes an empirical indicator." But it is important to realize that the latter weak construal of the entropic "definition" of "later than" is necessitated not by the merely statistical character of the thermodynamic anisotropy of time but by the existence of *non*-entropic irreversibility alongside entropic statistical irreversibility. This important consideration seems to have been overlooked by Carnap in setting forth a criticism of the entropic "definition" of "later than." Carnap writes:

> Reichenbach's definition, which is also accepted by Grünbaum,[1] appears to me very problematic. Reichenbach criticizes Boltzmann's definition by pointing out that the correlation be-

[1] My explicit statement of the differences between the "definitions" given by Reichenbach and myself respectively was not available to Carnap when he wrote the comment quoted here. But the reader will recall from the first part of this chapter that despite the obvious and substantial indebtedness of my entropic "definition" of "later than" to that of Reichenbach, there are important respects in which my "definition" does depart from his.

> tween the direction of time and the increase in entropy holds, not universally, but only with probability. I agree. But it seems to me that an analogous objection holds for Reichenbach's definition.[2]

In an unpublished amplification of the latter statement, which I now quote here with his kind permission, Professor Carnap wrote as follows:

> If we understand "earlier" in the customary sense of the term in physics, then for a single system the statement "If the entropy at the time point A is considerably lower than at the time point B, then A is earlier than B" holds not universally, but only with probability. But then it seems clear that the same statement with respect to a majority of the branch systems instead of a single branch system, holds likewise only with probability, although under certain conditions the probability may be overwhelmingly great. If the latter is the case, then the increase in entropy may certainly be taken as the basis for an inductive *inference* to the relation E [i.e., "earlier than"]. But it seems very doubtful whether it is legitimate to take a statistical correlation, however high it may be, as the basis for a theoretical *definition*. In other words, if a relation E′ is defined in this way, then there are cases, in which E and E′ do not agree, where E is understood in the customary sense.

It is quite correct, of course, that my entropic "definition" of "later than" is only statistical: this "definition" presupposes that in most space ensembles of branch systems, the vast majority of the members of the ensemble will increase their entropy in one of the two opposite time directions and decrease it in the other. And these two time directions had already been distinguished from one another to the extent that we utilized the relations of temporal o-betweenness to impose a real number coordinatization extrinsically without assuming at the outset that the entropy statistics of branch systems would turn out to be temporally asymmetric. Hence it is avowedly true that the direction of entropy increase of the majority of branch systems is not the

---

[2] R. Carnap: "Adolf Grünbaum on The Philosophy of Space and Time," in: P. A. Schilpp (ed.) *The Philosophy of Rudolf Carnap* (LaSalle: Open Court Publishing Co.; 1963), p. 954. The essentials of my entropic "definition" of "later than" were set forth in my contribution to the latter Schilpp volume entitled "Carnap's Views on the Foundations of Geometry," pp. 599–684.

same for all cosmic epochs which contain branch systems satisfying initial conditions of "randomness" but is the same for only the vast majority of such cosmic epochs. Indeed this fact prompted my speaking of the "statistical" anisotropy of time in this context. I deny altogether, however, Carnap's contention that, on the strength of being statistical, my entropic (and hence intrinsic) "definition" of "later than" (or correspondingly of "earlier than") is beset by the following difficulty charged against it by Carnap: (1) "there are cases, in which E [i.e., the customary "earlier than" of physics as used, for example, to plot the entropy of a permanently closed system against time] and E′ [i.e., the entropically "defined" relation of "earlier than"] do not agree," and that therefore (2) "it seems very doubtful whether it is legitimate to take a statistical correlation, however high it might be, as the basis for a theoretical *definition*." My denial of the vulnerability of my "definition" to this criticism of Carnap's is based on the fact that my "definition" of "later than" avowedly utilized the direction of entropy increase for a *typical representative* of the majority of cosmic epochs, so that the ascription of earlier-later relations to states belonging to entropically *atypical* cosmic epochs will be dictated by the fact that the *latter* epochs sustain relations of temporal o-betweenness to the entropically typical epochs. That there is no difficulty here of the kind alleged by Carnap is also apparent from my earlier account of the temporal description of fluctuation phenomena on the basis of my entropic "definition" of "later than": branch systems exhibiting cosmically "occasional" entropy *decreases* with *positive* time can be described as such, since these decreases are temporally *counter-directed* with respect to the entropy increase of the majority of branch systems.

Hence the statistical character of my "definition" of "later than" does not disqualify it. And the valid reason for the rejection by Reichenbach and myself of Boltzmann's attempt at a statistical definition was not that Boltzmann's definition was statistical. Carnap seems to have overlooked that instead, our reason for rejecting Boltzmann's attempt was that the relevant probabilities of Boltzmann's statistics were *completely time-symmetric*, these statistics being those of the long-term entropic behavior of a single, permanently closed system.

Chapter 9

THE ASYMMETRY OF RETRODICTABILITY AND PREDICTABILITY, THE COMPOSSIBILITY OF EXPLANATION OF THE PAST AND PREDICTION OF THE FUTURE, AND MECHANISM VS. TELEOLOGY

The temporally asymmetric character of the entropy statistics of branch systems has a number of important consequences which were not dealt with in Chapter Eight and to which we must now turn our attention. In particular, our conclusions regarding the entropy statistics of branch systems can now be used to elucidate (1) the conditions under which retrodiction of the past is feasible while prediction of the future is not,[1] (2) the relation of psychological time to physical time, (3) the consequence which the feasibility of retrodictability *without* corresponding predictability has for the compossibility of explainability of the past and the corresponding predictability of the future, and (4) the merits of the controversy between philosophical mechanism and teleology.

(A) CONDITIONS OF RETRODICTABILITY AND NON-PREDICTABILITY.

Suppose we encounter a beach whose sand forms a smooth surface except for one place where it is in the shape of a human

---

1 The reader will recall that conditions under which the *inverse* asymmetry obtains were discussed in Chapter Eight.

footprint. We know from our previous considerations with high probability that instead of having evolved *isolatedly* from a prior state of uniform smoothness into its present uneven configuration according to the statistical entropy principle for a permanently closed system, the beach was an *open* system in *interaction* with a stroller. And we are aware furthermore that if there is some quasi-closed wider system containing the beach and the stroller, as there often is, the beach achieved its ordered low entropy state of bearing the imprint or interaction-indicator at the expense of an at least compensatory entropy increase in that wider system comprising the stroller: the stroller increased the entropy of the wider system by scattering his energy reserves in making the footprint.

We see that the sandy footprint shape is a genuine indicator and not a randomly achieved form resulting from the unperturbed chance concatenations of the grains of sand. The imprint thus contains information in the sense of being a veridical indicator of an interaction. Now, in all probability the entropy of the imprint-bearing beach-system increases after the interaction with the stroller through the smoothing action of the wind. And this entropy increase is parallel, in all probability, to the direction of entropy increase of the majority of branch systems. Moreover, we saw that the production of the indicator by the interaction is likely to have involved an entropy increase in some wider system of which the indicator is a part. Hence, *in all probability the states of the interacting systems which do contain the indicators of the interaction are the relatively higher entropy states of the majority of branch systems, as compared to the interaction state. Hence the indicator states are the relatively later states as compared to the states of interaction which they attest.* And by being both *later* and indicators, these states have *retrodictive* significance, thereby being traces, records or "memories." And due to the high degree of retrodictive univocity of the low entropy states constituting the indicators, the latter are veridical to a high degree of *specificity*.

Confining our attention for the present to indicators whose *production* requires only the occurrence of the interaction which they attest, we therefore obtain the following conclusion. Apart from some kinds of *advance*-indicators requiring very special conditions for their production and constituting *exceptions,* it is the case that *with overwhelming probability, low entropy indicator-states can exist in systems whose interactions they attest only after and not before these interactions.*[2] If this conclusion is true (assuming that there are either no cases or not enough cases of bona fide precognition to disconfirm it), then, of course, it is not an a priori truth. And it would be very shallow indeed to seek to construe it as a trivial a priori truth in the following way: calling the indicator states "traces," "records," or "memories" and noting that it then becomes tautological to assert that traces and the like have only retrodictive and no predictive sigficance. But this transparent verbal gambit cannot make it true a priori that—apart from the exceptions to be dealt with below—interacting systems bear indicators attesting veridically only their earlier and not their later interactions with outside agencies.

Hence, the exceptions apart, we arrive at the fundamental asymmetry of recordability: *reliable indicators in interacting systems permit only retrodictive inferences concerning the interactions for which they vouch but no predictive inferences pertaining to corresponding later interactions.*

And the logical schema of these inductive inferences is roughly as follows: the premises assert (1) the presence of a certain relatively low entropy state in the system, and (2) the quasi-universal statistical law stating that most low entropy states are interaction-indicators *and* were *preceded* by the interactions for which they vouch. The conclusion from these premises is then the inductive retrodictive one that there was an earlier interaction of a certain kind.

As already mentioned, our affirmation of the temporal asymmetry of recordability of interactions must be qualified by dealing with two exceptional cases, the first of which is the pre-recordability of those interactions which are veridically predicted by human beings (or computers). For any event which

[2] Two of the exceptions, which we shall discuss in some detail below, are constituted by the following two classes of advance indicators: First, veridical predictions made and stored (recorded) by human (or other sentient, theory-using) beings, and physically registered, bona fide advance indicators produced by computers, and second, advance indicators (e.g., sudden barometric drops) which are produced by the very cause (pressure change) that also produces the future interaction (storm) indicated by them.

could be predicted by a scientist could also be "pre-recorded" by that scientist in various forms such as a written entry on paper asserting its occurrence at a certain later time, an advance drawing, or even an advance photograph based on the pre-drawing. By the same token, artifacts like computers can pre-record events which they can predict. A comparison between the written, drawn, or photographic pre-record (i.e., recorded prediction) of, say, the crash of a plane into a house and its post-record in the form of a caved-in house, and a like comparison of the corresponding pre- and post-records of the interaction of a foot with a beach will now enable us to formulate the essential differences in the conditions requisite to the respective production of pre-records and post-records as well as the usual differences in make-up between them.

The production of at least one retrodictive indicator or post-record of an interaction such as the plane's crash into the house requires only the occurrence of that interaction (as well as a moderate degree of durability of the record). The retrodictive indicator states in the system which interacted with an outside agency must, of course, be distinguished from the *epistemic use* which human beings may make of these physical indicator states. And our assertion of the sufficiency of the interaction for the production of a post-record allows, of course, that the *interpretation* of actual post-records by humans as bona fide documents of the past requires their use of theory and not just the occurrence of the interaction. In contrast to the sufficiency of an interaction itself for its (at least short-lived) post-recordability, no such sufficiency obtains in the case of the pre-recordability of an interaction: save for an overwhelmingly improbable freak occurrence, the production of even a single pre-record of the coupling of a system with an agency external to it requires, as a necessary condition, *either* (a) the use of an appropriate theory by symbol-using entities (humans, computers) having suitable information, *or* (b) the pre-record's being a partial effect of a cause that also produces the pre-recorded interaction, as in the barometric case to be dealt with below. And in contexts in which (a) is a necessary condition, we find the following: since pre-records are, by definition, veridical, this necessary condition cannot *generally* also be sufficient, unless the predictive theory employed is deterministic *and* the information available to the theory-using organism pertains to a closed system.

In addition to differing in regard to the conditions of their production, pre-records generally differ from post-records in the following further respect: unless the pre-record prepared by a human being (or computer) *happens* to be part of the interacting system to which it pertains, the pre-record will not be contained in states of the interacting system which it concerns but will be in some other system. Thus, a pre-record of the crash of a plane into a house in a heavy fog would generally *not* be a part of either the house or the plane, although it can happen to be. But in the case of *post*-recording, there will always be at least one post-record, however short-lived, in the interacting system itself to which that post-record pertains.

Our earlier example of the footprint on the beach will serve to illustrate more fully the asymmetry between the requirements for the production of a pre-record and of a post-record. The pre-recording of a *later* incursion of the beach by a stroller would require extensive information about the motivations and habits of people not now at the beach and also knowledge of the accessibility of the beach to prospective strollers. This is tantamount to knowledge of a large system which is *closed,* so that all relevant agencies can safely be presumed to have been included in it. For otherwise, we would be unable to guarantee, for example, that the *future* stroller will not be stopped enroute to the beach by some agency not included in the system, an eventuality whose occurrence would deprive our pre-record of its referent, thereby destroying its status as a veridical indicator. In short, in the case of the footprint, which is a post-record and not a pre-record of the interaction of a human foot with the beach, the interaction itself is *sufficient* for its post-recording (though not for the extended *durability* of the record once it exists) but *not* for its pre-recording and prediction. Since a future interaction of a potentially open system like the beach is not itself sufficient for its pre-recordability, open systems like beaches therefore do not themselves exhibit pre-records of their own future interactions. Instead—apart from the second species of pre-recordability to be considered presently—pre-recordability of interactions of potentially open systems requires the mediation of symbol and theory-using organisms or the operation of appropriate artifacts like computers. And such pre-recordability can obtain successfully only if the theory available to the pre-recording organism is deterministic and sufficiently comprehensive to include all the relevant laws and boundary conditions governing the pertinent closed systems.

The second species of exceptions to the asymmetry of recordability is exemplified by the fact that a sudden drop in the pressure reading of a barometer can be an advance-indicator or "pre-record" of a subsequent storm. To be sure, it is the immediately *prior* pressure change in the spatial vicinity of the barometer and only that particular prior change (i.e., the *past* interaction through pressure) which is recorded numerically by a given drop in the barometric reading, and not the pressure change that *will* exist at that same place at a *later* time: To make the predictions required for a *pre*-recording of the pressure changes which will exist at a given space point at later times (i.e., of the corresponding future interactions), comprehensive meteorological data pertaining to a large region would be essential. But it is possible in this case to base a rather reliable prediction of a future storm on the present sudden barometric drop. The latter drop, however, is, in fact, a bona fide advance indicator only because it is a partial effect of the very comprehensive cause which also produces (assures) the storm. Thus, it is the fulfillment of the *necessary condition* of having a causal ancestry that overlaps with that of the storm which is needed to confer the status of an advance indicator on the barometric drop. In contrast to the situation prevailing in the case of *post*-recordability, the existence of this necessary condition makes for the fact that the future occurrence of a storm is *not itself sufficient* for the existence of an advance indicator of that storm in the form of a sudden barometric drop at an earlier time.

An analogous account can be given of the following case, which Mr. F. Brian Skyrms has suggested to me for consideration: situations in which *human intentions* are highly reliable advance indicators of the events envisaged by these intentions. Thus, the desire for a glass of beer, coupled with the supposed presence of the conditions under which beer and a glass are obtainable, produces as a partial effect the intent to get it. And, if external conditions permit (the beer is available and accessible), and, furthermore, if the required internal conditions materialize (the person desiring the beer remains able to go and get it), then the intent will issue in the obtaining and drinking of the beer. But in contrast to the situation prevailing in the case of retrodictive indicators (post-records), the future consumption of the beer is not a sufficient condition for the existence of its probabilistic advance indicator in the form of an intention.

The consideration of some alleged counter-examples will serve to complete our statement of the temporal asymmetry of the recordability of interactions. These purported counter-examples are to the effect that there are pre-records not depending for their production on the use of predictive theory by symbol-using organisms or on the pre-record's being a partial effect of a cause that also produces the pre-recorded interaction.

In the first place, it might be argued that there are spontaneous pre-records as exemplified in the following two kinds of scientific contexts: first, in any essentially closed dynamical system such as the solar system, a dynamical state later than one occurring at a time t_0 is a sufficient condition for the occurrence of the state at time t_0, no less than is a state prior to t_0; hence the state at time t_0 can be regarded as a pre-record of the later state no less than it can be deemed a post-record of the earlier one, and second, a certain kind of death—say, the kind of death ensuing from leukemia—may be a sufficient condition for the existence of a pre-record of it in the form of the onset of active leukemia. But these examples violate the conditions on which our denial of spontaneous pre-recordability is predicated in the following essential respect: they involve later states which are not states of interaction with outside agencies entered into by an otherwise closed system, in the manner of our example of the beach.

In the second place, since the thesis of the temporal asymmetry of spontaneous recordability makes cases of bona fide precognition overwhelmingly improbable, it might be said that this thesis and the entropic considerations undergirding it are vulnerable to the discovery of a reasonable number of cases of genuine precognition, a discovery which is claimed by some to have already been made. To this I retort that if the purported occurrence of precognition turns out to become well authenticated, then I am, of course, prepared to envision such alterations in the body of current orthodox scientific theory as may be required.

The connection between low entropy states and retrodictive information which emerged in our discussion of the asymmetry of recordability throws light on the reason for the failure of Maxwell's well-known sorting demon: in substance, Maxwell's demon cannot violate the second law of thermodynamics, since the entropy decrease produced in the gas is more than balanced by the entropy increase in the mechanism procuring the informational data concerning individual gas molecules which are needed by the demon for making the sorting successful.[3]

We saw earlier how reliance on entropy enables us to ascertain which one of two causally connected events is *the* cause of the other because it is the earlier of the two. Our present entropic account of the circumstances under which the past can be inferred from the present while the future cannot, as well as our earlier statement in Chapter Eight of the circumstances when only the converse determination is possible enables us to specify the conditions of validity for the following statements by Reichenbach: "Only the totality of all causes permits an inference concerning the future, but the past is inferable from a partial effect alone" and "one can infer the total cause from a partial effect, but one cannot infer the total effect from a partial cause."[4] A partial effect produced in a system while it is open permits, on entropic grounds, an inference concerning the earlier interaction event which was its cause: even though we do not

[3] Cf. L. Brillouin: *Science and Information Theory, op. cit.;* E. C. Cherry: "The Communication of Information," *American Scientist,* Vol. XL (1952), p. 640; J. Rothstein: "Information, Measurement and Quantum Mechanics," *Science,* Vol. CXIV (1951), p. 171, and S. Watanabe: "Über die Anwendung Thermodynamischer Begriffe auf den Normalzustand des Atomkerns," *Zeitschrift für Physik,* Vol. CXIII (1939), pp. 482–513.

[4] H. Reichenbach: "Die Kausalstruktur der Welt und der Unterschied von Vergangenheit und Zukunft," *Berichte der Bayerischen Akademie München, Mathematisch-Naturwissenschaftliche Abteilung* (1925), p. 157, and "Les Fondements Logiques de la Mécanique des Quanta," *op. cit.*, p. 146. Cf. also C. F. von Weizsäcker: "Der Zweite Hauptsatz und der Unterschied von Vergangenheit und Zukunft," *op. cit.* By noting in his later publications (especially in *DT, op. cit.*, pp. 157–67) that the temporal asymmetry involved here has an entropic basis, Reichenbach abandoned his earlier view that it provides an *independent* criterion for the anisotropy of time. Thus, he has essentially admitted the validity of H. Bergmann's telling criticisms (*Der Kampf um das Kausalgesetz in der jüngsten Physik, op. cit.*, pp. 19–24) of his earlier view.

know the total present effect, we know that the part of it which is an ordered, low entropy state was (most probably) preceded by a still lower entropy state and that the diversity of the interactions associated with such a very low interaction entropy state is relatively small, thereby permitting a rather specific assertion about the past.

Thus, the asymmetry of inferability arises on the macro-level in the absence of knowledge of the microscopic state of the total (closed) system at a given time and is made possible by the relative retrodictive univocity of local low entropy states which result from interactions. We are therefore in possession of the answer to the question posed by J. J. C. Smart when he wrote: "Even on a Laplacian view, then, we still have the puzzling question: 'Why from a limited region of space can we deduce a great deal of the history of the past, whereas to predict similar facts about the future even a superhuman intelligence would have to consider initial conditions over a very wide region of space?' "[5]

(B) THE PHYSICAL BASIS FOR THE ANISOTROPY OF PSYCHOLOGICAL TIME.

We saw that the production of traces, records, or memories is usually accompanied by entropy increases in the branch systems. And therefore it follows that the direction of increase of stored information or "memories" either in inanimate recording-devices or in memory-gathering organisms like man must be the same as the direction of entropy increase in the majority of branch systems. But we noted in our discussion of Bridgman's critique of Eddington in Chapter Eight that the direction of the increase or accumulation of mental traces or memories is the *forward* direction of psychological time. Hence, what is *psychologically* later goes hand-in-hand with what is purely physically later on the basis of the entropic evolution of branch systems. And the anisotropy of psychological time *mirrors* the more fundamental anisotropy of physical time.

By way of concrete example, note that the vocal output of a lecturer increases the entropy of the lecture room and produces

[5] J. J. C. Smart: "The Temporal Asymmetry of the World," *op. cit.*, p. 81.

memories in his listeners at his physiological expense. At the start of his lecture, the lecturer had concentrated energy, and therefore the entropy of the lecture room was then relatively low. After his lecture much of that energy has been scattered in the form of sound waves, thereby increasing the entropy of the room and also registering memories in his listeners which enable them to tell that it is then later than it was when they first sat down for the lecture.

These considerations show that even if our physiological existence were possible in a state of total equilibrium—which it is not—we would then *not* be able to have the kind of temporal awareness that we do have. If we were unfortunate enough to be surviving while immersed in one vast cosmic equilibrium, it would then be unutterably dull to the point of loss of our psychological time sense.

(C) THE BEARING OF RETRODICTABILITY AND NON-PREDICTABILITY ON THE COMPOSSIBILITY OF EXPLAINABILITY AND PREDICTABILITY.

What is the bearing of the existence of the asymmetry between retrodiction and prediction on the following quite distinct question: Is there symmetry between the *explanation* of an event E on the basis of one or more *antecedents* of E, when E belongs to the *past* of the explaining scientist, on the one hand, and, on the other hand, the *prediction* of the same (kind of) E by reference to the same (kind of) *antecedent(s)* of E, when E belongs to the *future* of the scientist making the prediction? In short, what is the relevance of the retrodiction-prediction asymmetry to the thesis of symmetry (or structural equality) of explanation and prediction as put forward by writers such as C. G. Hempel?[6] Preparatory to answering this question, we shall clarify the temporal relations which are involved and will then represent the results on a diagram.

---

[6] I refer here to the original paper of C. G. Hempel and P. Oppenheim: "Studies in the Logic of Explanation," *Philosophy of Science,* Vol. XV (1948), p. 15. For Hempel's most recent statement of his account of scientific explanation, see his "Deductive Nomological vs. Statistical Explanation," in: H. Feigl and G. Maxwell (eds.) *Minnesota Studies in the Philosophy of Science* (Minneapolis: University of Minnesota Press; 1962), Vol. III, pp. 98–169.

For Hempel, the particular conditions C_i ($i = 1,2, \ldots n$) which, in conjunction with the relevant laws, account for the *explanandum*-event E, may be *earlier* than E in *both* explanation *and* prediction or the C_i may be *later* than E in *both* explanation *and* prediction. Thus, a case of *prediction* in which the C_i would be *later* than E would be one in astronomy, for example, in which a future E is accounted for by reference to C_i which are still further in the future than E. These assertions hold, since Hempel's criterion for an explanation as opposed to a prediction is that E belong to the scientist's *past* when he offers his account of it, and his criterion for a corresponding prediction is that E belong to the scientist's *future* when it is made.

However, in the retrodiction-prediction antithesis, a *retrodiction* is characterized by the fact that the C_i are *later* than E, while the C_i are *earlier* than E in the kind of *prediction* which is antithetical to retrodiction but *not* identical with Hempelian prediction.

In the accompanying diagram, the i, k, l, m, may each range over the values 1,2 . . . n.

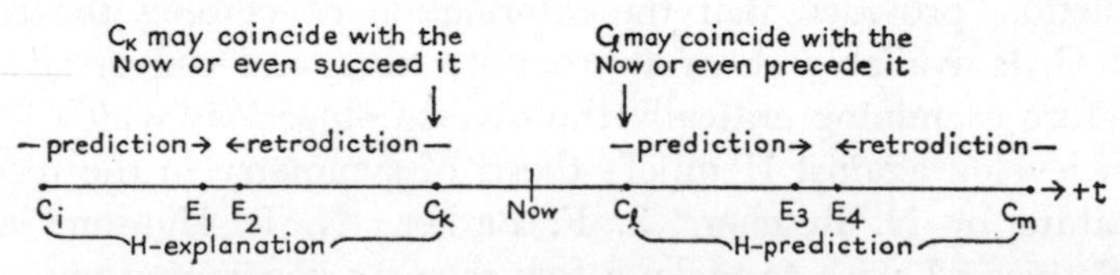

If we use the prefix "H" as an abbreviation for "Hempelian," then two consequences are apparent. Firstly, a retrodiction as well as a prediction can be an H-prediction, and a prediction as well as a retrodiction can be an H-explanation. Secondly, being an H-prediction rather than an H-explanation or conversely depends on the transient homocentric "now," but there is no such "now"-dependence in the case of being a retrodiction instead of a prediction, or conversely.

The passage in the Hempel-Oppenheim essay setting forth the symmetry thesis espoused by K. R. Popper and these authors reads as follows:

> the same formal analysis, including the four necessary conditions, applies to scientific prediction as well as to explanation. The difference between the two is of a pragmatic character. If E is given, i.e., if we know that the phenomenon described by

E has occurred, and a suitable set of statements $C_1, C_2, \ldots, C_k, L_1, L_2, \ldots, L_r$ is provided afterwards, we speak of an explanation of the phenomenon in question. If the latter statements are given and E is derived prior to the occurrence of the phenomenon it describes, we speak of a prediction. It may be said, therefore, that an explanation is not fully adequate unless its explanans, if taken account of in time, could have served as a basis for predicting the phenomenon under consideration.[7] Consequently, whatever will be said in this article concerning the logical characteristics of explanation or prediction will be applicable to either, even if only one of them should be mentioned.[8]

Hempel's thesis of symmetry or structural equality between H-explanation and H-prediction can therefore now be formulated in the following way: Any *prediction* which qualifies logically *and* methodologically as an H-explanation also qualifies as an H-prediction, provided that the scientist is in possession of the information concerning the C_i prior to the occurrence of E, and conversely. And any *retrodiction* which qualifies logically *and* methodologically as an H-explanation also qualifies as an H-prediction, provided that the information concerning the relevant C_i is available at an appropriate time, and conversely.

Before examining critically the diverse objections which have been leveled against Hempel's thesis of symmetry in the recent literature by N. Rescher,[9] S. F. Barker,[1] N. R. Hanson,[2] and M. Scriven,[3] I wish to make a few remarks concerning my con-

[7] "The logical similarity of explanation and prediction, and the fact that one is directed towards past occurrences, the other towards future ones, is well expressed in the terms 'postdictability' and 'predictability' used by Reichenbach in [Quantum Mechanics], p. 13."

[8] C. G. Hempel and P. Oppenheim: "Studies in the Logic of Explanation," *op. cit.*, §3.

[9] N. Rescher: "On Prediction and Explanation," *British Journal for the Philosophy of Science,* Vol. VIII (1958), p. 281.

[1] S. F. Barker: "The Role of Simplicity in Explanation," in: H. Feigl & G. Maxwell (eds.) *Current Issues in the Philosophy of Science, op. cit.,* pp. 265–86 and the Comments on this paper by W. Salmon, P. K. Feyerabend, and R. Rudner with Barker's Rejoinders.

[2] N. R. Hanson: "On the Symmetry Between Explanation and Prediction," *The Philosophical Review,* Vol. LXVIII (1959), p. 349.

[3] M. Scriven: "Explanation and Prediction in Evolutionary Theory," *Science,* Vol. CXXX (1959), pp. 477ff., excerpts from which are reprinted from *Science* by permission, and "Explanations, Predictions and Laws," in: H. Feigl & G. Maxwell (eds.) *Minnesota Studies in the Philosophy of Science, Vol. III, op. cit.,* pp. 170–230.

strual of both that thesis and of the philosophical task to whose fulfillment it pertains.

I take Hempel's affirmation of symmetry to pertain not to the *assertibility* per se of the *explanandum* but to the either deductive or inductive *inferability* of the *explanandum* from the *explanans*. Popper and Hempel say: To the extent that there is ever explanatory *inferability*, there is also predictive inferability and conversely. They do not claim that every time you are entitled to assert, on some grounds or other, that a certain kind of event did occur in the past, you are also entitled to say that the same kind of event will occur in the future. Being concerned with scientific understanding, Popper and Hempel said that there is temporal symmetry not of assertibility per se but of assertibility *on the strength of the explanans*. The scientific relevance of dealing with predictive arguments rather than mere predictive assertions can hardly be contested by claiming with Scriven that in this context "the crucial point is that, however achieved, a prediction is what it is simply because it is produced in advance of the event it predicts; it is *intrinsically* nothing but a bare description of that event."[4] For surely a soothsayer's unsupported prophecy that there will not be a third world war is not of scientific significance and ought not to command any scientific interest precisely because of the unreasoned manner of its achievement. Hence a scientifically warranted prediction of an event must be more than a mere pre-assertion of the event. And in any context which is to be scientifically relevant, the following two components can be distinguished in the meaning of the term "H-predict" no less than in the meaning of "H-explain" (or "post-explain"), and similarly for the corresponding nouns:

(1) the mere *assertion* of the *explanandum*, which *may* be based on grounds other than its scientific *explanans*,

(2) the logical *derivation* (deductive or inductive) of the *explanandum* from an *explanans*, the character of the content of the *explanans* remaining unspecified until later on in this section.

My attachment of the prefix "H" to the word "explain" (and to "explanation") and my use of "post-explain" as a synonym of "H-explain" will serve to remind us for the sake of clarity that this usage of "explain" results from a restriction to the past of *one*

[4] M. Scriven: "Explanations, Predictions and Laws," *op. cit.*, Sec. 3.4.

well-established usage which is *temporally neutral,* viz., "explain" in the sense of providing scientific understanding (or a scientific accounting) of why something did *or* will occur. But, to my mind, the philosophical task before us is not the ascertainment of how the words "explain" and "predict" are used, even assuming that there is enough consistency and precision in their usage to make this lexicographic task feasible. And hence the verdict on the correctness of Hempel's symmetry thesis cannot be made to depend on whether it holds for what is taken to be the actual or ordinary usage of these terms. Instead, in this context I conceive the philosophical task to be both the elucidation and examination of the provision of scientific understanding of an *explanandum* by an *explanans* as encountered in actual scientific theory. Accordingly, Hempel's symmetry thesis, which concerns the inferability of the *explananda* from a given kind of *explanans* and not their assertibility, must be assessed on the basis of a comparison of H-predictive with H-explanatory *arguments* with respect to the measure of scientific understanding afforded by them. Thus, the issue of the adequacy of the symmetry thesis will revolve around whether there is temporal symmetry in regard to the degree of entailment, as it were, characterizing the logical link between the *explanans* and the *explanandum.* Specifically, we shall need to answer both of the following questions: First, would the type of *argument* which yields a prediction of a future *explanandum*-event not furnish precisely the same amount of scientific understanding of a corresponding past event? And second, does an *explanans* explain an *explanandum* referring to a past event any more conclusively than this same kind of *explanans predictively implies* the *explanandum* pertaining to the corresponding future event?

We are now ready to turn to the appraisal of the criticisms of Hempel's symmetry thesis offered by Rescher, Barker, Hanson, and Scriven. In the light of my formulation of Hempel's thesis, it becomes clear that it does *not* assert, as Rescher supposes, that any set of C_i which permit a *predictive* inference also qualify for a corresponding *retrodictive* one, or that the converse is true. As Rescher notes correctly but irrelevantly, whether or not symmetry obtains between prediction and retrodiction in any given domain of empirical science is indeed not a purely logical question but depends on the content of the laws pertaining to the domain in question. We see therefore that Hempel was justified in claiming[5] that Rescher has confused H-explanation with retrodiction. And this confusion is also facilitated by one of Scriven's statements of the symmetry thesis, which reads: "to predict, we need a correlation between present events and future ones—to explain, between present ones and past ones."[6]

In agreement with Scheffler,[7] Rescher offers a further criticism of the Hempelian assertion of symmetry: "it is inconsistent with scientific custom and usage regarding the concepts of explanation and prediction," for, among other things, "Only true statements are proper objects for explanation, but clearly not so with prediction."[8] Thus, Rescher notes that we explain only phenomena which are known to have occurred, but we sometimes predict occurrences which do not come to pass. And in support of the latter claim of an "epistemological asymmetry," Rescher points to a large number of cases in which we have "virtually certain knowledge of the past on the basis of traces found in the present" but "merely probable knowledge of the future on the basis of knowledge of the present and/or the past."[9]

The question raised by Rescher's further objection is whether this *epistemological* asymmetry can be held to impugn the Hempelian thesis of symmetry. To deal with this question, it is fundamental to distinguish—as Rescher, Barker, Hanson, and Scriven unfortunately failed to do, much to the detriment of their theses —between the following two sets of ideas: First, an asymmetry between H-explanation and H-prediction both in regard to the grounds on which we claim to know that the *explanandum* is *true* and correlatively in regard to the *degree* of our *confidence* in the supposed truth of the *explanandum*, and second, an asymmetry, if any, between H-explanation and H-prediction with respect to the *logical relation* obtaining between the *explanans*

[5] C. G. Hempel: "Deductive-Nomological vs. Statistical Explanation," *op. cit.*, Sec. 6.

[6] M. Scriven: "Explanation and Prediction in Evolutionary Theory," *op. cit.*, p. 479.

[7] I. Scheffler: "Explanation, Prediction and Abstraction," *British Journal for the Philosophy of Science*, Vol. VII (1957), p. 293.

[8] N. Rescher: "On Prediction and Explanation," *op. cit.*, p. 282.

[9] *Ibid.*, p. 284.

and the *explanandum*. For the sake of brevity, we shall refer to the first asymmetry as pertaining to the "assertibility" of the *explanandum* while speaking of the second as an asymmetry in the "inferability" or "why" of the *explanandum*. In the light of this distinction, we shall be able to show that the existence of an epistemological asymmetry in regard to the assertibility of the *explanandum* cannot serve to impugn the Hempelian thesis of symmetry, which pertains to only the why of the *explanandum*.

If understood as pertaining to the *assertibility* per se of the *explanandum*, Rescher's contention of the existence of an epistemological asymmetry is indeed correct. For we saw in the first section of this chapter that there are *highly reliable* retrodictive indicators or records of past interactions but generally no spontaneously produced pre-indicators of corresponding future interactions. And this fact has the important consequence that while we can certify the *assertibility* or truth of an *explanandum* referring to a *past* interaction on the basis of a record without invoking the supposed truth of any (usual) *explanans* thereof, generally no pre-indicator but only the supposed truth of an appropriate *explanans* can be invoked to vouch for the assertibility or truth of the *explanandum* pertaining to a *future* interaction. And since the theory underlying our interpretations of records is confirmed better than are many of the theories used in an *explanans*, there is a very large class of cases in which an epistemological asymmetry or asymmetry of recordability does obtain with respect to the assertibility of the *explanandum*. But this asymmetry of *assertibility* cannot detract from the following *symmetry* affirmed by Popper and Hempel: To the same extent to which an *explanandum* referring to the past can be post-asserted *on the strength of its explanans* in an H-explanation, a corresponding *explanandum* referring to the future can be pre-asserted on the strength of the *same explanans* in an H-prediction. In other words, you can post-assert an *explanandum on the strength of its explanans* no better than you can pre-assert it.

The entire substance of both Barker's objection to Hempelian symmetry and of Hanson's (1959) critique of it is vitiated by the following fact: these authors adduced what they failed to recognize as a temporal asymmetry in the mere assertibility of the *explanandum* to claim against Hempel that there is a temporal asymmetry in the *why*. And they did so by citing cases in which they invoke a spurious contrast between the *non-assertibility* of an *explanandum* referring to the future and the *inductive inferability* of the corresponding *explanandum* pertaining to the past. Thus we find that Barker writes: "it can be correct to speak of explanation in many cases where specific prediction is not possible. Thus, for instance, if the patient shows all the symptoms of pneumonia, sickens and dies, I can then explain his death—I know what killed him—but I could not have definitely predicted in advance that he was going to die; for usually pneumonia fails to be fatal."[1] But all that Barker is entitled to here is the following claim, which is wholly compatible with Hempel's symmetry thesis: in many cases such as the pneumonia one, there obtains post-assertibility of the *explanandum* in the presence of the corpse but no corresponding pre-assertibility because of the asymmetry of spontaneous recordability. But this does not, of course, justify the contention that a past death, which did materialize and is reliably known from a record, can be explained by reference to earlier pneumonia any more conclusively than a future death can be inferred predictively on the basis of a present state of pneumonia. For the logical link between the *explanans* affirming a *past* state of pneumonia and the *explanandum* stating the recorded (known) death of a pneumonia patient is precisely the same inductive one as in the following case: the corresponding avowedly probabilistic *predictive* inference (H-prediction) of death on the basis of an *explanans* asserting a patient's *present* affliction with pneumonia.

It would seem that the commission of Barker's error of affirming an asymmetry in the *why* is facilitated by the following question-begging difference between the *explanans* used in his H-explanation of a death from pneumonia and the one used by him in the purportedly corresponding prediction: Barker's H-explanation of the past death employs an *explanans* asserting the onset of pneumonia at a past time as well as the sickening at a later past time, but the further condition of sickening is

[1] S. F. Barker: "The Role of Simplicity in Explanation," *op. cit.*, p. 271.

omitted from the antecedents of his corresponding H-prediction. Hence the spurious asymmetry of conclusiveness between the two cases.

It is now apparent that the valid core of Barker's statement is the commonplace that in the pneumonia case, as in others, post-assertibility of the *explanandum* does obtain even though pre-assertibility does not. And once it is recognized that the only relevant asymmetry which does obtain in cases of the pneumonia type is one of assertibility on grounds other than the usual *explanans,* the philosophical challenge of this asymmetry is to specify the complex reasons for it, as I have endeavored to do above. But no philosophical challenge is posed for Hempel's symmetry thesis.

An analogous confusion between the assertibility asymmetry and one in the *why* invalidates the paper by Hanson which Barker cites in support of his views. Suppose that a certain kind of past measurement yielded a particular Ψ-function which is then used in Schrödinger's equation for the H-explanation of a later past occurrence. And suppose also that the same kind of present measurement again yields the same Ψ-function for a like system and that this function is then used for the H-prediction of a correspondingly later future occurrence, which is of the same type as the past occurrence. It is patent that in quantum mechanics the *logical relation* between *explanans* (the function Ψ_1 and the associated set s_1 of probability distributions at the time t_1) and *explanandum* (the description of a particular micro-event falling within the range of one of the s_1 probability distributions) is no less statistical (inductive) in the case of H-explanation than in the case of H-prediction. And this *symmetry* in the *statistical why* is wholly compatible with the following asymmetry: the reliability of our knowledge *that* a specific kind of micro-event belonging to the range of one of the s_1 probability distributions has occurred in the past has no counterpart in our knowledge of the future occurrence of such an event, because only the results of past measurements (interactions) are available in records.

Hence it was wholly amiss for Hanson to have used the latter asymmetry of recordability as a basis for drawing a *pseudo-contrast* between the quantum mechanical *inferability* of a past micro-event—this inferability being logically identical with that of a future one—and the lack of *pre-assertibility* of the future occurrence of the micro-event. Says he: "any single quantum phenomenon P . . . can be completely *explained ex post facto;* one can *understand* fully just what kind of event occurred, in terms of the well-established laws of the . . . quantum theory. . . . But it is, of course, the most fundamental feature of these laws that the *prediction* of such a phenomenon P is, as a matter of theoretical principle, quite impossible."[2] Hanson overlooks that the asymmetry between pre-assertibility and post-assertibility obtaining in quantum mechanics in no way makes for an asymmetry between H-explanation and H-prediction with respect to the relation of the *explanandum* to its quantum mechanical *explanans.* And the statistical character of quantum mechanics enters only in the following sense: when coupled with the recordability asymmetry of classical physics, it makes for a temporal asymmetry in the assertibility of the *explanandum.*

We see that the statistical character of the quantum mechanical account of micro-phenomena is no less compatible with the symmetry between H-explanation and H-prediction than is the *deterministic* character of Newton's mechanics. And this result renders untenable what Hanson regards as the upshot of his 1959 paper on the symmetry issue, viz., "that there is a most intimate connection between Hempel's account of the symmetry between explanation and prediction and the logic of Newton's *Principia.*"[3]

It remains to deal in some detail with Scriven's extensive critique of Hempel's thesis. Scriven argues that (1) evolutionary explanations and explanations like that of the past occurrence of paresis due to syphilis fail to meet the symmetry requirement by not allowing corresponding predictions, (2) predictions based on mere *indicators* (rather than causes) such as the prediction of a storm from a sudden barometric drop are not matched by corresponding explanations, since indicators are not explanatory though they may serve to predict (or, in other cases, to retrodict). And, according to Scriven, these indicator-based predictions show that the mere inferability of an *explanandum* does not

[2] N. R. Hanson: "On the Symmetry Between Explanation and Prediction," *op. cit.*, pp. 353–54.

[3] *Ibid.*, p. 357.

guarantee scientific understanding of it, so that symmetry of inferability does not assure symmetry of scientific understanding between explanation and prediction.

I shall now examine several of the paradigm cases adduced by Scriven in support of these contentions.

I. *Evolutionary Theory.*

He cites evolutionary theory with the aim of showing that "Satisfactory explanation of the past is possible even when prediction of the future is impossible."[4]

Evolutionary theory does indeed afford valid examples of the epistemological asymmetry of assertibility. And this for the following two reasons growing out of Section A of the present chapter: First, the ubiquitous role of interactions in evolution brings the recordability asymmetry into play. And that asymmetry enters not only into the assertibility of the *explanandum.* For in cases of an H-prediction based on an *explanans* containing an *antecedent* referring to a *future* interaction, there is also an asymmetry of assertibility between H-prediction and H-explanation in regard to the *explanans.* And second, the existence of biological properties which are *emergent* in the sense that even if all the laws were strictly *deterministic,* the occurrence of these properties could not have been *predicted* on the basis of any and all laws which could possibly have been discovered by humans in advance of the first known occurrence of the respective properties in question. Thus, evolutionary theory makes us familiar with past biological changes which were induced by prior past interactions, the latter being post-assertible on the basis of present records. And these past interactions can serve to explain the evolutionary changes in question. But the logical relation between *explanans* and *explanandum* furnishing this explanation is completely *time-symmetric.* Hence this situation makes for asymmetry only in the following innocuous sense: since corresponding future interactions cannot be rationally preasserted—there being no advance records of them—there is no corresponding pre-assertibility of those future evolutionary changes that will be effected by future interactions.

[4] M. Scriven: "Explanation and Prediction in Evolutionary Theory," *op. cit.*, p. 477.

In an endeavor to establish the existence of an asymmetry damaging to Hempel's thesis on the basis of the account of a case of non-survival given in evolutionary theory, Scriven writes:

> there are . . . good grounds [of inherent unpredictability] for saying that even in principle explanation and prediction do *not* have the same form. Finally, it is not in general possible to list all the exceptions to a claim about, for example, the fatal effects of a lava flow, so we have to leave it in probability form; this has the result of eliminating the very degree of certainty from the prediction that the explanation has, when we find the fossils in the lava.[5]

But all that the lava case entitled Scriven to conclude is that the merely *probabilistic* connection between the occurrence of a lava flow and the extinction of certain organisms has the result of depriving pre-*assertibility* of the very degree of certainty possessed by post-assertions here. Scriven is not at all justified in supposing that *predictive inferability* in this case lacks even an iota of the certainty that can be ascribed to the corresponding post-explanatory inferability. For wherein does the greater degree of certainty of the post-explanation reside? I answer: only in the assertibility of the *explanandum,* not in the character of the logical relation between the *explanans* (the lava flow) and the *explanandum* (fatalities on the part of certain organisms). What then must be the verdict on Scriven's contention of an asymmetry in the *certainty* of prediction and post-explanation in this context? We see now that this contention is vitiated by a confusion between the following two radically distinct kinds of asymmetry: First, a difference in the degree of certainty (categoricity) of our knowledge of the truth of the *explanandum* and of the claim of environmental unfitness made by the *explanans,* and second, a difference in the "degree of entailment," as it were, linking the *explanandum* to the *explanans*.

Very similar difficulties beset Scriven's analysis of a case of biological survival which is accounted for on the basis of environmental fitness. He says:

> It is fairly obvious that no characteristics can be identified as contributing to "fitness" in all environments. . . . we cannot

[5] *Ibid.*, p. 480.

> predict which organisms will survive except in so far as we can predict the environmental changes. But we are very poorly equipped to do this with much precision.[6] . . . However, these difficulties of prediction do not mean that the idea of fitness as a factor in survival loses all of its *explanatory* power. . . . animals which happen to be able to swim are better fitted for surviving a sudden and unprecedented inundation of their arid habitat, and in some such cases it is just this factor which explains their survival. Naturally we could have said in advance that *if* a flood occurred, they would be likely to survive; let us call this a hypothetical probability prediction. But hypothetical predictions do not have any value for actual predictions except in so far as the conditions mentioned in the hypothesis are predictable. . . . hence there will be cases where we can *explain why* certain animals and plants survived even when we could not have *predicted that* they would.[7]

There would, of course, be complete agreement with Scriven, if he had been content to point out in this context, as he does, that there are cases in which we can "explain why" but not "predict that." But he combines this correct formulation with the incorrect supposition that cases of post-explaining survival on the basis of fitness constitute grounds for an indictment of Hempel's thesis of symmetry. Let me therefore state the points of agreement and disagreement in regard to this case as follows. Once we recognize the ubiquitous role of *interactions* we can formulate the valid upshot of Scriven's observations by saying: insofar as future fitness and survival depend on future interactions which cannot be predicted from given information, whereas past fitness and survival depended on past interactions which *can* be retrodicted from that same information, there is

---

[6] The environmental changes which Scriven goes on to cite are all of the nature of *interactions* of a potentially open system. And it is this common property of theirs which makes for their role in precluding the predictability of survival.

[7] *Ibid.*, p. 478. In a recent paper "Cause and Effect in Biology" (*Science*, Vol. CXXXIV [1961], p. 1504), the zoologist E. Mayr overlooks the fallacy in Scriven's statement which we are about to point out and credits Scriven with having "emphasized quite correctly that one of the most important contributions to philosophy made by the evolutionary theory is that it has demonstrated the independence of explanation from prediction." And Mayr rests this conclusion among other things on the contention that "The theory of natural selection can describe and explain phenomena with considerable precision, but it cannot make reliable predictions."

an epistemological asymmetry between H-explanation and H-prediction in regard to the following: the assertibility both of the antecedent fitness affirmed in the *explanans* and of the *explanandum* claiming survival.

This having been granted as both true and illuminating, we must go on to say at once that the following considerations—which Scriven can grant only on pain of inconsistency with his account of asymmetry in the lava case—are no less true: the scientific inferability from a cause and hence our understanding of the *why* of survival furnished by an *explanans* which *does* contain the antecedent condition that the given animals are able to swim during a sudden, unprecedented inundation of their arid habitat is *not* one iota more probabilistic (i.e., *less* conclusive) in the case of a *future* inundation and survival than in the case of a *past* one. For if the logical nerve of intelligibility linking the *explanans* (fitness under specified kinds of inundational conditions) with the *explanandum* (survival) is only probabilistic in the future case, how could it possibly be any less probabilistic in the past case? It is evident that post-explanatory inductive inferability is entirely on a par here with predictive inferability from fitness as a cause. Why then does Scriven feel entitled to speak of "probability prediction" of *future* survival *without also* speaking of "probability explanation" of past survival? It would seem that his reason is none other than the *pseudo-contrast* between the *lack* of pre-assertibility of the *explanandum* (which is conveyed by the term "probability" in "probability prediction") with the obtaining of post-explanatory inductive inferability of the *explanandum.* And this pseudo-contrast derives its plausibility from the tacit appeal to the bona fide asymmetry between the pre-assertibility and post-assertibility of the *explanandum,* an asymmetry which cannot score against the Popper-Hempel thesis.

II. *The Paresis Case.*

In a further endeavor to justify his repudiation of Hempel's thesis, Scriven says:

> we can explain but not predict whenever we have a proposition of the form "The only cause of X is A" (I)—for example, "The only cause of paresis is syphilis." Notice that this is perfectly compatible with the statement that A is often not followed by

> X–in fact, very few syphilitics develop paresis (II). Hence, when A is observed, we can predict that X is *more* likely to occur than without A, but still extremely unlikely. So, we must, on the evidence, still predict that it will *not* occur. But if it does, we can appeal to (I) to provide and guarantee our explanation. . . . Hence an event which cannot be predicted from a certain set of well-confirmed propositions can, if it occurs, be explained by appeal to them.[8]

In short, Scriven's argument is that although a past case of paresis can be explained by noting that syphilis was its cause, one cannot predict the future occurrence of paresis from syphilis as the cause. And he adds to this the following further comment:

> Suppose for the moment we include the justification of an explanation or a prediction in the explanation or prediction, as Hempel does. From a general law and antecedent conditions we are then entitled to deduce that a certain event will occur in the future. This is the deduction of a prediction. From one of the propositions of the form the only possible cause of y is x and a statement that y has occurred we are able to deduce, not only that x must have occurred, but also the proposition the cause of y in this instance was x. I take this to be a perfectly sound example of deducing and explanation. Notice however, that what we have deduced is not at all a description of the event to be explained, that is we have not got an *explanandum* of the kind that Hempel and Oppenheim envisage. On the contrary, we have a specific causal claim. This is a neat way of making clear one of the differences between an explanation and a prediction; by showing the different kinds of proposition that they often are. When explaining Y, we do not have to be able to deduce that Y occurs, for we typically know this already. What we have to be able to deduce (if deduction is in any way appropriate) is that Y occurred *as a result of* a certain X, and of course this needs a very different kind of general law from the sort of general law that is required for prediction.[9]

---

[8] M. Scriven: "Explanation and Prediction in Evolutionary Theory," *op. cit.*, p. 480.

[9] M. Scriven: "Comments on Professor Grünbaum's Remarks at The Wesleyan Meeting," *Philosophy of Science,* Vol. XXIX (1962), pp. 173–74. For Scriven's most recent criticism of my views, see his "The Temporal Asymmetry of Explanations and Predictions," in: B. Baumrin (ed.) *Philosophy of Science* (New York: John Wiley and Sons; 1963), Vol. I, pp. 97–105.

I shall now show that Scriven's treatment of such cases as post-explaining paresis on the basis of syphilis suffers from the same defect as his analysis of the evolutionary cases: *Insofar as there is an asymmetry, Scriven has failed to discern its precise locus, and having thus failed, he is led to suppose erroneously that Hempel's thesis is invalidated by such asymmetry as does obtain.*

Given a particular case of paresis as well as the proposition that the only cause of paresis is syphilis—where a "cause" is understood here with Scriven as a "contingently necessary condition"—what can be inferred? Scriven maintains correctly that what follows is that both the paretic concerned had syphilis and that in his particular case, syphilis was the cause in the specified sense of "cause." And then Scriven goes on to maintain that his case against Hempel is established by the fact that we are able to assert that syphilis *did cause* paresis while *not* also being entitled to say that syphilis *will cause* paresis. But Scriven seems to have completely overlooked that our not being able to make both of these assertions does not at all suffice to discredit Hempel's thesis, which concerns the time-symmetry of the *inferability* of the *explanandum* from the *explanans*. The inadequacy of Scriven's argument becomes evident the moment one becomes aware of the reason for not being entitled to say that syphilis "will cause" paresis though being warranted in saying that it "did cause" paresis.

The sentences containing "did cause" and "will cause" respectively each make two affirmations as follows: First, the assertion of the *explanandum* (paresis) per se, and second, the affirmation of the obtaining of a causal relation (in the sense of being a contingently necessary condition) between the *explanans* (syphilis) and the *explanandum* (paresis). Thus, for our purposes, the statement "Syphilis *will cause* person Z to have paresis" should be made in the form "Person Z will have paresis *and* it will have been caused by syphilis," and the statement "Syphilis *did cause* person K to have paresis" becomes "Person K has (or had) paresis *and* it was caused by syphilis." And the decisive point is that in so far as a *past* occurrence of paresis can be inductively *inferred from* prior syphilis, so also a future occurrence of paresis can be. For the causal relation or connection between syphilis and paresis is incontestably time-symmetric:

precisely in the way and to the extent that syphilis *was* a necessary condition for paresis, it also *will* be! Hence the only bona fide asymmetry here is the record-based but innocuous one in the *assertibility* of the *explanandum* per se, but there is no asymmetry of *inferability* of paresis from syphilis. The former innocuous asymmetry is the one that interdicts our making the predictive assertion "will cause" while allowing us to make the corresponding post-explanatory assertion "did cause." And it is this fact which destroys the basis of Scriven's indictment of Hempel's thesis. For Hempel and Oppenheim did not maintain that an *explanandum* which can be post-asserted can always also be pre-asserted; what they did maintain was only that the *explanans* never post-explains any better or more conclusively than it implies predictively, there being complete symmetry between post-explanatory *inferability* and predictive *inferability* from a given *explanans*. They and Popper were therefore fully justified in testing the adequacy of a proffered *explanans* in the social sciences on the basis of whether the post-explanatory inferability of the *explanandum* which was claimed for it was matched by a corresponding predictive inferability, either inductive or deductive as the case may be.

What is the force of the following comment by Scriven: in the post-explanation of paresis we do not need to infer the *explanandum* from the *explanans* à la Hempel and Oppenheim, because we know this already from prior records (observations) of one kind or another; what we do need to infer instead is that the *explanandum*-event occurred *as a result of* the cause (necessary condition) given by the *explanans,* an inference which does not allow us to *predict* (i.e., pre-assert) the *explanandum*-event? This comment of Scriven's proves only that here there is record-based post-assertibility of paresis but no corresponding pre-assertibility.

In short, Scriven's invocation of the paresis case, just like his citation of the cases from evolutionary theory, founders on the fact that he has confused an *epistemological* asymmetry with a *logical* one. To this charge, Scriven has replied irrelevantly that he has been at great pains in his writings—as for example in his discussion of the barometer case which I shall discuss below—to distinguish valid arguments based on true premises which do qualify as scientific explanations from those which do not so qualify. This reply is irrelevant, since Scriven's caveat against identifying (confusing) arguments based on true premises which are both valid and explanatory with those which are valid without being explanatory does not at all show that he made the following crucial distinction here at issue: the distinction between (1) a difference (asymmetry) in the assertibility of either a conclusion (*explanandum*) or a premise (*explanans*), and (2) a difference (asymmetry) in the inferability of the *explanandum* from its *explanans.* Although the distinction which Scriven does make cannot serve to mitigate the confusion with which I have charged him, his distinction merits examination in its own right.

To deal with it, I shall first consider examples given by him which involve *non*-predictive valid deductive arguments to which he denies the status of being explanatory arguments. And I shall then conclude my refutation of Scriven's critique of Hempel's thesis by discussing the following paradigm case of his: the deductively valid *predictive* inference of a storm from a sudden barometric drop, which he adduces in an endeavor to show that such a valid deductive inference could not possibly qualify as a post-*explanation* of a storm.

It would be agreed on all sides, I take it, that no *scientific understanding* is afforded by the deduction of an *explanandum* from itself even though such a deduction is a species of valid inference. Hence it can surely be granted that the class of valid deductive arguments whose conclusion is an *explanandum* referring to some event or other is wider than the class of valid deductive arguments affording scientific understanding of the *explanandum*-event. But it is a quite different matter to claim, as Scriven does, that no *scientific understanding* is provided by those valid deductive arguments which ordinary usage would not allow us to call "explanations." For example, Scriven cites the following case suggested by S. Bromberger and discussed by Hempel:[1] the height of a flagpole is deducible from the length of its shadow and a measurement of the angle of the sun taken in conjunction with the principles of geometrical optics, but the height of the flagpole could not thereby be said to have been

[1] Cf. C. G. Hempel: "Deductive-Nomological vs. Statistical Explanation," *op. cit.,* Sec. 4.

"explained." Or take the case of a rectilinear triangle in physical space for which Euclidean geometry is presumed to hold, and let it be given that two of the angles are 37° and 59° respectively. Then it can be deductively inferred that the third angle is one of 84°, but according to Scriven, this would not constitute an explanation of the magnitude of the third angle.

Exactly what is shown by the flagpole and angle cases concerning the relation between valid deductive arguments which furnish scientific understanding and those which, according to ordinary usage, would qualify as "explanations"? I maintain that while differing in one respect from what are usually called "explanations," the aforementioned valid deductive arguments yielding the height of the flagpole and the magnitude of the third angle provide scientific understanding no less than "explanations" do. And my reasons for this contention are the following.

In the flagpole case, for example, the *explanandum* (stating the height of the flagpole) can be deduced from two different kinds of premises: First, an *explanans* of the type familiar from geometrical optics and involving laws of coexistence rather than laws of succession, antecedent events playing no role in the *explanans,* and second, an *explanans* involving causally antecedent events and laws of succession and referring to the temporal genesis of the flagpole as an artifact. But is this difference between the kinds of premises from which the *explanandum* is deducible a basis for claiming that the coexistence-law type of *explanans* provides less *scientific understanding* than does the law-of-succession type of *explanans*? I reply: certainly not. And I hasten to point out that the difference between *pre*-axiomatized and axiomatized geometry conveys the measure of the scientific understanding provided by the geometrical account given in the flagpole and angle cases on the basis of laws of coexistence. But is it not true after all that ordinary usage countenances the use of the term "explanation" only in cases employing causal antecedents and laws of succession in the *explanans*? To this I say: this *terminological* fact is as unavailing here as it is philosophically unedifying.

Finally, we turn to Scriven's citation of cases of deductively valid predictive inferences which, in his view, invalidate Hempel's thesis because they could not possibly also qualify as post-explanations.

III. *The Barometer Case.*

Scriven writes:

> What we are trying to provide when making a prediction is simply a claim that, at a certain time, an event or state of affairs will occur. In explanation we are looking for a cause, an event that not only occurred earlier but stands in a special relation to the other event. Roughly speaking, the prediction requires only a correlation, the explanation more. This difference has as one consequence the possibility of making predictions from indicators other than causes—for example, predicting a storm from a sudden drop in the barometric pressure. Clearly we could not say that the drop in pressure in our house caused the storm: it merely presaged it. So we can sometimes predict what we cannot explain.[2]

Other cases of the barometer type are cases such as the presaging of mumps by its symptoms and the presaging of a weather change by rheumatic pains.

When we make a predictive inference of a storm from a sudden barometric drop, we are inferring an effect of a particular cause from another (earlier) presumed effect of that same cause. Hence the inference to the storm is not from a cause of the storm but only from an indicator of it. And the law connecting sudden barometric drops to storms is therefore a law affirming only an indicator type of connection rather than a causal connection.

The crux of the issue here is whether we have no scientific understanding of phenomena on the strength of their deductive inferability from indicator laws (in conjunction with a suitable antecedent condition), scientific understanding allegedly being provided only by an *explanans* making reference to one or more causes. If that were so, then Scriven could claim that although the mere inferability of particular storms from specific sudden barometric drops is admittedly time-symmetric, there is no time-symmetry in positive scientific understanding. It is clear from the discussion of the flagpole case that the terminological practice of restricting the term "explanation" though not the term "prediction" to cases in which the *explanans* makes reference to a partial or total cause rather than to a mere indicator cannot

[2] M. Scriven: "Explanation and Prediction in Evolutionary Theory," *op. cit.*, p. 480.

settle the questions at issue, which are: would the type of argument which yields a *prediction* of a future *explanandum*-event (storm) from an indicator-type of premise furnish any scientific understanding, and, if so, does this type of argument provide the same positive amount of scientific understanding of a corresponding *past* event (storm)?

These questions are, of course, not answered in the negative by pointing out correctly that the law connecting the cause of the storm with the storm can serve as a *reason* for the weaker indicator-law. For this fact shows only that the causal law can account for both the storm and the indicator law, but it does not show that the indicator law cannot provide any scientific understanding of the occurrence of particular storms. To get at the heart of the matter, we must ask what distinguishes a causal law from an indicator law such that one might be led to claim, as Scriven does, that subsumption under indicator laws provides no scientific understanding at all, whereas subsumption under causal laws does.

Let it be noted that a causal law which is used in an *explanans* and is not itself derived from some wider causal law is fully as *logically contingent* as a mere indicator law which is likewise not derived from a causal law but is used as a premise for the deduction of an *explanandum* (either predictively or post-dictively, i.e., H-explanatorily). Why then prefer (predictive or post-dictive) subsumption of an *explanandum* under a causal law to subsumption under a mere indicator law? The justification for this preference would seem to lie not merely in the greater generality of the causal law; it also appears to rest on the much larger variety of empirical contingencies which must be ruled out in the *ceteris paribus* clause specifying the relevant conditions under which the indicator law holds, as compared to the variety of such contingencies pertaining to the corresponding causal law.[3] But this difference in both generality and in the variety of contingencies does not show that the indicator-law

[3] For example, in addition to all the things that can interfere with the materializing of a storm when its avowed total cause (a pressure drop over a wide area) is presumed to be present, there are a variety of further contingencies such that the occurrence of any one of them will eventuate in the *absence* of a storm when we observe a sudden barometric drop at one place: this sudden drop may arise from a *narrowly local* pressure-lowering cause (device) in the immediate spatial vicinity of the barometer. And

provides no scientific understanding of particular phenomena subsumable under it; it shows only, so far as I can see, that one might significantly speak of degrees of scientific understanding. And this conclusion is entirely compatible with the contention required by the symmetry thesis that the barometric indicator law furnishes the same positive amount of scientific understanding of a past storm as of a future one predicted by it.

I believe to have shown, therefore, that with respect to the symmetry thesis, *Hempel ab omni naevo vindicatus.*[4]

(D) THE CONTROVERSY BETWEEN MECHANISM AND TELEOLOGY.

The results of our discussion of the temporal asymmetry of recordability have a decisive bearing on the controversy between mechanism and teleology.

---

hence, in that case, the sudden barometric drop would not all betoken the presence of a pressure drop over an area sufficiently large to eventuate in a storm. Similarly, the occurrence of what appear to be symptoms of mumps may not betoken the presence of the filter-passing virus which can cause this disease; instead these presumed symptoms may arise from any one of a number of other kinds of causes, no one of which would issue in mumps. These other kinds of causes must be ruled out, *in addition* to those entering the *ceteris paribus* clause of the law relating the mumps to *its cause,* if the *symptoms* of mumps are to serve as a reliable basis for inferring the subsequent onset of this disease.

Although the distinction between a causal law (C-law) and an indicator law (I-law) seems to be sufficiently clear in the examples adduced by Scriven, Professor Richard Rudner has suggested in private correspondence that, in view of the well-known difficulties besetting the characterization of a C-law as such, we should be leery of supposing that a generally clear and tenable distinction between C-laws and I-laws has been drawn. In the barometer and mumps cases, we distinguish the initial total cause from a mere indicator by pointing out that the indicator is itself a partial effect of the initial total cause. If the difficulties besetting the general characterization of a C-law are actually such that this distinguishing criterion will not do and such that no other viable criterion seems to be in sight, then my remarks about our preference for subsumption of an *explanandum* under a C-law rather than an I-law would lose generality and would have to be restricted to specific examples like those cited by Scriven. It is clear that Scriven's argument here depends for its very statement on the tenability of the distinction between C-laws and I-laws. And if that distinction were indeed fundamentally untenable, this fact alone would suffice to refute Scriven's argument.

[4] Believing (incorrectly) to have cleansed Euclid of all blemish, G. Saccheri (1667–1733) published a book in 1733 under the title *Euclides ab omni naevo vindicatus.*

By mechanism we understand the philosophical thesis that all explanation must be *only a tergo,* i.e., that occurrences at a time t can be explained *only* by reference to *earlier* occurrences and *not also* by reference to later ones.[5] And by teleology we understand a thesis which is the contrary rather than the contradictory of mechanism: all phenomena belonging to a specified domain and occurring at a time t are to be understood by reference to *later* occurrences only. We note that, thus understood, mechanism and teleology can both be false.

During our post-Newtonian epoch there is a misleading incongruity in using the term "mechanism" for the thesis of the monopoly of *a tergo* explanations. For in the context of the *time-symmetric* laws of Newton's mechanics, the given state of a closed mechanical system at a time t can be inferred from a state *later* than t (i.e., *retrodicted*) no less than the given state can be inferred from a state *earlier* than t (i.e., *predicted*). Instead of furnishing the prototype for mechanistic explanation in the philosophical sense, the phenomena described by the time-symmetric laws of Newton's mechanics constitute a domain with respect to which both mechanism and teleology are false, thereby making the controversy between them a pseudo-issue. More generally, that controversy is a pseudo-issue with respect to any domain of phenomena constituted by the evolution of closed systems obeying time-symmetric laws, be they deterministic or statistical.

But there is indeed a wide class of phenomena with respect to which mechanism is true. And one may presume that tacit reference to this particular class of phenomena has conferred plausability on the thesis of the *unrestricted* validity of mechanism: traces or marks of interaction existing in a system which is essentially closed at a time t are accounted for scientifically by *earlier interactions* or *perturbations* of that system—which are called "causes"—and not by later interactions of the system. Thus, we

[5] A weaker version of mechanism might be the thesis that all phenomena can be understood by reference to earlier occurrences, a thesis which *allows* that understanding might be provided by reference to later occurrences as well. This weaker and less influential version of mechanism, though likewise incompatible with teleology, is not, however, the one whose assessment is illuminated by the asymmetry of recordability. Hence I discuss only the stronger, though more vulnerable, version of mechanism here.

explain a scar on a person by noting that he did sustain an injury in the past, not by claiming that he will suffer injury in the future.

In view of the demonstrated *restricted* validity of mechanism, we must therefore deem the following statement by H. Reichenbach as too strong: "We conclude: If we define the direction of time in the usual sense, there is no finality, and only causality is accepted as constituting explanation."[6]

---

[6] H. Reichenbach: *DT, op. cit.*, p. 154.

Chapter 10

IS THERE A "FLOW" OF TIME OR TEMPORAL "BECOMING"?

It is clear that the anisotropy of time resulting from the existence of irreversible processes consists in the mere structural differences between the two opposite senses of time but provides no basis at all for singling out *one* of the two opposite senses as "*the* direction" of time. Hence the assertion that irreversible processes render time anisotropic is *not at all* equivalent to such statements as "time flows one way."

Thus we must clearly distinguish the *anisotropy* of physical time from the feature of common sense (psychological) time which is rendered by such terms as "the transiency of the Now" or "becoming" and by such metaphors as the "flux," "flow" or "passage" of time. And our concern in the present chapter is to determine the factual credentials, if any, of the concepts which are represented by these terms.

Since the instants of anisotropic time are ordered by the relation "earlier than" no less than by the converse relation "later than," the anisotropy of time provides no warrant at all for singling out the "later than" sense as "the" direction of time. Instead, the inspiration for speaking about "the" direction of time derives from the supposition that there is a transient "now" or "present" which can be claimed to shift so as to single out the future direction of time as the sense of its "advance."

The transient or shifting division of the time continuum into the *past* and the *future* depends on the transient Now and is *not* furnished solely by the "static" relation of "earlier than" or its converse "later than" with respect to which we formulated the anisotropy of time. The obtaining of the relation "earlier than" or "later than" between two physical events or states does not, of course, depend at all on the transient Now.[1] Thus the year 1910 is earlier than the year 1920, no less than 1950 is earlier than 1970, and than 1970, in turn, is earlier than 2850. And yet, at this writing in 1962, which is at the focus of my immediate experience, the events of 1970 and 2850 belong to the future while those of the other years we mentioned belong to the past such that the Now within 1962 on which this classification depends is not an arbitrarily chosen time of reference. For the past is the class of events earlier than those events which constitute the Now in the sense that the past is constituted by the events which "no longer exist" while those of the Now or present "actually exist." And the future is correspondingly the set of events that are later than now in the sense that they are yet to "acquire existence," as it were, or to "come *into* being." Thus, writing about himself in 1925, Reichenbach speaks of having "the feeling that *my* existence is a reality, while *Plato's* life merely still casts its shadows onto reality" and declares that we are unable to escape "the compulsion which distinguishes for us a Now-point in an absolute way as the experience of the divide between past and future."[2]

The transience of the Now is a feature of psychological (and common sense) time in the sense that there is a *diversity* of the Now-contents of immediate awareness. Hence it is a matter of fact that the Now "shifts" in conscious awareness to the extent that there is a *diversity* of the Now-contents, and it is likewise a fact that the Now-contents are temporally *ordered.* But since these diverse Now-contents are ordered with respect to the relation "earlier than" no less than with respect to its converse "later

[1] This independence *cannot,* of course, be gainsaid by the following consideration from the special theory of relativity (cf. Chapter Twelve): in the case of those and only those particular pairs of events which are *not* linkable by causal chains, the usual choice of a definition of simultaneity made in that theory leads to a dependence *on the inertial system* of the conventional ascription of the relations "later than" and "simultaneous with."

[2] H. Reichenbach: "Die Kausalstruktur der Welt und der Unterschied von Vergangenheit und Zukunft," *op. cit.,* pp. 133–75, esp. p. 140.

than," it is a mere *tautology* to say that the Now shifts *from* earlier *to* later. For this metaphorical affirmation of shifting *in the future direction along the time-axis* tells us no more than that later Nows are later than earlier ones, just as earlier Nows are earlier than later ones! By the same token, the assertion that the "flow" of time is *unidirectional* is a tautology, as is the claim that time "flows" from the past to the future. The factual emptiness of the latter formulations must not tempt us into overlooking the following: a *non*-directional or directionally *neutral* claim of the transiency of the Now and of the temporal order of the various "Now"-contents does codify factual truths pertaining to psychological (common sense) time. And we shall see in some detail later on that because of its inherent dependence on consciousness, the transiency of the Now is not also a feature of physical time.

These considerations now enable us to evaluate the following contentions by Reichenbach:

> We have to distinguish here between two problems. First, the procedure described leads to an *order* of time, in the same sense in which the points on a line are ordered. Such a series of points has two directions, neither of which has any distinguishing characteristic. Temporal order, too, has two directions, the direction to earlier and the direction to later events, but in this case one of the two directions has a distinguishing characteristic: time flows from the earlier to the later event. Time therefore represents not only an ordered series generated by an asymmetrical relation, but is also *unidirectional*. This fact is usually ignored. We often say simply: the direction from earlier to later events, from cause to effect, is the direction of the progress of time. However, in this form the assertion is empty unless we specify what "progress of time" means. In the same fashion we could say that the points on a line progress from left to right; but this assertion is empty, since the progress of points means here nothing but the progressing in the selected direction. When we speak about the progress of time, in contrast, we intend to make a synthetic assertion which refers both to an immediate experience and to physical reality. This particular problem can only be solved if we can formulate the content of the assertion more precisely. We shall leave this problem for the time being and content ourselves with the con-

> clusion that the direction which we have defined as earlier-later is the same direction as that of the progress of time. For the problems dealt with in the theory of relativity it suffices that there exists a serial *order* of time, i.e., that we can distinguish between two directions which are opposite to each other.[3]

We saw that the anisotropy of time lies in the fact that *each* of the two opposite directions of time has a distinguishing characteristic. But Reichenbach believes that the transiency of the Now entitles him to maintain the following: "one of the two directions has a distinguishing characteristic: time flows from the earlier to the later event" and is "*uni*directional" in this sense. Discerning clearly that the charge of tautology might be leveled against the assertion that the direction from earlier to later is "the direction of the progress of time," Reichenbach contends that this assertion can be synthetic. And his reason is that the assertion renders the content of "an immediate experience" as well as an objective feature of "physical reality." Deferring until later our appraisal of the relevance of the transient Now to *physical* as distinct from common sense (psychological) time, we see that Reichenbach overlooked a fundamental point here: it is only the obtaining of a *diversity* of Now-contents which is a matter of fact, but not the allegedly "unidirectional" character of "the progress of time." For, as we saw earlier, the factuality of the diversity of the Now-contents does not suffice to give synthetic content to the assertion that time progresses unidirectionally from earlier to later. And Reichenbach's allegation here of the synthetic character of this assertion is avowedly left unsubstantiated by him here.

We shall now turn to a critical analysis of those of his writings in which he attempted to justify his cited contention that a Now and its associated transient division between the past and the future is a feature of *physical* time or "physical reality" no less than of psychological (common sense) time.

Having noted that the concept of the purported unidirectional progress of time has not found a place within the theory of relativity because "it suffices that there exists a serial *order* of time"[4]

---

[3] H. Reichenbach: *PST, op. cit.*, pp. 138–39. Reprinted through permission of the publisher.

[4] *Ibid.*, p. 139.

for the problems dealt with in that theory, Reichenbach deems that theory's Minkowskian world picture incomplete as follows:

> A topology of time can be constructed in which the basic concepts "earlier," "later" and "simultaneous" are defined. But what could *not* be solved in this way so far is the problem of the "now."
>
> What does "now" mean? Plato lived before me, and Napoleon VII will live after me. But which one of these three lives *now*? I undoubtedly have a clear feeling that *I* live now. But does this assertion have an objective significance beyond my subjective experience?[5]
>
> . . . In the condition of the world, a cross-section called the present is distinguished; the "now" has an objective significance. Even when no human being is alive any longer, there is a "now"; the "present state of the planetary system" is then just as much a determinate specification as "the state of the planetary system at the time of the birth of Christ."
>
> In the four-dimensional picture of the world, such as used by the theory of relativity, there is no such distinguished cross-section. But this is due only to the fact that an essential content is omitted from this picture.[6]

That the theory of relativity does not make any allowance for the *transient* Now of common sense time is indeed correct. For the "Now" in the "Here-Now" of the Minkowski diagram designates no more than a kind of *arbitrary* zero or origin of temporal coordinates: we can make use of the Minkowski diagram at noon on June 1, 1962 to let "Here-Now" designate a certain event occurring after the extinction of the sun. And the "Absolute Past" and the "Absolute Future" are no more than the set of events respectively absolutely earlier and later than the event *arbitrarily* designated as the "Here-Now." Instead of allowing for the transient division of time into the past and future by the shifting Now of experienced time, the theory of relativity conceives of events as simply being and sustaining relations of earlier and later, but not as "coming *into* being": we conscious organisms then "come across" them by "entering" into their absolute future,

---

[5] H. Reichenbach: "Die Kausalstruktur der Welt und der Unterschied von Vergangenheit und Zukunft," *op. cit.*, p. 139.

[6] *Ibid.*, p. 141.

as it were. And upon experiencing their immediate effects, we regard them as "taking place" or "coming into being."[7]

Like Eddington[8] before him and G. J. Whitrow[9] after him, Reichenbach supposed that the exclusion of the concept of the unidirectional progress of time by a physical theory is attributable to the *deterministic* character of that theory. And believing—quite erroneously as we shall see—that an *indeterministic* physics can provide a *physical* basis for the transient Now, he attempted to find an objective physical basis for the present or Now in his paper of 1925[1] as follows: he gave a *probabilistic* interpretation of causality according to which the past is "objectively determined," while the future is "objectively undetermined" in virtue of the non-existence of a complete set of partial causes, a knowledge of which would render our predictions certain. In this context he conceived the present *without* the inadmissible use of absolute simultaneity as the class of events not causally connectible with the particular "now." But (as is evident from the world-line diagram in Chapter 12, p. 354), this characterization yields a conception of the present which differs from the "Now" of conscious experience in the following essential respect: a given event E_2 at a point P_2 in space will remain simultaneous for an observer at the distant point P_1 *throughout* a time interval constituted by a continuum of events having a space-like separation from E_2! It is because Reichenbach maintained that on the basis of determinism, "the morrow has already occurred today in the same sense as yesterday has," thus purportedly making nonsense of all our planning, that he rejected determinism in 1925 just before the full advent of quantum mechanics and sought a physical basis for the present and thereby for becoming. Just before his death in 1953, he argued[2]

[7] Cf. A. S. Eddington: *The Nature of the Physical World, op. cit.*, p. 68, and *Space, Time and Gravitation, op. cit.*, p. 51. See also E. Cassirer: *Zur Einsteinschen Relativitätstheorie* (Berlin: Bruno Cassirer Verlag; 1921), pp. 120–21.

[8] A. S. Eddington: *Space, Time and Gravitation, op. cit.*, p. 51.

[9] G. J. Whitrow: *The Natural Philosophy of Time, op. cit.*, p. 295.

[1] H. Reichenbach: "Die Kausalstruktur der Welt und der Unterschied von Vergangenheit und Zukunft," *op. cit.*, pp. 141–43.

[2] H. Reichenbach: *DT, op. cit.*, pp. 211–24 and "Les Fondements Logiques de la Mécanique des Quanta," *op. cit.*, pp. 154–57.

that the micro-indeterminism of quantum mechanics is not an ephemeral *pis-aller* of present-day physical theory and extended his early views by attempting to utilize the indeterminacies of quantum mechanics. He writes:

> Let us suppose that consecutive measurements are made alternately of two noncommuting [i.e., complementary] quantities. One will obtain a series of macroscopic events which one cannot predict, but which one can record. This series provides us with a clear distinction between the past and the future: the past is determined, but the future is not. . . . The analysis of classical physics has shown us that one can record the past but not the future. The combination of this result with Heisenberg's uncertainty leads us to the consequence that one can know the past but that one cannot predict the future. . . .
>
> . . . Modern science . . . furnishes us with precisely the difference between the past and the future, which Laplace's physics could not recognize.
>
> To be sure, Boltzmann's physics, if coupled with the hypothesis of the branch structure, yields a certain structural difference between the past and the future. . . . But while this difference enabled us to distinguish between the past and the future, it was not associated with a difference in determination: although one cannot record the future, one could predict it on the basis of the totality of causes. Thus, one cannot call the future undetermined. . . .
>
> It is no longer that way in quantum physics. . . . Here is the difference: there are future facts which cannot possibly be predicted, whereas there are no past facts which it would be impossible to know. In principle, they can always be recorded. . . . The distinction between the indeterminateness ("l'indéterminisme") of the future and the determinateness ("détermination") of the past has, in the final analysis, found expression in the laws of physics. . . . The concept of "becoming" acquires significance in physics: the present, which separates the future from the past, is the moment at which that which was undetermined becomes determined, and "becoming" has the same meaning as "becoming determined." . . .
>
> . . . The term "determination" denotes a relation between two situations A and B; the situation A does or does not determine the situation B. It is meaningless to say that the situation B, considered by itself, is determined. If we say that the past is determined or that the future is undetermined, it is tacitly

> understood that we are relating this to the present situation; it is with respect to "now" that the past is determined and that the future is not.[3]

Reichenbach's view here is shared by Eddington, who writes:

> The division into past and future (a feature of time-order which has no analogy in space-order) is closely associated with our ideas of causation and free-will. In a perfectly determinate scheme the past and future may be regarded as lying mapped out—as much available to present exploration as the distant parts of space. Events do not happen; they are just there, and we come across them. "The formality of taking place" is merely the indication that the observer has on his voyage of exploration passed into the absolute future of the event in question; and it has no important significance.[4]

In the same vein, the astronomer H. Bondi contends that "In a theory with indeterminacy . . . the passage of time transforms statistical expectation into real events."[5] And G. J. Whitrow claims[6] that "There is indeed a profound connection between the reality of time and the existence of an incalculable element in the universe." If Reichenbach, Eddington, Bondi, Whitrow and others had merely maintained that indeterminacy makes for our human inability to know in advance of their actual occurrence what particular kinds of events will in fact materialize, then, of course, there could be no objection. But I take them to have claimed that the *existential* status of future events in an indeterministic world is that of *coming into being* with time, whereas in a deterministic world it is one of simply being.

I believe that the issue of determinism *vs.* indeterminism is *totally irrelevant* to whether becoming is a significant attribute of the time of physical nature independently of human consciousness. And I wish to explain now why I regard the thesis of Reichenbach, Eddington, Bondi, Whitrow, and of many others that indeterminism confers flux onto physical time as untenable. I have given my reasons for likewise rejecting Reichenbach's

[3] H. Reichenbach: "Les Fondements Logiques de la Mécanique des Quanta," *op. cit.*, pp. 154–57.

[4] A. S. Eddington: *Space, Time and Gravitation, op. cit.*, p. 51.

[5] H. Bondi: "Relativity and Indeterminacy," *Nature*, Vol. CLXIX (1952), p. 660.

[6] G. J. Whitrow: *The Natural Philosophy of Time, op. cit.*, p. 295.

further claim that "The paradox of determinism and planned action is a genuine one"[7] in other publications.[8]

In the indeterministic quantum world, the relations between the sets of measurable values of the state variables characterizing a physical system at different times are, in principle, not the one-to-one relations linking the states of classically behaving closed systems. But this holds for a given state of a physical system and its absolute future quite independently of whether that state occurs at midnight on December 31, 1800 or at noon on March 1, 1984. Moreover, if we consider *any one* of the temporally successive regions of space-time, we can assert the following: the events belonging to its particular absolute past could be (more or less) uniquely specified in records which are a part of that region, whereas its particular absolute future is thence quantum mechanically unpredictable. Accordingly, every "now," be it the "now" of Plato's birth or that of Reichenbach's, always constitutes a divide in Reichenbach's sense between its own recordable past and its unpredictable future, thereby satisfying Reichenbach's definition of the "present." But this fact is fatal to his avowed aim of providing a physical basis for a "unique," transient "now" and thus for "becoming."[9] Reichenbach's recent characterization of the determinacy of the past as recordability as opposed to the quantum mechanical indeterminacy of the future can therefore not serve to vindicate his conception of becoming any more than did his paper of 1925,[1] which was penetratingly criticized by Hugo Bergmann as follows:

> Thus, according to Reichenbach, a cross-section in the state of the world is distinguished from all others; the now has an

---

[7] H. Reichenbach: *DT*, *op. cit.*, p. 12.

[8] Cf. A. Grünbaum: "Causality and the Science of Human Behavior," *American Scientist*, Vol. XL (1952), pp. 665–76 [reprinted in: H. Feigl and M. Brodbeck (eds.) *Readings in the Philosophy of Science* (New York: Appleton-Century-Crofts, Inc.; 1953), pp. 766–78]; "Das Zeitproblem," *Archiv für Philosophie*, Vol. VII (1957), pp. 203–206; "Complementarity in Quantum Physics and its Philosophical Generalization," *The Journal of Philosophy*, Vol. LIV (1957), pp. 724–27, and "Science and Man," *Perspectives in Biology and Medicine*, Vol. V (1962), pp. 483–502. See also J. J. C. Smart's telling criticisms [*Philosophical Quarterly*, Vol. VIII (1958), esp. p. 76] of Reichenbach's contention that we can "change the future" but not the past.

[9] This aim is stated by him in "Die Kausalstruktur der Welt und der Unterschied von Vergangenheit und Zukunft," *op. cit.*, pp. 139–42.

[1] *Ibid.*

objective significance. Even when no man is alive any longer, there is a now. "The present state of the planetary system" would even then be just as precise a descriptive phrase as "the state of the planetary system in the year 1000."

Concerning this definition one must ask: Which now is intended, if one says: the present state of the planetary system? That of the year 1800 or 2000 or which other one? Reichenbach's reply is: the now is the threshold of the transition from the state of indeterminacy to that of determinacy. But (if Reichenbach's indeterminism holds) this transition has *always* occurred and will always occur. And if the rejoinder would be: the indeterminacy of the year 1800 has already been transformed into a determinacy, then one must ask: For whom? Evidently for us, for the present, for our now. Accordingly, this definition by Reichenbach seems to refer after all to a now which it must first define. What is the objective difference between the now of the year 1800 and the now of the present instant? The answer must be: now is the instant of the transition from indeterminacy to determinacy, that is, one explains the present now . . . by reference to itself.

. . . Reichenbach writes: The problem can be formulated as the question concerning the difference between the past and the future. For determinism, there is no such difference. . . . But the reproach which Reichenbach directs at determinism here should be aimed not at it but at the world view of physics, which does not take cognizance of any psychological categories, for which there is no "I," . . . a concept which is inextricably intertwined with the concept "now." Even those who regard the supplanting of determinism by indeterminism as admissible, as we do, will not be willing to admit that the concept of "now" can be assigned a legitimate place within indeterministic physics. Even if one assumes—as we wish to do along with Reichenbach—that the future is not uniquely determined by a temporal cross-section, one can say only that this indeterminacy prevails just as much for Plato as for myself and that I cannot decide by physical means who is living "now." For the difference is a psychological one.

. . . "Now" is the temporal mode of the experiencing ego.[2]

[2] H. Bergmann: *Der Kampf um das Kausalgesetz in der jüngsten Physik*, *op. cit.*, pp. 27–28. Wilfrid Sellars has independently developed the basis for similar criticisms of the alleged connection between indeterminism and becoming as part of his penetrating study of a complex of related issues: cf. W. Sellars's "Time and The World Order," in *Minnesota Studies in the Philosophy of Science, Vol. III, op. cit.*, pp. 527–616.

I maintain with Bergmann that the transient now with respect to which the distinction between the past and the future of common sense and psychological time acquires meaning has no relevance at all to the time of physical events, because it has no significance at all apart from the egocentric perspectives of a *conscious* (human) organism and from the immediate experiences of that organism. If this contention is correct, then *both* in an indeterministic *and* in a deterministic world, *the coming into being or becoming of an event, as distinct from its merely being, is thus no more than the entry of its effect(s) into the immediate awareness of a sentient organism (man).* For what is the difference between these two worlds in regard to the determinateness of future events? The difference concerns only the type of functional connection linking the attributes of the future events to those of present or past events. But this difference does *not* make for a precipitation of future events into existence in a way in which determinism does not. Nor does indeterminacy make for any difference whatever at any time in regard to the *attribute-specificity* of the future events themselves. For in either kind of universe, it is a fact of logic that what will be, will be! The result of a future quantum mechanical measurement may not be definite prior to its occurrence in relation to earlier states, and thus our prior knowledge of it correspondingly cannot be definite. But as an event, it is as fully attribute-definite and occurs just as a measurement made in a deterministic world does. The belief that in an indeterministic world, the future events come into being or become actual or real with the passage of time would appear to confuse two quite different things: (1) the *epistemological* precipitation of the actual event-properties of future events out of the wider matrix of the possible properties allowed by the quantum-mechanical probabilities, and (2) an existential coming into being or becoming actual or real. Only the *epistemological* precipitation is affected by the passage of time through the transformation of a statistical expectation into a definite piece of information. But this does *not* show that in an indeterministic world there is any kind of *precipitation into existence* or *coming into being* with the passage of time. And even in a deterministic world, the effects of physical events come into our awareness at a certain time and *in that sense* can be thought of as coming into being.

Bergmann's demonstration above that an indeterminist universe fails to define a *non*-psychological objective transient now can be extended in the following sense to justify his contention that the concept "now" involves features peculiar to consciousness: the "flux of time" or transiency of the "now" has a meaning only in the context of the egocentric perspectives of *sentient* organisms and does *not also* have relevance to the relations between purely inanimate individual recording instruments and the environmental physical events they register, as Reichenbach claims. For what can be said of every state of the universe can also be said, *mutatis mutandis,* of every state of a given inanimate recorder. Moreover, the irrelevance of the transient now to the accretion of time-tagged marks or *traces* on an *inanimate* recording tape also emerges from William James's and Hans Driesch's correct observation that a simple isomorphism between a succession of *brain traces* and a succession of states of awareness does not explain the now-contents of such psychological phenomena as melody awareness. For the hypothesis of isomorphism of traces and states of awareness renders only the succession of states of awareness but not the *instantaneous awareness of succession,*[3] which is an essential ingredient of the meaning of "now": the now-content, when viewed as such in awareness, includes an awareness of the order of succession of events in which the occurrence of that awareness constitutes a *distinguished element.* And the transiency of the now or the "flux" of time arises from the *diversity* of the now-contents having the latter attributes: there are striking differences in the *membership* of the set of remembered (recorded) and/or forgotten events of which we have instantaneous awareness.

I cannot see, therefore, that Reichenbach is justified in considering the accretion of time-tagged marks or traces on an *inanimate* recording tape so as to form an expanding spatial series as illustrating the "flux" of time. Thus, Bergmann's exclusively psychologistic conception of this flux or becoming must be upheld against Reichenbach: the flux depends for its very *existence* on the perspectival role of consciousness, since the coming *into* being (or becoming) of an event is no more than

[3] Cf. W. James: *The Principles of Psychology, op. cit.,* pp. 628–29, and H. Driesch: *Philosophische Gegenwartsfragen* (Leipzig: E. Reinicke; 1933), pp. 96–103.

the entry of its effect(s) into the immediate awareness of a sentient organism (man).

We saw earlier that the locution "Here-Now" of the relativistic Minkowski diagram does not commit that entirely non-psychological theory to the transient now encountered in common sense time. Hence the purely physical character of the special theory of relativity *cannot* be adduced to show that the transient now is relevant to physical time, i.e., it cannot be adduced to refute our claim of the dependence of the "now" and, correlatively, of the transient division of the time continuum into "past" and "future," on the perspectival role of consciousness.[4]

It was none other than the false assumption that "flux" must be a feature of physical no less than of psychological (common sense) time that inspired Henri Bergson's misconceived polemic against the mathematical treatment of motion, which he unfoundedly charged with having erroneously *spatialized* time by a description which leaves out the flux of becoming and renders only the "static" relations of earlier and later.[5]

Hermann Weyl has given a metaphorical rendition of the dependence of coming into being on consciousness by writing:[6] "The objective world simply *is*, it does not *happen*. Only to the gaze of my consciousness, crawling[7] upward along the life- [i.e., world-] line of my body, does a section of this world come to life as a fleeting image in space which continuously changes in time." This poetic but sound declaration has given rise to serious misunderstandings, as shown by the following objection from Max Black:

---

[4] Neither does Minkowski's use of the locution "Here-Now" show conversely that the special theory of relativity makes essential use of psychological temporal categories in its assertive content (as distinct from the pragmatics of its verification by us humans).

[5] Cf. H. Bergson: *Creative Evolution* (New York: Random House, Inc.; 1944) and *Matière et Mémoire* (Geneva: A. Skira; 1946). Related criticisms of Bergson's treatment of other aspects of time are given in A. Grünbaum: "Relativity and the Atomicity of Becoming," *op. cit.*, pp. 144–55.

[6] H. Weyl: *Philosophy of Mathematics and Natural Science, op. cit.*, p. 116.

[7] The metaphor "crawling" must not, of course, be taken to suggest the "metaphysical error" charged against it by J. J. C. Smart ["Spatializing Time," *Mind*, Vol. CXIV (1955), p. 240] that psychologically time itself "flows" spatially at a certain rate measured in some non-existent hypertime. We shall see shortly that the concept of the "forward" shifting now does not involve this logical blunder.

> But this picture of a "block universe," composed of a timeless web of "world-lines" in a four-dimensional space, however strongly suggested by the theory of relativity, is a piece of gratuitous metaphysics. Since the concept of change, of something happening, is an inseparable component of the common-sense concept of time and a necessary component of the scientist's view of reality, it is quite out of the question that theoretical physics should require us to hold the Eleatic view that nothing happens in "the objective world." Here, as so often in the philosophy of science, a useful limitation in the form of representation is mistaken for a deficiency of the universe.[8]

But contrary to Black, Weyl's claim that the time of inanimate nature is devoid of *happening* in the sense of *becoming* is not at all tantamount to the Eleatic doctrine that *change* is an illusion of the human mind! It is of the essence of the relativistic account of the inanimate world as embodied in the Minkowski representation that there is change in the sense that different kinds of events can (do) occur at different times: the attributes and relations of an object associated with any given world-line may be different at different times (e.g., its world-line may intersect with different world-lines at different times). Consequently, the total states of the world (when referred to the simultaneity criterion of a particular Galilean frame) are correspondingly different at different times, i.e., they change with time. It is Black's own misidentification of mere change with becoming ("happening") which leads him to the astonishing and grotesque supposition that Weyl's mentalistic account of becoming bespeaks Weyl's unawareness that "the concept of change . . . is . . . a necessary component of the scientist's view of reality." Black refers to the web of earlier-later relations represented by the world-lines as "timeless" just because they do not make provision for becoming. And he suggests that Weyl conceives of them as forming a four-"space" in the sense in which physical space *excludes* the system of temporal relations obtaining with respect to "earlier than." It is apparent that Black's use of the terms "timeless" and "space" in this context is misleading to the point of conveying question-begging falsehoods. Weyl's thesis is that coming *into* being ("happening"), as con-

[8] M. Black: "Review of G. J. Whitrow's *The Natural Philosophy of Time*," *Scientific American*, Vol. CCVI (April, 1962), pp. 181–82.

trasted with simply being, is only coming into the present awareness of a sentient organism. And that thesis is not vulnerable to Black's charge of having mistaken "a useful limitation in the form of representation" for "a deficiency of the universe," all the less so, since Weyl makes a point of the difference between space and time by speaking of the world as "a $(3 + 1)$ – dimensional metrical manifold"[9] rather than as a 4-dimensional one.

In an endeavor to erect a *reductio ad absurdum* of Weyl's thesis, M. Capek has given an even more grotesque account of that thesis than Black did. Capek writes:

> although the world scheme of Minkowski eliminates succession in the physical world, it recognizes at least the *movement of our consciousness* to the future. Thus arises an absurd dualism of the timeless physical world and temporal consciousness, that is, a dualism of two altogether disparate realms whose correlation becomes completely unintelligible. . . . in such a view . . . we are already dead without realizing it now; but our consciousness creeping along the world line of its own body will certainly reach any pre-existing and nominally future event which in its completeness *waits* to be finally reached by our awareness. . . . To such strange consequences do both spatialization of time and strict determinism lead.[1]

But Capek states a careless and question-begging falsehood by declaring that on Weyl's view the physical world is "timeless." For what Weyl is contending is only that the physical world is devoid of becoming, while fully granting that the states of physical systems are ordered by an "earlier than" relation which is isomorphic, in important respects, with its counterpart in consciousness. Capek's claim of the unintelligibility of the correlation between physical and psychological time within Weyl's framework is therefore untenable, especially in the absence of an articulation of the kind (degree) of correlation which Capek requires and also of a *justification* of that requirement. More unfortunate still is the grievous mishandling of the meaning of Weyl's metaphor in Capek's attempt at a *reductio ad absurdum* of Weyl's view, when Capek speaks of our "already" being dead

[9] H. Weyl: *Space-Time-Matter, op. cit.*, p. 283.

[1] M. Capek: *The Philosophical Impact of Contemporary Physics* (Princeton: D. Van Nostrand Company, Inc.; 1961), p. 165.

without realizing it now and of our completed future death *waiting* to be finally "reached" by our awareness. This gross distortion of Weyl's metaphorical rendition of the thesis that coming *into* being is only *coming into present awareness* rests on a singularly careless abuse of the temporal and/or kinematic components of the meanings of the words "already," "completed," "wait," "reach," etc.

I have argued that in the psychological (common sense) context to which the transient now does have relevance, it is tautological to assert that time "flows" from the past to the future or from earlier to later. But I wish to conclude by noting that it is unjustified to charge the latter assertion with a breach of logical grammar.

We saw that the concept of the transiency of the now or of the flow of time is a qualitative concept without any metrical ingredients. Hence the entirely *non*-metrical concept represented by the metaphor "forward flow" is not at all vulnerable to the metrical *reductio ad absurdum* offered by J. J. C. Smart[2] in the following form: "The concept of the flow of time or of the advance of consciousness is, however, an illusion. How fast does time flow or consciousness advance? In what units is the rate of flow or advance to be measured? Seconds per — ?" Max Black[3] likewise unwarrantedly asks the *metrical* question "How fast does time flow?" and then goes on to claim quite mistakenly that a non-existent super-time would be needed to give meaning to the flow of psychological time. Furthermore Black supposes incorrectly that the metaphor "forward flow" commits one to saying that *time* is "changing"[4] and to the contention that it makes sense to speak of the cessation of the flow of psychological time. The absurdities which Black is then able to derive from the literal interpretation of these latter assertions can therefore *not* serve to discredit the concept of the transiency of the Now as understood in this book.

---

[2] J. J. C. Smart: "The Temporal Asymmetry of the World," *op. cit.*, p. 81.
[3] M. Black: "The 'Direction' of Time," *op. cit.*, p. 57.
[4] *Ibid.*

Chapter II

EMPIRICISM AND THE THREE-DIMENSIONALITY OF SPACE

The success of empiricism in accounting for our knowledge of the tri-dimensionality of the physical world is intimately connected with its ability to refute Kant's claim that the existence of such *similar* but *incongruent* counterparts as the left and right hands constitutes evidence for his transcendental a priori of space.[1] Since the reasons for the untenability of *this particular* Kantian contention are not given even in Reichenbach's definitive empiricist critique of the transcendental idealist theory of space[2] and are not sufficiently known to the philosophical public, I shall give a brief statement of them.

If we take two arbitrarily (irregularly) shaped objects in a Euclidean plane, which are metrically symmetric or "reflected" about a straight line in that plane, it will be seen that so long as we confine these two objects to that plane, they cannot be brought into congruence such that the points of one coincide with their respective image points in the other. But such congruence *can* be achieved, if we are allowed to rotate one of these reflected two-dimensional objects about the axis of symmetry, thereby making use of the next higher (third) dimension. G. Lechalas credits Delboeuf with discovering that, in

[1] I. Kant: *Werke,* edited by E. Cassirer (Berlin: Bruno Cassirer Verlag; 1912), Vol. II, pp. 393–400 and Vol. IV, §13, pp. 34–36.

[2] H. Reichenbach: "Kant und die Naturwissenschaft," *Die Naturwissenschaften,* Vol. XXI (1933), pp. 601–606 and 624–26; *Relativitätstheorie und Erkenntnis Apriori* (Berlin: J. Springer; 1920), and *PST, op. cit.*

general, given two (n−1)-dimensional objects, metrically symmetric about some (n−2)-dimensional object, then to achieve congruence such that the points of the one coincide with their respective image points in the other, a continuous rotation in n-dimensional space is necessary.[3] It should be pointed out, however, that the requirement of rotation through a hyper-space does not hold unrestrictedly: it does hold for spaces whose topology is Euclidean or spherical, but it fails to hold for the Möbius strip two-space or for a one-dimensional space whose topology is that of the numeral 8.

Accordingly, the three-dimensional right-hand cannot be brought into the stated sort of congruence with the three-dimensional left-hand by a continuous rigid motion, because of the *empirical* fact that the four-dimensional space needed for the required kind of rotation is physically unavailable! This same fact enables us to infer the *three*-dimensionality as opposed to the two-dimensionality of optically active molecules from their dextro-rotary or levo-rotary behavior. For if they were only two-dimensional, then it would be possible to convert a given dextro-rotary molecule into a levo-rotary one by merely flipping it over. But this cannot be done.[4]

Contrary to Kant, the specific structural difference between the right and left hands can be given a conceptual rather than *only* a denotatively intuitive characterization as follows:[5] the group of Euclidean rigid motions is only a *proper* sub-group of the group of length-preserving ("non-enlarging") similarity mappings. For the determinant of the coefficients of the particular linear transformations constituting the latter type of similarity mappings must have either the value +1 or the value −1. But only those similarity transformations whose determinant

[3] G. Lechalas: "L'Axiome de libre Mobilité," *Revue de Métaphysique et de Morale,* Vol. VI (1898), p. 754. This property of reflections had already been pointed out by Möbius in his *Der Barycentrische Calcul* (Leipzig: Barth; 1827), p. 184.

[4] Cf. John Read: *A Direct Entry to Organic Chemistry* (London: Methuen & Company, Ltd.; 1948), Chapter vii.

[5] Cf. F. Klein: *Elementary Mathematics From An Advanced Standpoint,* (New York: Dover Publications, Inc.; 1939), Vol. II, pp. 39–42; H. Weyl: *Philosophy of Mathematics and Natural Science, op. cit.,* pp. 79–85. For a more elementary account, see O. Hölder: *Die Mathematische Methode, op. cit.,* pp. 387–89.

("Jacobian")[6] is +1 form the group of Euclidean rigid motions, the remainder being the *reflections* whose Jacobian is −1 and which *include* the case of *Kant's left and right hands.*[7] But the very existence of the latter pairs of physical objects in our physical three-space which realize a formal transformation whose Jacobian is −1 is an *empirical* fact, as is the *non*-existence of a *physical* four-dimensional hyper-space through which they could otherwise be rotated so as to be brought into congruence.

Since the tri-dimensionality of physical space has turned out to be a logically contingent empirical fact, one naturally wonders whether it is an autonomous, irreducible empirical fact or not with respect to the standard formulations of physical theory. Huyghens's principle in optics tells us that if a single spherical light wave is produced by a disturbance at a point which lasts for a very short time between $t = t_0 - \epsilon$ and $t = t_0$, then the effect at a point P at a distance cT (where c = the velocity of light) is null until the instant $t = t_0 - \epsilon + T$ *and* is null again *after* the instant $t = t_0 + T$. And thus, according to Huyghens's principle, a single spherical wave would leave no residual after-effect at a point P. Now, J. Hadamard has shown[8] that this requirement of Huyghens's principle is satisfied only by wave equations having an *even* number of independent variables. Since the time variable in conjunction with the space variables constitute the independent variables of these equations, Hadamard's result

---

[6] Let F_1 and F_2 be two functions of u and v. Then

$$J = \begin{vmatrix} \frac{\partial F_1}{\partial u} & \frac{\partial F_1}{\partial v} \\ \frac{\partial F_2}{\partial u} & \frac{\partial F_2}{\partial v} \end{vmatrix} = \frac{\partial F_1}{\partial u} \cdot \frac{\partial F_2}{\partial v} - \frac{\partial F_1}{\partial v} \cdot \frac{\partial F_2}{\partial u} \text{ is the functional}$$

determinant or "*Jacobian*" of F_1 and F_2 with respect to u and v. In the three-dimensional case relevant to the incongruent hands, we are dealing with three functions F_1, F_2, and F_3 of u, v, and w, and the corresponding Jacobian is a determinant of three rows and three columns.

[7] A requirement for the preservation of the dimensionality under a transformation is that the Jacobian *not* be equal to zero, the latter condition being necessary and sufficient that the transformation be one-to-one. Cf R. S. Burington and C. C. Torrance: *Higher Mathematics* (New York: McGraw-Hill Book Company, Inc.; 1939), pp. 132–42.

[8] J. Hadamard: *Lectures on Cauchy's Problem in Linear Partial Differential Equations* (New Haven: Yale University Press; 1923), pp. 53–54, 175–77, and 235–36.

shows that Huyghens's principle holds only for cases in which the number of *space* dimensions is *odd,* as in the case of the three-dimensional physical space of our world.[9] Sharing the view of Aristotle and Galileo that the tri-dimensionality of physical space might be explainable as a consequence of other, more comprehensive empirical principles, H. Weyl suggests[1] that the difference between spaces of even and odd numbers of dimensions in regard to the transmission of waves may be one clue to the required explanation.[2]

Contrary to a widely held interpretation, Poincaré's account of the status of the tri-dimensionality of space is not incompatible with the empirico-realistic conception of that attribute of the physical world just set forth. In his rejoinder to Russell, we find Poincaré saying without elaboration: "I consider the axiom of three dimensions as conventional in the same way as those of Euclid."[3] But in his posthumous book,[4] he tells us that since his earlier treatment of this axiom was "very compressed," he now wishes to clarify it. Then, after explaining that in classifying the elements of a manifold as the same in some respect, we use the "basic convention" of abstracting from other qualitative differences among them, he notes that the three-dimensionality of the perceptual localizations of physical events is obtained upon abstracting from a variety of qualitative *non*-positional differences between them. This sense of "convention," however, hardly renders three-dimensionality non-objective any more than the reference to kinds of events makes particular causal statements true by convention. That Poincaré was entirely clear on this is apparent from the following assertion by

[9] For an explanation of this result by reference to the special case of three and two dimensions, cf. B. Baker and E. T. Copson: *The Mathematical Theory of Huyghens's Principle* (Oxford: Oxford University Press; 1939), pp. 46–47.

[1] H. Weyl: *Philosophy of Mathematics and Natural Science, op. cit.,* p. 136.

[2] Cf. also P. Ehrenfest: "In What Way Does It Become Manifest in the Fundamental Laws of Physics That Space Has Three Dimensions?" *Proceedings of the Amsterdam Academy,* Vol. XX (1917), pp. 200–209, or, in German translation: "Welche Rolle spielt die Dreidimensionalität des Raumes in den Grundgesetzen der Physik," *Annalen der Physik,* Vol. LXI (1920), pp. 440–46.

[3] H. Poincaré: "Sur les Principes de la Géométrie," *op. cit.,* p. 73.

[4] H. Poincaré: *Letzte Gedanken, op. cit.,* p. 59.

him: "We see on the basis of this brief explanation what experimental facts lead us to ascribe three dimensions to space. As a consequence of these facts, it would be more convenient for us to attribute three dimensions to it than four or two; but the term 'convenient' is perhaps not strong enough; a being which had attributed two or four dimensions to space would be handicapped in a world like ours in the struggle for existence."[5] After exhibiting how that handicap would arise from an interpretation of space as two- or four-dimensional, he shows on the basis of group-theoretical arguments[6] that in the context of causality, physical facts lead to the tri-dimensionality of physical space just as the structure of perceptual data had. And he concludes by saying that since we have the capacity to construct mathematically a continuum of an arbitrary number of dimensions, "this capacity would . . . permit us to construct a space of four dimensions just as well as one of [only] three dimensions. *It is the external world, experience,* which determines our developing our ideas more in one of these directions than in the other."[7]

The mathematical continuity which is attributed to physical space is a *topological* property just as its tri-dimensionality. We shall therefore now conclude our consideration of philosophical problems of the *topology* of time and space in Part II by inquiring whether continuity can be held to have any kind of *empirical* status.

We argued in Chapter One that the continuity postulated for physical space and time issues in the metric amorphousness of these manifolds and thus makes for the conventionality of congruence, much as the conventionality of non-local metrical simultaneity in special relativity is a consequence of the postulational *fact* that light is the fastest causal chain *in vacuo* and that clocks do not define an absolute metrical simultaneity under transport. But it has been objected that there is an important epistemological difference between the latter Einstein postulate and the postulational ascription of continuity (in the mathematical sense) to physical space and time instead of some discontinuous structure: the continuity postulate, it is held, cannot be regarded,

---

[5] *Ibid.*, p. 86.
[6] *Ibid.*, Chapter iii, §5, pp. 87–94.
[7] *Ibid.*, p. 99, my italics. Cf. also O. Hölder: *Die Mathematische Methode, op. cit.*, p. 393.

even in principle, as a *factual* assertion in the sense of being either true or false. For according to this objection, there can be no *empirical* grounds for accepting a geometry postulating continuous intervals in preference to one which postulates discontinuous intervals consisting of, say, only the algebraic or only the rational points. The objector therefore contends that the rejection of the latter kind of denumerable geometry in favor of a non-denumerable one affirming continuity has no kind of factual warrant but is based solely on considerations of *arithmetic convenience* within the analytic part of geometry. And hence the *topology* is claimed to be no less infested with features springing from conventional choice than is the metric. Accordingly, the exponent of this criticism concludes that it is not only a misleading emphasis but outright incorrect for us to discern conventional geometrical elements only in the *metrization* of the topological substratum while deeming the *continuity* of that substratum to be *factual*.

This plea for a conventionalist conception of continuity is not convincing, however. Admittedly, the justification for regarding continuity as a broadly *inductive framework-principle* of physical geometry cannot be found in the direct verdicts of measuring rods, which could hardly disclose the super-denumerability of the points on the line. And prima facie there is a good deal of plausibility in the contention that the postulation of a super-denumerable infinity of irrational points in addition to a denumerable set of rational ones is dictated solely by the desire for such arithmetical convenience as having closure under operations like taking the square root, etc. But, even disregarding the Zenonian difficulties which may vitiate denumerable geometries logically, as we saw in Chapter Six, these considerations lose much of their force, it would seem, as soon as one applies the *acid test* of a convention to the conventionalist conception of continuity in physical geometry: the availability or demonstrated feasibility of one or more *alternate* formulations dispensing with the particular alleged convention and yet permitting the successful theoretical rendition of the same total body of experiential findings, such as in the case of the choice of a particular system of *units* of measurement.

Upon applying this test, what do we find? No mathematically discontinuous alternative set of theories have been shown to be as viable as those which are based on the continuum by demonstrating that these two kinds of theories must be, *in principle,* empirically indistinguishable from one another.[8] In the absence of such a demonstration, let alone of the actual, full-fledged elaboration of an empirically adequate discontinuous alternative theory, the charge that in a geometry the topological component of continuity is no less conventional than the metric itself is *unfounded.* And pending the elaboration of a successful alternative to the continuum, the following suspicion insinuates itself: the empirical facts codified in terms of the classical mathematical apparatus in our most sophisticated and best-confirmed physical theories support continuity in a ***broadly inductive*** sense as a *framework principle* to the exclusion of the prima facie rivals of the continuum.

It would be an error to believe that this conclusion requires serious qualification in the light of recent suggestions of space (or time) quantization. For as H. Weyl has noted

> so far it [i.e., the atomistic theory of space] has always remained mere speculation and has never achieved sufficient contact with reality. How should one understand the metric relations in space on the basis of this idea? If a square is built up of miniature tiles, then there are as many tiles along the diagonal as there are along the side; thus the diagonal should be equal in length to the side.[9]

And Einstein has remarked that:

> From the quantum phenomena it appears to follow with certainty that a finite system of finite energy can be completely described by a finite set of numbers (quantum numbers). This does not seem to be in accordance with a continuum theory, and must lead to an attempt to find a purely algebraic theory for the description of reality. But nobody knows how to obtain the basis of such a theory.[1]

---

[8] P. K. Feyerabend has suggested ["Comments on Grünbaum's 'Law and Convention in Physical Theory,'" in: H. Feigl and G. Maxwell (eds.) *Current Issues in the Philosophy of Science, op. cit.,* pp. 160–61] that a crucial experiment between continuous and discontinuous descriptions of nature is logically possible by giving a sketch of such an experiment.

[9] H. Weyl: *Philosophy of Mathematics and Natural Science, op. cit.,* p. 43.

[1] A. Einstein: *The Meaning of Relativity* (Princeton: Princeton University Press; 1955), pp. 165–66.

By claiming to have established the *unfoundedness* of the conventionalist conception of continuity, I do not, of course, maintain that I have demonstrated its falsity. P. K. Feyerabend has asserted[2] that the absence at this time of a discontinuous alternative to current physical theory "does not prove anything," since "after all, the full theory of continuity was not developed before the second half of the nineteenth century." But Feyerabend overlooked here that my argument does establish the *gratuitousness* of the conventionalist conception of continuity by showing that, at best, its advocates are offering only a *program.*

Can my argument be strengthened by the failure of neo-intuitionist attempts to dispense with the continuum of geometry and analysis via the provision of substitutes for all of classical mathematics? These neo-intuitionistic endeavors to base classical mathematics on more restrictive foundations failed by involving truncations of mathematical physics whose range A. A. Fraenkel has characterized as follows:

> intuitionistic restriction of the concept of continuum and of its handling in analysis and geometry, though carried out in quite different ways by various intuitionistic schools, always goes as far as to exclude vital parts of those two domains. (This is not altered by Brouwer's peculiar way of admitting the continuum *per se* as a "medium of free growth.")[3]

But, as Feyerabend has pointed out correctly, the failures of the neo-intuitionists cannot be regarded as strengthening my argument:

> For what we want is not repetition, in new terms, of the *whole* of classical mathematics; what we want is a new mathematical system that may be adequate for the description of the universe which is also articulate enough for a crucial experiment to be possible.[4]

---

[2] P. K. Feyerabend: "Comments on Grünbaum's 'Law and Convention in Physical Theory,'" *op. cit.,* p. 160.

[3] A. A. Fraenkel and Y. Bar-Hillel: *Foundations of Set Theory* (Amsterdam: North Holland Publishing Company; 1958), p. 200. Chapter iv of this work gives an admirably comprehensive and lucid survey of the respects in which neo-intuitionist restrictions involve mutilations of the system of classical mathematics.

[4] P. K. Feyerabend: "Comments on Grünbaum's 'Law and Convention in Physical Theory,'" *op. cit.,* p. 161.

PART III

Philosophical Issues in the Theory of Relativity

Chapter 12

PHILOSOPHICAL FOUNDATIONS OF THE SPECIAL THEORY OF RELATIVITY, AND THEIR BEARING ON ITS HISTORY

(A) INTRODUCTION

Since the publication of Hans Reichenbach's definitive books on the philosophy of the theory of relativity during the nineteen twenties,[1] the literature on the philosophy and history of Einstein's special theory of relativity (hereafter called STR) has been enlarged by contributions which call for critical evaluation. In this chapter, I shall give such an evaluation in the course of presenting: (a) an up-to-date analysis of the intertwined philosophical and empirical foundations of the kinematics of the STR, with attention to neglected issues and prevalent misconceptions, and (b) a demonstration that a rigorous grasp of the philosophical conceptions underlying the fully evolved STR and distinguishing it from its ancestors is decisively prerequisite to (1) the very posing of well-conceived, searching historical questions in regard to the STR, and hence to (2) the provision of a historically sound and illuminating account of its genesis.

Specifically, it will be useful to interlace the presentations of (a) and (b) in sections as follows:

(B) Einstein's conception of simultaneity, its prevalent misrepresentations, and its history,

[1] H. Reichenbach: *Axiomatik der relativistischen Raum-Zeit-Lehre, op. cit.*, Vol. 72 of *Die Wissenschaft*, and *PST, op. cit.*

(C) The history of Einstein's enunciation of the limiting character of the velocity of light *in vacuo*,
(D) The principle of the constancy of the speed of light, and the falsity of the aether-theoretic Lorentz-Fitzgerald contraction hypothesis,
(E) The experimental confirmation of the kinematics of the STR, and
(F) The philosophical issue between Einstein and his aether-theoretic precursors; its bearing on E. T. Whittaker's history of the STR.

Although the historical portions of this chapter will emphasize the indispensability of philosophical mastery of the STR to the unraveling of its history, I do not wish to deny that the history of the STR, in its turn, can be relevant to the understanding of the philosophical ingredients of that theory. On the contrary, historical inquiry may be illuminating philosophically by disclosing, for example, the vicissitudes in Einstein's own philosophical orientation, thereby explaining, as G. Holton has pointed out, why advocates of contending schools of philosophic thought can each "find some part of Einstein's work to nail to his mast as a battle flag against the others."[2] Despite the capability of the history of a theory to serve as a propaedeutic to the analysis of its philosophical foundations, it happens that what is known reliably so far about the history of the STR has failed signally, as we shall see, to contribute to the clarification of the more subtle questions concerning its *epistemological* basis. And it will become apparent that some of the historical accounts which have been given are beset by serious and puzzling contradictions.

(B) EINSTEIN'S CONCEPTION OF SIMULTANEITY, ITS PREVALENT MISREPRESENTATIONS, AND ITS HISTORY.

Einstein discusses the problem of simultaneity epistemologically in Section 1 of his 1905 paper[3] well before showing at the

[2] G. Holton: "On the Origins of the Special Theory of Relativity," *American Journal of Physics,* Vol. XXVIII (1960), p. 627.

[3] A. Einstein: "On the Electrodynamics of Moving Bodies," *op. cit.,* pp. 38–40. The cited collection containing this paper will be cited hereafter as "*PR.*"

end of Section 2 that the two fundamental postulates of his STR entail discordant judgments of simultaneity as between relatively moving Galilean frames. And he stresses in Section 1, though only rather concisely, that, contrary to Newtonian theory, the metrical simultaneity of two spatially separated events involves a convention *within* any given inertial system. Nevertheless, numerous expositions of the STR have nurtured the misunderstanding that the philosophical repudiation of Newton's absolute simultaneity by the STR occurs, *in the first instance,* in the context of the theory of the relative motion of different inertial systems. We shall see in detail why it is an error to regard the discordant judgments of simultaneity made by different Galilean frames as the *primary* locus of the conceptual innovation wrought by Einstein in regard to the concept of simultaneity. This misunderstanding has the following consequences: First, it obscures the fact that Einstein's philosophical supplanting of Newton's conception of simultaneity is *presupposed* by rather than first derived from his enunciation of the fundamental postulate of the constancy of the speed of light, and second, it precludes awareness that Einstein's philosophical characterization of distant simultaneity as *conventional* rests on specifiable *physical assumptions,* thereby preventing the recognition of the logical role played by these particular physical assumptions in the very foundations of the STR.

The important bearing of the philosophical misconstrual of Einstein's conception of simultaneity on the investigation of the history of the STR is therefore the following: the lack of philosophical mastery on the part of the historian will conceal from him that there exists the historical problem as to the grounds on which Einstein felt entitled in 1905 to make the particular physical assumptions undergirding his philosophical doctrine of the conventionality of the simultaneity of spatially separated events.

This doctrine is set forth *very concisely* in Section 1 of his 1905 paper, which is entitled "Definition of Simultaneity." There he writes: "We have not defined a common 'time' for [the spatially separated points] A and B, for the latter cannot be defined at all unless we establish *by definition* that the 'time' required by light to travel from A to B equals the 'time' it requires to travel from B to A."[4] This conception of simultaneity rests on a number of philosophical and physical assumptions which we shall now discuss.

It is clear that the sense in which two spatially separated events can be simultaneous is different from the one in which two quasi-coinciding events can be regarded as simultaneous: in the latter case of quasi-coincidence, local or contiguous simultaneity obtains, and not simultaneity at a distance. It might be thought that distant simultaneity can be readily grounded on local simultaneity as follows. Suppose that two spatially separated events E_1 and E_2 produce effects which intersect at a sentient observer so as to produce the experience of sensed (intuitive) simultaneity at himself. Then this local simultaneity of the effects of E_1 and E_2 permits us to infer that the distant events occurred simultaneously, if the influence chains that emanated from them had appropriate *one-way velocities* as they traversed their respective distances to the location of the sentient observer. But this procedure is unavailing for the purpose of first characterizing the conditions under which the separated events E_1 and E_2 can be held to be simultaneous. For the *one-way* velocities invoked by this procedure presuppose *one-way transit times* which are furnished by *synchronized* clocks at the locations of E_1 and E_2 respectively. And the conditions for the synchronism of two spatially separated clocks, in turn, presuppose a criterion for the distant simultaneity of the events at these clocks.[5]

These considerations enable us to demonstrate the complete failure of A. N. Whitehead's attempt to ground the concept of distant simultaneity of physical theory on the sensed coincidences experienced by sentient observers.[6] In the first place, there are inconsistencies between the sensed simultaneity verdicts concerning a given pair of separated events when furnished

[4] A. Einstein: *PR*, p. 40; *italics in the original.* But, as has been correctly noted by C. Scribner ("Mistranslation of a Passage in Einstein's Original Paper on Relativity," *American Journal of Physics* [1963], Vol. XXXI, p. 398), the middle part of this sentence was mistranslated into English by Perrett and Jeffery and should read instead as follows: "the latter time can now be defined in establishing *by definition* that the 'time,' etc."

[5] H. Reichenbach: *Axiomatik der relativistischen Raum-Zeit-Lehre, op. cit.*, pp. 12–17, and *PST, op. cit.*, pp. 123–26.

[6] A. N. Whitehead: *The Concept of Nature, op. cit.*, pp. 53 and 56.

by sentient observers who are stationed at different space points of the same inertial system: if influences emanating from two spatially separated events E_1 and E_2 intersect at a point P_1 so as to produce sensed coincidence in the awareness of a sentient observer stationed at P_1, then there will be *other* points P_2, P_3, . . ., P_n in the *same* inertial system at which these same events E_1 and E_2 will *not* produce sensed coincidence in the respective stationary sentient observers.[7] And if Whitehead is to avoid these *inconsistencies* between the diverse verdicts as to sensed simultaneity in a manner coherent with physical theory, he must *restrict* the ascription of simultaneity on the basis of sensed simultaneity to only the following kinds of separated events: Those whose simultaneity would be compatible with the physical one-way velocities of the influence chains emanating from these events and meeting at the sentient observer after traversing the respective given distances. But the information concerning one-way velocities which Whitehead requires in order to be safe in interpreting the sensed coincidence as signifying simultaneity presupposes a prior criterion of simultaneity or clock-synchronization, which sensed coincidence is manifestly incompetent to furnish without involving Whitehead in a vicious circle.

One must therefore endorse Einstein's rejection of sensed simultaneity as the basis for distant physical simultaneity. His constructive alternative to Whitehead's unsuccessful attempt can be understood on the basis of his repudiation of both philosophical and physical assumptions made by the Newtonian theory as follows.

In opposition to the absolutistic conception of space and time ingredient in the Newtonian theory, the STR rests on the following conceptions of the identity of events and of their temporal order: first, prior to the construction of any system of coordinates, physical things and events first define, by their own identity, points and instants which can then constitute the spatial and temporal loci of other physical objects and events; and second, the temporal order of physical events is first constituted by their properties and relations qua physical events. By maintaining that the very *existence* of *temporal* relations between non-coinciding

[7] Cf. F. S. C. Northrop: "Whitehead's Philosophy of Science," *op. cit.*, p. 200.

events depends on the obtaining of some *physical* relations between them, Einstein espoused a conception of time (and space) which is *relational* by regarding them as systems of relations between physical events and things. Since time relations are first constituted by the system of physical relations obtaining among events, the character of the temporal order will be determined by the physical attributes in virtue of which events will be held to sustain relations of "simultaneous with," "earlier than," or "later than." In particular, it is a question of physical fact whether these attributes are of the kind to define temporal relations *uniquely*—i.e., such that on the strength of the obtaining of these attributes, *every* pair of events is *unambiguously* ordered with respect to one of the relations "earlier than," "later than," or "simultaneous with." If the temporal order were to have the latter character, it would be "absolute" in the compound sense that the time relation between any two events would be a *uniquely* obtaining *factual* relation between them which is wholly independent of any particular reference system and hence the same in every reference system. Were it to obtain in the case of simultaneity, it would have the consequence that *as a matter of physical fact, one and only one event* at a point Q would be simultaneous with a given event occurring at a point P elsewhere in space, a state of affairs which I have characterized by saying for brevity that simultaneity would be a "*uniquely* obtaining factual relation."

But is there in actual fact a physical basis for relations of *absolute* simultaneity among spatially separated events? If the behavior of transported clocks were of the kind assumed by the Newtonian theory, then indeed absolute simultaneity could be said to obtain merely in virtue of the coincidence of physical events with suitable readings of transported clocks of identical constitution. But the STR denies the existence of precisely this physical-clock basis for absolute simultaneity. For it makes the following assumption, to which we shall refer as "assumption (i)": within the class of physical events, material clocks do not define relations of absolute simultaneity under transport, because the relations of simultaneity which are defined by transported clocks *do* depend on the particular clock that is used, in the following sense: If two clocks U_1 and U_2 are initially synchronized at the same place A and then transported via paths of *different lengths* to a different place B such that their arrivals at B coincide, then U_1 and U_2 will no longer be synchronized at B. And if U_1 and U_2 were brought to B via the same path (or via different paths which are of equal length) such that their arrivals do *not* coincide, then too their initial synchronization would be destroyed. Thus, depending upon which one of the discordant clocks would serve as the standard, a given pair of events at A and B *would* or *would not* be held to be simultaneous. Hence, this dependence on the particular clock used prevents transported clocks from defining relations of absolute simultaneity within the class of physical events. And it *also* prompted Einstein to *deny* that even *within* a single inertial system, the physical basis of the simultaneity of two spatially separated events E and E* can be constituted, in part, by the readings produced by clocks as follows: two clocks U_1 and U_2 are transported to separate places from a common point in space at which they had identical readings, and then event E occurs at the location of clock U_1, and event E* at the place of clock U_2 such that the readings of U_1 and U_2 are the same.

The failure of transported clocks to constitute relations of absolute simultaneity among spatially separated events does not, of course, show that there is also no other physical relatedness among events which would make for such relations. But non-coinciding and *spatially separated* events can sustain physical relations of one kind or another only in virtue of the presence or absence of some kind of actual (or physically possible) physical linkage between them. And it would seem that the only direct linkage connecting the latter kinds of events can be constituted by *causal chains* of which they are the termini.

Let us consider, therefore, under what conditions a pair of spatially separated events can be simultaneous, if the relations of temporal order between such events depend on the obtaining or non-obtaining of a physical relation of *causal connectibility* between them. But we recall from Chapters Seven and Eight that (1) the characterization of *one* of two causally connected (or connectible) and genidentical events as "the" (possible partial or total) cause of the other presupposed the anisotropy of time, and (2) our account of the anisotropy of time, i.e., of the physical basis of the relations "earlier than" and "later than," in turn, presupposed distant simultaneity. Hence our impending statement of the bearing of the non-obtaining of causal connectibility on simultaneity will invoke only the symmetric causal relation of Chapter Seven, which is non-committal as to whether there are physical criteria for singling out one of two causally connected events as *the* (partial or total) cause of the other. And only after the concept of simultaneity ingredient in the account of the anisotropy of time thus becomes logically available will we invoke the latter to characterize one of two causally connected events as the earlier of the two.

Accordingly, we now assert: if and only if two *non*-coinciding events sustain the symmetric relation of genidentical (or quasi-genidentical) causal *connectibility* discussed in Chapter Seven will these two events be said to sustain the relation of *temporal separation*, i.e., the relation of being either earlier or later. And, therefore, two events will be said to sustain the relation of being neither earlier nor later, i.e., the relation of being *topologically simultaneous* or of having a *space*like separation, if and only if the specified kind of causal connectibility does *not* obtain between them.

If the physical basis for the relation of topological simultaneity among events lies in the impossibility of their being the termini of influence chains, we ask: Are the physically possible causal chains of nature such as to define a *unique* set of temporal relations in which every pair of events has an unambiguous place with respect to the relations of temporal separation and topological simultaneity? To answer this question, consider four events E_1, E_2, E_3, and E which satisfy the following conditions: First, E_1, E_2, and E_3 are genidentically connectible *by a light ray* such that E_2 is temporally o-between E_1 and E_3, and second, E_1, E, and E_3 are genidentically connectible other than by a light ray such that E is temporally o-between E_1 and E_3, and all three of these events occur at the *same* space point P_1 of an inertial system, while E_2 occurs at a different space point P_2 of that same inertial system. Furthermore, let E_x, E_y, and E_z be events temporally o-between E_1 and E such that E_x is o-between E_1 and E_y, while E_y is o-between E_x and E_z. And finally, let E_α,

E_β, and E_γ be o-between E and E_3 as shown in the following diagram, which is a world-line diagram but devoid of any of the arrows that might invoke the anisotropy of time as a basis for an (intrinsic) characterization of the positive time direction. As will be recalled from Chapters Seven and Eight, the latter direction could, of course, be introduced purely extrinsically by a mere coordinatization as effected through standard clocks. But we shall postpone the introduction of such a coordinatization, since it will be instructive to see that Einstein's thesis of the conventionality of metrical simultaneity can be formulated without it.

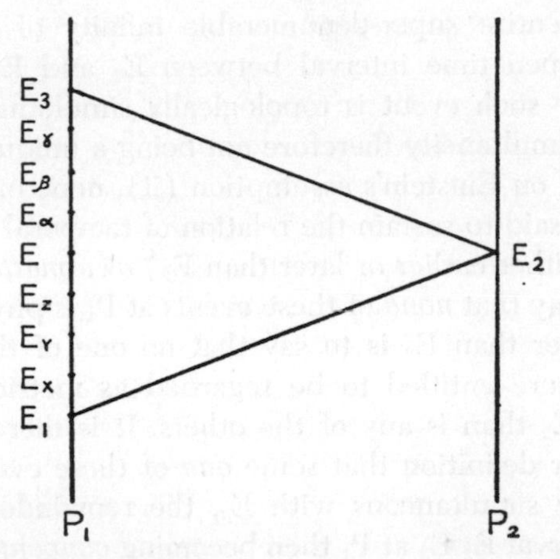

It is clear that the given relations that E_2 is temporally o-between E_1 and E_3, and that E, E_x, E_y, E_z, E_α, E_β, and E_γ are each o-between E_1 and E_3 as specified do not furnish any basis for relations of temporal separation or topological simultaneity between E_2 and any one of the events at P_1 lying within the open interval between E_1 and E_3. The existence of the latter relations will therefore depend on whether or not it is physically possible that there be direct causal chains which could link E_2 to a particular event in the open interval at P_1.

Though admitting that no light ray or other electromagnetic causal chains could provide a link between E_2 and E_z, or E_2 and E_α, Newtonian physics adduces its second law of motion to assert the physical possibility of other causal chains (e.g., moving particles) which would indeed do so.[8] But precisely this latter possibility is *denied* by Einstein in the STR, since he affirms the following topological postulate to which we shall refer as "assumption (ii)"[9]: it is physically impossible that *in vacuo*[1] there be genidentical (or other) causal chains which would link E_2 to any two events one or both of which lies *within* the open time interval $\overline{E_1 E_3}$ such that E_2 would be temporally o-between the two events in question.

Einstein's topological affirmation in assumption (ii) of the limiting role of light propagation *in vacuo* within the class of causal chains has the following fundamental consequence: each one of the entire super-denumerable infinity of events at P_1 within the open time interval between E_1 and E_3 rather than only a single such event is topologically simultaneous with E_2, topological simultaneity therefore not being a uniquely obtaining relation. For, on Einstein's assumption (ii), none of these events at P_1 can be said to sustain the relation of temporal separation to E_2 (i.e., be either earlier or later than E_2) *as a matter of physical fact*. But to say that *none* of these events at P_1 is physically either earlier or later than E_2 is to say that no one of them is objectively any more entitled to be regarded as metrically simultaneous with E_2 than is any of the others. It is therefore only by *convention* or definition that some *one* of these events comes to be metrically simultaneous with E_2, the remainder of them in the open interval $\overline{E_1 E_3}$ at P_1 then becoming *conventionally* either earlier or later than E_2.

---

[8] In the context of the Newtonian assumption that transported clocks can furnish a coordinatization of time by serving as indicators of absolute simultaneity, the relevant content of the second law of motion can be stated as follows: no matter what the velocity, the ratio of the force to the acceleration is constant, and particles can therefore be brought to arbitrarily high speeds in excess of the speed c of light by appropriately large forces acting for a sufficient time.

[9] A. Einstein: "Autobiographical Notes," in: P. A. Schilpp (ed.) *Albert Einstein: Philosopher-Scientist* (New York: Tudor Publishing Company; 1949), p. 53.

[1] An account of the reason for the qualification *in vacuo* can be found in I. E. Tamm: "Radiation of Particles with Speeds Greater than that of Light," *American Scientist*, Vol. XLVII (1959), p. 169, and "General Characteristics of Vavilov-Cherenkov Radiation," *Science*, Vol. CXXXI (1960), p. 206.

Unlike the Newtonian situation, in which there was only a single event E which could be significantly held to be simultaneous with E_2, the physical facts postulated by relativity require the introduction, *within a single inertial frame* S, of a convention stipulating which *particular* pair of topologically simultaneous events at P_1 and P_2 will be chosen to be metrically simultaneous. We see incidentally that *topological* simultaneity or having a *space*like separation is *not* a transitive relation: while E_x is topologically simultaneous with E_2, and E_2 is, in turn, topologically simultaneous with E, the events E_x and E are *not* topologically simultaneous but temporally separated. To this it has been objected that the "ordinary usage" of the term "simultaneous" in common sense discourse is such as to entail the transitivity of the simultaneity relation.

My reply to this objection is that it is no more incumbent upon Einstein to use the technical theoretical term "topologically simultaneous" so as to name a transitive relation on the strength of the "ordinary usage" of "simultaneous" than it is obligatory in physics to use such terms as "temperature" and "work" in their ordinary sense rather than as homonyms of their "ordinary language" counterparts. For the ordinary usage of temporal (and spatial) terms reflects the conceptual commitments of "ordinary users." And these commitments are frequently circumscribed by scientific and philosophical ignorance. In particular, the present-day advocacy of the imposition of ordinary language restrictions on the term "simultaneous" when used in its topological (i.e., *non*-metrical) sense constitutes an insistence on the retention of Newtonian beliefs and can therefore serve as an impediment to the exposition and understanding of the theory of relativity. Small wonder then that within the philosophy of science, ordinary language analysis has been almost entirely a pernicious obscurantism in so far as it has found adherents. And to the extent that ordinary language analysis has been claimed to be the sole task of philosophy, it has most unfortunately served to convince some scientists that no scientifically relevant conceptual clarification can be expected from any professional philosophers. For the bulk of the practitioners of ordinary language analysis themselves, who have enthroned the "ordinary man" to be the intellectual arbiter of the world, this state of affairs has provided the comforting assurance that scientific ignorance is not a handicap for a philosopher.

In the light of our foregoing considerations, Einstein's conceptual innovation regarding simultaneity can therefore be summarized somewhat as follows: The time relations among events having been assumed as first constituted by physical relations obtaining between them, these physical relations turned out to be such that topological simultaneity is not a uniquely obtaining relation and hence cannot serve, as it stands, as a metrical synchronization rule for clocks at the spatially separated points P_1 and P_2. Metrical simultaneity having thus been left indeterminate by both topological simultaneity and by the relativistic behavior of transported clocks, a supplementary *convention* and not merely relevant physical facts must be invoked to assert that an event at P_2 sustains a uniquely obtaining equality relation of metrical simultaneity to an event at P_1. In short, Einstein's physical innovation is that the physical relatedness which makes for the very existence of the temporal order has the kind of gaps that issue in the non-existence of absolute simultaneity, ascriptions of temporal order *within* these gaps therefore being conventional as follows: they rest on a conventional choice of a *unique* pair of events at P_1 and at P_2 as metrically simultaneous from within the class of pairs of events at P_1 and P_2 which are topologically simultaneous. It is evident that the failure of human measuring operations to disclose relations of absolute simultaneity is therefore only the epistemic consequence of the primary non-existence of these relations.

Before giving mathematical expression to the conventionality of metrical simultaneity, I wish to note that we formulated the conventionality of metrical simultaneity without any reference whatever to the relative motion of different Galilean frames and hence without any commitment whatever to there being a *discordance* or *relativity* of simultaneity as between such frames. And we shall see below in concrete detail that the repudiation of Newtonian simultaneity by the thesis of the conventionality of simultaneity does not depend at all for its validity upon there being disagreement among different Galilean observers as to the simultaneity of pairs of non-coinciding events. On the contrary, it will become apparent that the conventionality of simultaneity provides the logical framework within which the *relativity* of simultaneity can first be understood: if *each* Galilean observer adopts the particular metrical synchronization rule adopted by Einstein in Section 1 of his fundamental paper[2] and if the spatial separation of P_1 and P_2 has a component along the line of the relative motion of the Galilean frames, then that relative motion issues in their choosing as metrically simultaneous *different pairs of events* from within the class of topologically simultaneous events at P_1 and P_2, a result embodied in the familiar Minkowski diagram.

Now let the events at P_1 be assigned time numbers by means of a clock stationed there as follows: t_1 is the time of E_1, t_3 the time of E_3 and $\frac{1}{2}(t_3 + t_1)$ the time of E. The conventionality of metrical simultaneity then expresses itself in the fact that even *within* the given inertial system S, the obtaining of metrical simultaneity in the system depends on a choice—*not* dictated by any facts—of a particular numerical value between t_1 and t_3 as the temporal name to be assigned to E_2 at P_2 by an appropriate setting of a like clock stationed there in S. Using Reichenbach's notation,[3] we can say, therefore, that depending on the particular event at P_1 that is chosen to be simultaneous with E_2, upon the occurrence of E_2 we set the clock at P_2 to read

$$t_2 = t_1 + \epsilon\,(t_3 - t_1), \tag{1}$$

where ϵ has the particular value between 0 and 1 appropriate to the choice we have made. If, for example, we choose ϵ such that E_y—in our now arrow-equipped world-line diagram below —becomes simultaneous with E_2, then all of the events between E_1 and E_y become definitionally (i.e., not objectively) earlier than E_2, while all of the events between E_y and E_3 become definitionally later than E_2. Clearly, we could alternatively choose ϵ such that instead of becoming simultaneous with E_2, E_y would become earlier than E_2 or later in the same frame S. This freedom to decree definitionally the relations of temporal sequence merely expresses the objective indeterminateness of unique time relations between *causally non-connectible* events; and, of course, such

[2] A. Einstein: *PR*, *op. cit.*, p. 40.
[3] Cf. H. Reichenbach: *PST*, *op. cit.*, p. 127.

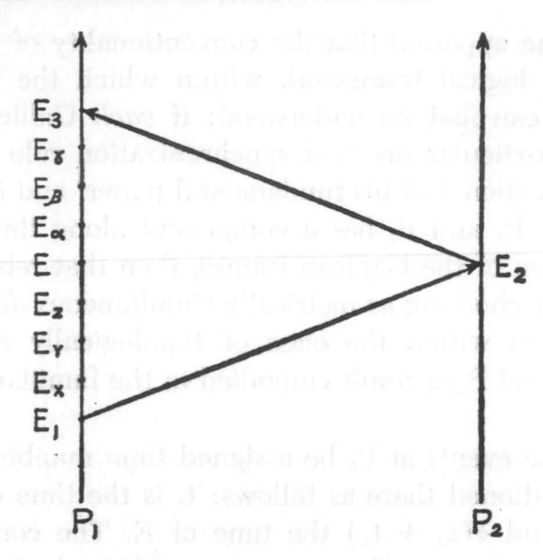

WORLD-LINE DIAGRAM

freedom can be exercised only with respect to such pairs of events.[4] For, as we saw, in the case of causally connectible events the relation of temporal separation has an objective physical basis. And once a criterion of metrical simultaneity has been chosen as outlined, the ensuing time coordinatization makes it unambiguous which one of two objectively time-separated events is the earlier of the two, and which one the later.

It is apparent that no fact of nature found in the objective temporal relations of physical events precludes our choosing a value of ϵ between 0 and 1 which *differs* from 1/2. But it is only for the value 1/2 that the velocity of light in the direction P_1P_2 becomes equal to the velocity in the opposite direction P_2P_1 in virtue of the then resulting equality of the respective transit times $t_2 - t_1$ and $t_3 - t_2$. A choice of $\epsilon = 3/4$, for instance, would make the velocity in the direction P_1P_2 only one-third as great as the return velocity. More generally, since the total round-trip time $t_3 - t_1$ is split up into the outgoing one-way time $t_2 - t_1$

---

[4] In his posthumously published work *A Sophisticate's Primer of the Special Theory of Relativity* (Middletown, Connecticut: Wesleyan University Press; 1962 [Chapter ii]), P. W. Bridgman has maintained that there is no need to ground the theory of time order of the STR on causal connectibility and non-connectibility, since time order can allegedly be based in the STR on philosophically self-contained alternative physical foundations. But in my Epilogue to this book by Bridgman, I have argued (*ibid.*, pp. 181–90) that Bridgman's contention is unsound.

and the return one-way time $t_3 - t_2$, the ratio of these two times is given by

$$\frac{t_2 - t_1}{t_3 - t_2} = \frac{\epsilon\,(t_3 - t_1)}{(1 - \epsilon)(t_3 - t_1)} = \frac{\epsilon}{1 - \epsilon}.$$

And the corresponding ratio of the one-way outgoing and return *velocities* is therefore $\frac{1 - \epsilon}{\epsilon}$. Couldn't one argue, therefore, that the choice of a value of ϵ is not a matter of convention after all, the ground being that only the value $\epsilon = 1/2$ accords with such physical facts as the isotropy of space and with the methodological requirement of inductive simplicity, which enjoins us to account for observed facts by minimum postulational commitments? And, in any case, must we not admit (at least among friends!) that the value 1/2 is more "true"? The contention that either the isotropy of space or Occam's "razor" is relevant here is profoundly in error, and its advocacy arises from a failure to understand the import of Einstein's statement that "we establish *by definition* that the 'time' required by light to travel from A to B equals the 'time' it requires to travel from B to A."[5] For, in the first place, since no statement concerning a one-way transit time or one-way velocity derives its meaning from mere facts but also requires a prior *stipulation* of the criterion of clock synchronization,[6] a choice of $\epsilon \neq 1/2$, which renders the transit times (velocities) of light in opposite directions *un*equal, cannot possibly conflict with such physical isotropies and symmetries as prevail *independently* of our descriptive conventions.[7]

---

[5] A. Einstein: *PR, op. cit.*, p. 40.

[6] In the case of some one-way measurements of the velocity of light such as the one by Roemer, the synchronization rule is introduced sufficiently tacitly not to be obviously present: cf. H. Reichenbach's account of Roemer's determination in "Planetenuhr und Einsteinsche Gleichzeitigkeit," *Zeitschrift für Physik,* Vol. XXXIII (1925), pp. 628–31.

[7] An example of such an independent kind of isotropy is the fact, discovered by Fizeau, that if, in a given inertial system, the emission of a light beam traversing a *closed* polygon in a given direction coincides with the emission of a beam traveling in the opposite direction, then the returns of these beams to their common point of emission will likewise coincide. We know from the experiments by Sagnac and by Michelson and Gale [cf. A. A. Michelson and H. G. Gale: *Astrophysical Journal,* Vol. LXI (1925), p. 140, and M. G. Trocheris: *Philosophical Magazine,* Vol. XL (1949), p. 1143] that in a *rotating* system this directional symmetry will *not* obtain.

And, in the second place, the canon of simplicity which we are pledged to observe in all of the inductive sciences is not implemented any better by the choice $\epsilon = 1/2$ than by any one of the other allowed fractional values; for no distinct hypothesis concerning physical facts is made by the choice $\epsilon = 1/2$ as against one of the other permissible values. There can therefore be no question of having propounded a hypothesis making lesser postulational commitments by that special choice. On the contrary, it is the postulated fact that light is the fastest signal which assures that *each one* of the permissible values of ϵ will be equally compatible with all possible matters of fact which are independent of how we decide to set the clock at P_2. Thus, the value $\epsilon = 1/2$ is not simpler than the other values in the inductive sense of assuming less in order to account for our observational data, but only in the descriptive sense of providing a *symbolically* simpler representation of the theory explaining these data. But this greater symbolic simplicity arising from the value 1/2 expresses itself not only in the equality of the velocities of light in opposite directions, the ratio of the outgoing and return velocities being in general $(1-\epsilon)/\epsilon$; the value $\epsilon = 1/2$ also shows its unique descriptive advantages by assuring that synchronism will be both a *symmetric* and a *transitive* relation upon using *different* clocks in the same system: the synchronism based on $\epsilon = 1/2$ is *symmetric* as between two clocks A and B, because the setting of clock B from A in accord with $\epsilon = 1/2$ issues automatically in the fact that clock A will have the readings required by a setting from clock B in accord with $\epsilon = 1/2$, and, *mutatis mutandis* for the transitivity of the synchronism of three clocks A, B, and C.[8] For values of $\epsilon \neq 1/2$, the symmetry and transitivity of synchronism generally do *not* obtain.

We have already refuted the ordinary language objection against using the term "topologically simultaneous" to name a relation which is not transitive. It is obvious that precisely the same refutation applies to such a linguistic objection against the philosophical admissibility of synchronization rules based on $\epsilon \neq 1/2$, which issue in a synchronism that is neither symmetric nor transitive.

I shall now rebut a final objection to the philosophical ad-

[8] Cf. H. Reichenbach: *Axiomatik der relativistischen Raum-Zeit-Lehre, op. cit.*, pp. 34–35 and 38–39.

missibility of values $\epsilon \neq 1/2$ before showing the following by reference to Einstein's lightning bolt experiment: if we avail ourselves of the latitude for synchronization rules given us by the philosophical admissibility of values $\epsilon \neq 1/2$, then the physical facts postulated by the STR do not dictate discordant judgments by different Galilean frames concerning the simultaneity of given events.

To state the objection whose untenability I wish to demonstrate, I first call attention to the following results. It is a presumed empirical fact of optics in the STR that if in an inertial system, the *round*-trip time of light for an open path of one-way length l is 2T, then the round-trip time for a *closed* path of total length nl is n/2 times 2T independently of direction along the closed path. Thus, for example, in the equilateral triangle ABC whose sides are each of length l, the time increment on the clock at A will be 3T for a counter-clockwise or a clockwise closed path journey ABCA (or ACBA) of a light ray which begins and ends at A.

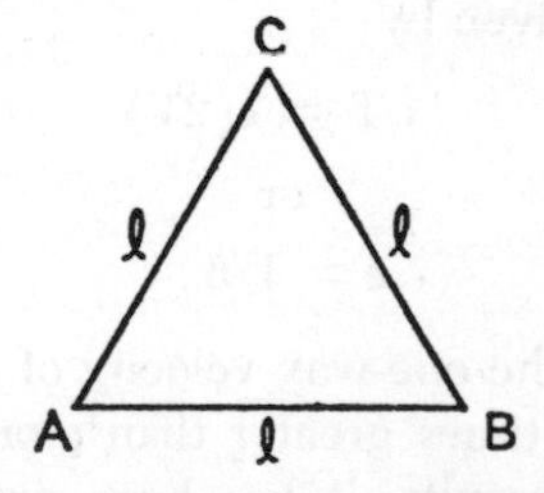

Assume that a light ray departs from A when the clock at A reads $t = 0$. Then that clock will of course read $t = 3T$ upon the return of that light ray to A quite independently of the way in which we have synchronized (set) the clocks at B and at C. Now suppose that we use $\epsilon = 1/2$ to synchronize the B clock with the one at A, and also to synchronize the C clock with the B clock. In that case, a light ray leaving A at $t = 0$ and traveling in the counter-clockwise direction ABCA will reach the B clock when the latter reads the time

$$t_2 = t_1 + \epsilon\ (t_3 - t_1) = 0 + \tfrac{1}{2}\ (2T) = T.$$

And, by the same token, this same ray will reach C upon a reading of 2T on the C clock. We have noted that the A clock reads 3T upon the arrival there of our counter-clockwise light ray.

Accordingly we can see that *if* we wish to use $\epsilon = 1/2$ again to synchronize the A clock with the C clock, then the A clock is automatically already synchronized: a light-ray departing from the C clock when it reads 2T requires a one-way transit time of *one half* of 2T, or T, to reach the point A, if $\epsilon = 1/2$ is used to synchronize the A clock with the one at C.

But now suppose that instead of $\epsilon = 1/2$, we had used $\epsilon = 2/3$ to synchronize the B clock with the one at A, and the C clock with the one at B. In that case, the arrival time of the counter-clockwise light ray at B would have been

$$t_2 = t_1 + \epsilon\,(t_3 - t_1) = 0 + \tfrac{2}{3}\,(2T) = \tfrac{4}{3}\,T.$$

And, by the same token, the arrival time of the ray at C would have been $\tfrac{8}{3}$ T. But since the previous setting of the A clock issues in a reading of 3T or $\tfrac{9}{3}$ T on it upon the return of the light ray to A, it is clear that the one-way transit time for the ray's journey from C to A was only $\tfrac{1}{3}$ T. And this means that the value of ϵ which was used to make the A clock synchronous with the one at C is given by

$$\tfrac{1}{3}\,T = \epsilon(2T)$$

or

$$\epsilon = 1/6\,.$$

Clearly, in that case the one-way velocity of light in the direction from C to A is four times greater than along either AB or BC.

In view of these results, it has been argued that the latter "compensatory" increase in the one-way velocity of light along CA is inductively highly improbable, thereby allegedly showing that on *factual* grounds of truth and falsity the value $\epsilon = 2/3$ is impermissible for setting the clocks at B and at C while the value $\epsilon = 1/2$ is uniquely warranted factually. But this argument is totally unsound, since the concept of inductive improbability of a compensation or adjustment in the one-way velocity of light along CA has no relevance here at all. For the argument is predicated on the *false* supposition that there is a factually true simultaneity of events at C and A such that there is a corresponding factually "true" synchronization of clocks at these points and a corresponding factual one-way velocity behavior of light along CA. But the one-way velocity of light depends on the criterion of simultaneity, which is conventional. And hence it is a sheer *petitio principii* to offer a *reductio ad absurdum* of the conventionality of simultaneity on the alleged ground that a particular one-way velocity of light along CA is inductively improbable. The "compensation" in the one-way velocity of light along the final leg of the closed path journey of the light ray can therefore be no more inductively improbable than the following: the hundredfold increase in the numbers representing the extensions of all bodies when we shift from meters to centimeters as units of length! It can be no more false or inductively improbable that the one-way velocities of light along AB, BC, and CA correspond to the values $\epsilon = 2/3$, $\epsilon = 2/3$, and $\epsilon = 1/6$ respectively rather than to $\epsilon = 1/2$ than it is that the lengths of bodies are those corresponding to centimeters rather than to meters! Just as the convention of using centimeters after having used meters guarantees the hundredfold increase of all length numbers independently of any inductive probabilities (or improbabilities), so also the dependence of the one-way velocity along CA on the ϵ-setting of the clock at A *automatically assures* precisely the one-way velocity appropriate to that setting!

Since clock synchronizations employing values $\epsilon \neq 1/2$ are just as admissible philosophically as those employing $\epsilon = 1/2$, I shall now show that the physical facts postulated by the STR do not dictate *discordant* judgments by different Galilean frames concerning the simultaneity of given events. To furnish the required proof, consider Einstein's familiar lightning bolt experiment in which two vertical bolts of lightning strike a moving train at points A′ and B′ respectively, and the ground at points A and B respectively, A and A′ being to the left of B and B′, respectively. Let the distance AB in the ground system be 2d. If we now define simultaneity in the ground system by the choice $\epsilon = 1/2$, then the velocity of light traveling from A to the midpoint C of AB becomes equal to the velocity of light traveling in the opposite direction BC.[9] And if the arrivals of these two oppositely directed light pulses coincide at C, then the ground observer will say that the bolt at AA′ struck simultaneously with the bolt

[9] The choice of $\epsilon = 1/2$ is *not* the only choice which would assure the equality of the velocities in the opposite directions AC and BC. The same result could be achieved by using any other *equal* values of ϵ $(0 < \epsilon < 1)$ to synchronize the clocks at A and B with the clock at C, although, in that case, the velocity along CA would *not* equal the velocity along AC, and similarly for the velocities along CB and BC.

at BB′, say at $t = 0$. Will the train observer's observations of the lightning flashes *compel* him to say that the bolts did *not* strike simultaneously? Decidedly not! To be sure, since the train is moving, say, to the left relatively to the ground, it will be impossible that the two horizontal pulses which meet at the midpoint C in the ground system also meet at the midpoint C′ of the train system. For C′ was adjacent to the point C when the ground-system clock at C read $t = 0$ and the train clock at C′ read $t' = 0$, C and C′ being the respective origins of the spatial coordinates.[1] Hence the flash from A′ will arrive at C′ *earlier than* the one from B′. But this fact will hardly require the train observer to say that the point A′ was struck by lightning earlier than the point B′, *unless* the train observer *too* has chosen *equal* values of ϵ such as 1/2 to set the clocks at A′ and at B′ to be each synchronized with the clock at C′ in his own system, a choice which then commits him to say that the time required by light from the left to traverse the distance A′C′ is *the same* as the transit time of light from the right for an equal distance B′C′.[2] But, *apart from descriptive convenience,* what is to prevent the train observer from choosing suitably *unequal* values of ϵ $(0 < \epsilon < 1)$ such that he too will say that the lightning bolts struck simultaneously, just as the ground observer did?

As I shall now show, the answer to this question is: nothing at all. To justify my contention in concrete detail, I shall demonstrate the following:

(i) the time-difference $\Delta t'_{C'}$ on the local clock at C′ between the arrivals there of the light pulses from A′ and B′ is

$$\Delta t'_{C'} = \frac{2d \cdot \beta}{c\sqrt{1 - \beta^2}}.$$

(ii) If in the train system we choose values $\epsilon_{A'}$ and $\epsilon_{B'}$ $(0 < \epsilon < 1)$ that satisfy the equation

$$\epsilon_{A'} - \epsilon_{B'} = \beta \qquad \left(\text{where } \beta \equiv \frac{v}{c}\right)$$

[1] A statement of the reason for postulating this adjacence of C and C′ can be found in P. Bergmann: *Introduction to the Theory of Relativity* (New York: Prentice Hall; 1946), p. 31, n. 1 and p. 33.

[2] If a train observer moving to the *right* relatively to the ground chooses the same ϵ for synchronizing the clocks at A′ and at B′ with the one at C′, he will, of course, say that point A′ was struck *later* than B′.

to synchronize the clocks at A′ and at B′ respectively with the clock at C′, then the lightning flashes at A′ and B′ will be simultaneous in the train system while *also* being simultaneous in the ground system.

(i) The time difference $\Delta t'_{C'}$ on the *one* clock at C′ is an objective fact independent of any choices of ϵ which we may make to synchronize the A′ and B′ clocks with the one at C′. Now the Lorentz transformation equations of the STR are based on the choice of $\epsilon = 1/2$ for all synchronizations in any given frame. But in view of the independence of $\Delta t'_{C'}$ from ϵ, we may use these Lorentz equations to compute the value of $\Delta t'_{C'}$ *without* any prejudice to our own subsequent choice of ϵ values other than 1/2 to synchronize a particular train system clock with the one at C′ in part (ii) of our present proof.

Accordingly, looking at the diagram,

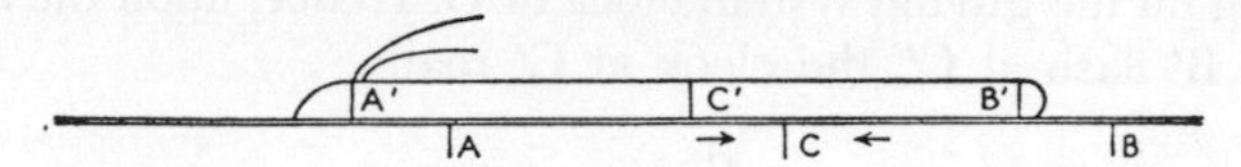

we have the following results. When the pulse from A′ reaches C′ from the left, the ground system clock at a point P adjacent to C′ reads a time t given by the distance equation

$$ct + vt = d$$

or

$$t = \frac{d}{c+v}.$$

But, in virtue of the relativistic clock retardation given by the Lorentz transformations, the train clock at C′ is slow by a factor of $\sqrt{1-\beta^2}$ as compared to the ground clock at P. Hence, upon the arrival of the A′ flash at C′, the clock at C′ reads the time

$$t'_{C'} = \frac{d}{c+v}\sqrt{1-\beta^2}.$$

To find the train time on the C′ clock when the flash from B′ reaches C′, we first calculate the time reading for that event on a ground system clock at the point Q which is then adjacent to C′.

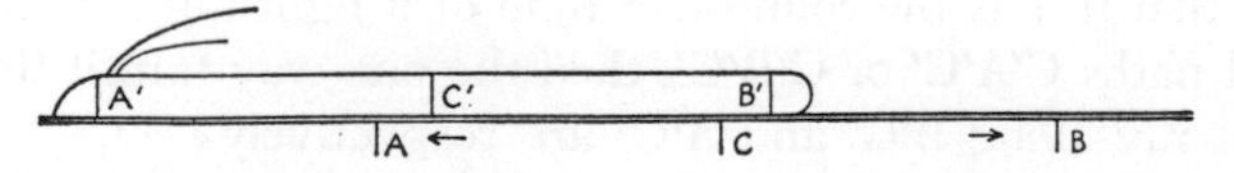

This time t is given by the following simultaneous equations, which express the fact that a light ray departing from the point B at $t = 0$ with velocity c on the ground will catch up at a distance x from B with the point C′, which departed from C at $t = 0$ in the same direction:

$$x = ct$$

and

$$x = d + vt,$$

so that

$$ct = d + vt,$$

or

$$t = \frac{d}{c - v}.$$

But again, the C′ clock reading will be slow with respect to this reading on the ground system clock at Q. Hence, upon the arrival of the B′ flash at C′, the clock at C′ reads

$$t'_{C'} = \frac{d}{c - v}\sqrt{1 - \beta^2}.$$

Accordingly, the difference on the train clock at C′ between the earlier arrival of the flash from A′ and the later arrival of the one from B′ is given by

$$\Delta t'_{C'} = \frac{d}{c - v}\sqrt{1 - \beta^2} - \frac{d}{c + v}\sqrt{1 - \beta^2} = \frac{2d \cdot \beta}{c\sqrt{1 - \beta^2}},$$

where $\beta \equiv v/c$, which is less than 1.

(ii) We must now ask: what are the respective values of $\epsilon_{A'}$ and $\epsilon_{B'}$ which must be used to synchronize the respective clocks at A′ and B′ from C′ such that in the train system the two bolts struck *simultaneously* at A′ and B′? Clearly, the required values of ϵ are those which will yield the required difference $\Delta t'_{C'}$ in the arrivals at C′ by yielding suitably *unequal* one-way transit times for the respective flash transits from A′ to C′ and from B′ to C′.

To compute the required values $\epsilon_{A'}$ and $\epsilon_{B'}$, we must first recall that if T is the round-trip time of a light ray for the equal closed paths C′A′C′ or C′B′C′, then the one-way transit times for a light ray along B′C′ and A′C′ are respectively

$$(1 - \epsilon_{B'})T$$

and

$$(1 - \epsilon_{A'})T.$$

Now let τ be the time on the A′ clock and *also* the time on the B′ clock at which the two bolts struck *simultaneously* in the train system. Then the one-way transit time for the flash along B′C′ is also given by the difference between the departure time τ and the arrival time $\frac{d}{c - v}\sqrt{1 - \beta^2}$ at C′, while the corresponding one-way transit time for the light ray journey along A′C′ is given by the difference between τ and the arrival time $\frac{d}{c + v}\sqrt{1 - \beta^2}$ at C′. We therefore have

$$\frac{d}{c - v}\sqrt{1 - \beta^2} - \tau = (1 - \epsilon_{B'})T$$

for the light ray journey along B′C′, and

$$\frac{d}{c + v}\sqrt{1 - \beta^2} - \tau = (1 - \epsilon_{A'})T$$

for the light ray journey along A′C′. Subtraction of the second of these equations from the first eliminates the quantity τ on the left-hand side, leaving only the difference $\Delta t'_{C'}$ on that side, so that we have

$$\frac{2d \cdot \beta}{c\sqrt{1 - \beta^2}} = (\epsilon_{A'} - \epsilon_{B'})T.$$

But since the equal round-trip distances C′A′C′ and C′B′C′ are each $\frac{2d}{\sqrt{1 - \beta^2}}$, as we know from the Lorentz transformations, the corresponding round-trip time T must be $T = \frac{2d}{c\sqrt{1 - \beta^2}}$. Substituting this result in our preceding equation, we obtain

$$\beta = \epsilon_{A'} - \epsilon_{B'} \qquad \textit{Q.E.D.}$$

And using any positive fractional values of ϵ compatible with the latter equation as well as the stated value of T, either of our earlier equations containing the time τ of the *simultaneous* occurrence of the bolts will permit us to solve for the value of τ

appropriate to our particular choices of ϵ. Thus, if $\beta < 1/2$ and we choose $\epsilon_{B'} = 1/2$, then $\epsilon_{A'}$ will have the fractional value $\frac{1}{2} + \beta$, and we obtain

$$\tau = \frac{d \cdot v}{c^2 \sqrt{1 - \beta^2}}$$

for the time of the *simultaneous* origination of the lightning bolt flashes at A′ and B′ in the train system. We see that, on the indicated *unequal* choices of $\epsilon_{A'}$ and $\epsilon_{B'}$, the train observer at C′ can judge the bolts at A′ and B′ to have struck simultaneously, because he would account for the time difference $\dfrac{2d \cdot \beta}{c\sqrt{1 - \beta^2}}$ between the arrivals of the light pulses from them at C′ by the *corresponding difference* between the *one-way velocities* of the pulses along A′C′ and B′C′ respectively.

It is worthy of notice that this judgment of simultaneity in the train system need not be confined to a train observer at the midpoint C′ of A′B′. To see that this is so, we point out first that since by hypothesis, the light pulses originating at the lightning flashes at A′ and at B′ will meet at the midpoint C in the ground system, they will also meet at a point D′ on the train adjacent to C and lying between C′ and B′, as shown on the diagram.

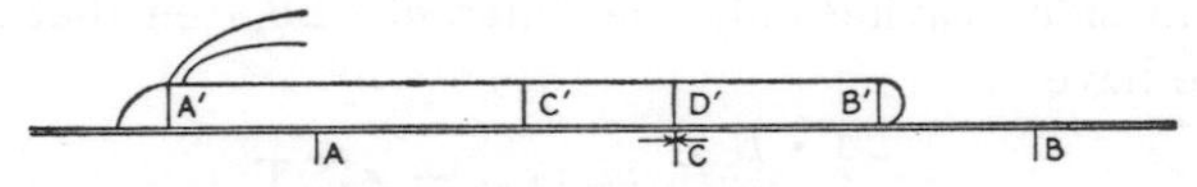

Let us now compute the distance of the point of encounter D′ from both B′ and A′ from the relevant Lorentz transformation equation

$$x' = \frac{x + vt}{\sqrt{1 - \beta^2}}.$$

Now the x-coordinate of D′ is that of the origin at C, and the ground system time t of encounter of the two flashes at C and D′ is given by $t = d/c$. Hence, inserting this value of t and $x = 0$ in the Lorentz equation, we obtain for the distance $\overline{C'D'}$

$$\overline{C'D'} = x' = \frac{vd}{c\sqrt{1 - \beta^2}}.$$

By the Lorentz transformations, the distance C′B′ on the train is

$$\overline{C'B'} = \frac{d}{\sqrt{1 - \beta^2}}.$$

Accordingly, subtracting $\overline{C'D'}$ from $\overline{C'B'}$ and remembering that $\beta \equiv v/c$, we obtain for the distance $\overline{B'D'}$

$$\overline{B'D'} = \frac{d}{\sqrt{1 - \beta^2}} - \frac{v}{c}\frac{d}{\sqrt{1 - \beta^2}} = d\sqrt{\frac{c - v}{c + v}}.$$

Now the distance $\overline{A'B'}$ is simply twice that of $\overline{C'B'}$, and upon subtracting from $\overline{A'B'}$ the distance $\overline{B'D'}$, we find that the distance $\overline{A'D'}$ is

$$\overline{A'D'} = \frac{2d}{\sqrt{1 - \beta^2}} - d\sqrt{\frac{c - v}{c + v}} = d\sqrt{\frac{c + v}{c - v}}.$$

Accordingly, the *ratio* of the distance $\overline{A'D'}$ to $\overline{B'D'}$ is

$$(c + v)/(c - v).$$

To assure the *simultaneity* of the flashes at A′ and B′ from the standpoint of a train observer at the non-midpoint D′, we need only stipulate that the ratio of the one-way velocity of light along A′D′ to the one-way velocity of light along B′D′ is the *same* as the ratio $(c + v)/(c - v)$ of the distances $\overline{A'D'}$ and $\overline{B'D'}$, thereby guaranteeing that the flashes required equal transit times to traverse these unequal paths. And the required one-way velocities along A′D′ and B′D′ can be assured by merely making appropriate choices of $\epsilon_{A'}$ and $\epsilon_{B'}$ for synchronizing the clocks at A′ and at B′ with the clock at D′. This can be done in an infinitude of different ways to each of which corresponds, for a given setting of the clock at D′, a particular time of the *simultaneous* occurrence of the two bolts as judged by the train observer at D′. Thus, if we again choose $\epsilon_{B'} = 1/2$ (as we did when synchronizing the B′ clock with the one at C′), then the one-way velocity of light along B′D′ is c and hence the one-way transit time along B′D′ is

$$\Delta t'_{B'D'} = \frac{d}{c}\sqrt{\frac{c - v}{c + v}}.$$

Since the round-trip time of light for the closed path D′A′D′ is $\frac{2d}{c}\sqrt{\frac{c+v}{c-v}}$, the one-way transit time $\Delta t'_{A'D'}$ along A′D′ is given by

$$\Delta t'_{A'D'} = (1 - \epsilon_{A'}) \frac{2d}{c}\sqrt{\frac{c+v}{c-v}}.$$

Since we wish to choose a value of $\epsilon_{A'}$ which will result in $\Delta t'_{A'D'} = \Delta t'_{B'D'}$, we have:

$$(1 - \epsilon_{A'}) \frac{2d}{c}\sqrt{\frac{c+v}{c-v}} = \frac{d}{c}\sqrt{\frac{c-v}{c+v}}.$$

Solving for $\epsilon_{A'}$, we find

$$\epsilon_{A'} = \frac{c + 3v}{2\,(c + v)}.$$

To see that this value of $\epsilon_{A'}$ meets the requirement of being between 0 and 1 in the relativistic context of $v < c$, we simply rewrite $\epsilon_{A'}$ in the form

$$\epsilon_{A'} = \frac{c + 2v + v}{c + 2v + c}.$$

If the clock at D′ is set in accord with the Lorentz transformations to read a time $t' = \frac{d}{c\sqrt{1 - \beta^2}}$ upon the joint arrival of the flashes from A′ and B′, then the stated choices of $\epsilon_{A'}$ and $\epsilon_{B'}$ by the train observer at D′ will have the following result: just as the train observer at C′, so also the train observer at D′ will synchronize the A′ and B′ clocks with his own so as to assign the time $t' = \frac{d \cdot v}{c^2\sqrt{1 - \beta^2}}$ to *each* of the lightning strokes at A′ and B′, thereby judging them to be simultaneous. So much for the possibility of having *concordant* judgments of simultaneity as between the ground system and the train system.

The flash at AA′ is causally absolutely non-connectible with any of those events at point B in the ground system which are members of the open time interval I of events bisected by the flash at BB′ and of duration 4d/c units of time in the ground system. It is this absolute causal non-connectibility that enables the ground observer to say that the BB′ flash was simultaneous with the AA′ flash, as required by his choice of $\epsilon = 1/2$, and *also* enables the train observer, if he wishes, to choose $\epsilon = 1/2$ for himself as well and to accept the following *discordance* of simultaneity judgments as a consequence of his own separate choice of $\epsilon = 1/2$: First, it is *not* the flash at BB′ that is simultaneous with the AA′ flash, as the ground-system observer maintains by his assignment of $t = 0$ to both events, but a different I-interval event E_e at B which is absolutely earlier than the BB′ flash, the ground-system clock at B reading the time $t = -2vd/c^2$ upon the occurrence of E_e; and second, the train-system time of the AA′ flash and of the event E_e is

$$t' = -\frac{vd}{c^2\sqrt{1-\beta^2}},$$

but the BB′ flash occurs at the later train-system time

$$t' = +\frac{vd}{c^2\sqrt{1-\beta^2}}.$$

It follows that the conventionality of simultaneity within each inertial system entailed by the aforementioned absolute causal nonconnectibility allows each Galilean observer to choose his own values of ϵ *either* so as to agree with the other observers on simultaneity *or* so as to disagree. If each separate Galilean frame does choose the value 1/2—which, as we saw, Einstein assumed in the formulation of the theory for the sake of the resulting descriptive simplicity—then the relative motion of these frames indeed makes disagreement in their judgments of simultaneity unavoidable, except in regard to pairs of events lying in the planes perpendicular to the direction of their relative motion.[3] But, in the first instance, it is not the relative motion of inertial systems but the limiting character of the velocity of light and the behavior of clocks under transport which give rise to the relativity of simultaneity. For the Newtonian theory affirmed absolute simultaneity while countenancing the relative motion of inertial systems.

Moreover the presumed physical facts which give rise to the

---

[3] Since $t' = \dfrac{t - \frac{v}{c^2}x}{\sqrt{1-\beta^2}}$, spatially separated events occurring at the same time t in system S will occur at different times t′ in system S′, unless they lie in the planes given by $x =$ constant. Cf. H. Reichenbach: *Axiomatik der relativistischen Raum-Zeit-Lehre*, *op. cit.*, pp. 55–56.

conventionality and relativity of simultaneity (as distinct from man's discovery of these facts) are evidently quite independent of man's presence in the cosmos and of his measuring activities. Hence, the relativity of simultaneity is not first generated by our inability to carry out those measuring operations that would yield relations of absolute simultaneity in the Newtonian sense. Instead, the failure of human signaling and measuring operations to disclose relations of absolute simultaneity is only the epistemic consequence of the primary nonexistence of these relations. It is therefore erroneous or at best highly misleading to give an epistemic or homocentrically operational twist to Einstein's conception of simultaneity by emphasizing the role of signals as a means of human knowledge. Such a homocentrically operational twist falsely portrays what Einstein construes, in the first instance, as an ontological feature of the temporal relatedness of physical events. For the homocentrically operational account suggests that Einstein's repudiation of Newton's absolute simultaneity rests on a mere *epistemic* limitation on the *ascertainability* of the existence of relations of absolute simultaneity. To be sure, human operations of measurement are indispensable for *discovering* or *knowing* the physical relations and thereby the time relations sustained by particular events. But these relations are or are not sustained by physical macro-events quite apart from *our* actual or hypothetical measuring operations and are *not* first *conferred* on nature by our operations. In short, it is because no relations of absolute simultaneity *exist* to be measured that measurement cannot disclose them; it is not the mere failure of measurement to disclose them that *constitutes* their nonexistence, much as that failure is *evidence* for their nonexistence.

Only a philosophical obfuscation of this state of affairs can make plausible the view that the relativity of simultaneity (or, for that matter, any of the other philosophical innovations of relativity theory) lends support to the subjectivism of homocentric operationism or of phenomenalistic positivism.[4]

---

[4] For further details on why I deny that the STR supports an operational account of scientific concepts in P. W. Bridgman's homocentric sense, see A. Grünbaum: "Operationism and Relativity," *The Scientific Monthly*, Vol. LXXIX (1954), p. 228; reprinted in *The Validation of Scientific Theories*, *op. cit.*, pp. 84–94.

(C) HISTORY OF EINSTEIN'S ENUNCIATION OF THE LIMITING CHARACTER OF THE VELOCITY OF LIGHT IN VACUO.

We can restate what we have called assumption (ii) above as follows:

(ii) light is the fastest signal *in vacuo* in the following topological sense: no kind of causal chain (moving particles, radiation) emitted *in vacuo* at a given point A together with a light pulse can reach any other point B earlier, as judged by a local clock at B which merely orders events there in a metrically arbitrary fashion, than this light pulse.[5]

We saw that this assumption (ii)—along with Einstein's postulate that clock transport does *not* define relations of absolute simultaneity, which we called assumption (i)—is fundamental to Einstein's conception of simultaneity. It therefore behooves us to inquire in the present section into the grounds which Einstein had in 1905 for making assumption (ii). And we note, incidentally, that our very awareness of the fundamental historical problem we have just posed depends upon a clear prior philosophical comprehension of Einstein's conception of simultaneity, free from the prevalent misunderstandings which we have criticized.

As will presently be evident, Einstein himself gives us tantalizingly incomplete explicit information concerning the grounds for his original confidence in his intuition that assumption (ii)

[5] It must be remembered that, in its metrical form, this postulate does not confine purely "geometric" velocities which are not the velocities of causal chains to the value c. For example, within any given inertial system K, special relativity allows us to subtract vectorially (as against by means of the Einstein velocity addition formula) a velocity $v_1 = .9c$ of body A in K and a velocity $v_2 = -.3c$ of body B in K to obtain 1.2c as the relative velocity of separation of A and B as judged by the K-observer [though not by an observer attached to either A or B!]. For no body or disturbance is thereby asserted to be traveling relative to any other at the velocity 1.2c of separation, and since the process of separation has no direction of propagation, it cannot be a causal chain. See Max von Laue: *Die Relativitätstheorie* (Braunschweig: F. Vieweg; 1952), Vol. I, pp. 40–41; L. Silberstein: *The Theory of Relativity* (New York: The Macmillan Company; 1914), p. 164n; A. Sommerfeld: Notes in *PR*, a collection of original memoirs (New York: Dover Publications, Inc.; reprint), p. 94; A. Grünbaum: *Philosophy of Science*, Vol. XXII (1955), p. 53; H. P. Robertson: *Mathematical Reviews*, Vol. XVI (1955), p. 1166; G. D. Birkhoff: *The Origin, Nature and Influence of Relativity* (New York: The Macmillan Company; 1925), p. 104, and J. Weber: *American Journal of Physics*, Vol. XXII (1954), p. 618.

is true. But even if we did possess full clarity on that score, we still confront the same question in regard to assumption (i) and are compelled to try to answer it with even less assistance from Einstein himself, as we shall see.

The importance of also understanding the grounds on which Einstein thought he could safely make assumption (i) can be gauged by the following basic fact: if, in accord with Newtonian conceptions, assumption (i) had been thought to be *false,* then the belief in the truth of (ii) alone would *not* have warranted the abandonment of the received Newtonian doctrine of absolute simultaneity. And, in that eventuality, the members of the scientific community to whom Einstein addressed his paper of 1905 would have been fully entitled to *reject* his *conventionalist* conception of *one*-way transit times and velocities. But the denial of absolute simultaneity by the latter conception is *crucial* for his principle of the constancy of the speed of light, as is evident from his statement of this principle in Section 2 of his 1905 paper. Specifically, Einstein emphasizes there that if the *one-way* velocity of light is also to have the numerical value c, then "time interval is to be taken in the sense of the [convention or] definition [of simultaneity] in Section 1."[6] The indispensability of assumption (i) for Einstein's conception of simultaneity is further apparent from the fact that a Newtonian physicist quite naturally regards *not* signal connectibility but the readings of suitably transported clocks as the fundamental indicators of temporal order. The Newtonian recognizes, of course, that the truth of (ii) compels such far-reaching revisions in his theoretical edifice as the repudiation of his second law of motion, which allows particles to attain arbitrarily high velocities through accelerations of appropriate durations. But he stoutly and rightly maintains that *if* (i) is *false,* absolute simultaneity remains intact, unencumbered by the truth of (ii).

It is of both historical and logical importance to note that in "The Principles of Mathematical Physics," an address delivered in St. Louis on September 24, 1904, and published in *The Monist,* Vol. XV (1905), pp. 1-24,[7] H. Poincaré had already envisioned

[6] Cf. A. Einstein: *PR, op. cit.,* p. 41.

[7] This address by Poincaré is reprinted in the *Scientific Monthly,* Vol. LXXXII (1956), p. 165.

the construction of a new mechanics in which the velocity of light would play a limiting role. Although he concluded this paper with the remark that "as yet nothing proves that the (old) principles will not come forth from the combat victorious and intact,"[8] Poincaré did prophesy that:

> From all these results, if they are confirmed, would arise an entirely new mechanics, which would be, above all, characterized by this fact, that no velocity could surpass that of light, any more than any temperature, could fall below the zero absolute, because bodies would oppose an increasing inertia to the causes, which would tend to accelerate their motion; and this inertia would become infinite when one approached the velocity of light.[9]

But, unlike Einstein, Poincaré failed to see the import of this new limiting postulate for the meaning of simultaneity: in the very same paper,[1] he speaks of spatially separated clocks as marking *the same hour at the same physical instant* (my italics). And he distinguishes between watches which mark "the true time" and those that mark only the "local time," a distinction which was also invoked, as we shall see, by Larmor and Lorentz. But Poincaré does note that a Galilean observer will not be able to detect whether his frame is one in which the clocks mark the allegedly spurious local time. Accordingly, he affirms "The principle of relativity, according to which the laws of physical phenomena should be the same, whether for an observer fixed, or for an observer carried along in a uniform movement of translation; so that we have not any means of discerning whether or not we are carried along in such a motion."[2]

Specifically what does Einstein himself tell us about his original grounds for assuming the truth of assumption (ii)? At this point, it is essential to quote him *in extenso*. He writes:

> By and by I despaired of the possibility of discovering the true laws by means of constructive efforts based on known

[8] H. Poincaré: "The Principles of Mathematical Physics," *The Monist*, Vol. XV (1905), p. 24.

[9] *Ibid.*, p. 16.

[1] *Ibid.*, p. 11.

[2] *Ibid.*, p. 5.

For a useful, brief account of these pre-Einsteinian conceptions, cf. P. Bergmann: "Fifty Years of Relativity," *Science*, Vol. CXXIII (1956), p. 123.

> facts. The longer and the more despairingly I tried, the more I came to the conviction that only the discovery of a universal formal principle could lead us to assured results. The example I saw before me was thermodynamics. The general principle was there given in the theorem: the laws of nature are such that it is impossible to construct a *perpetuum mobile* (of the first and second kind). How, then could such a universal principle be found? After ten years of reflection such a principle resulted from a paradox upon which I had already hit at the age of sixteen: If I pursue a beam of light with a velocity c (velocity of light in a vacuum), I should observe such a beam of light as a spatially oscillatory electromagnetic field at rest. However, there seems to be no such thing, whether on the basis of experience or according to Maxwell's equations. From the very beginning it appeared to me intuitively clear that, judged from the standpoint of such an observer, everything would have to happen according to the same laws as for an observer who, relative to the earth, was at rest. For how, otherwise, should the first observer know, i.e., be able to determine, that he is in a state of fast uniform motion?
>
> One sees that in this paradox the germ of the special relativity theory is already contained. Today everyone knows, of course, that all attempts to clarify this paradox satisfactorily were condemned to failure as long as the axiom of the absolute character of time, viz., of simultaneity, unrecognizedly was anchored in the unconscious. Clearly to recognize this axiom and its arbitrary character really implies already the solution of the problem. The type of critical reasoning which was required for the discovery of this central point was decisively furthered, in my case, especially by the reading of David Hume's and Ernst Mach's philosophical writings.[3]

We see that Einstein gives essentially three reasons for his original belief in assumption (ii), noting that, like two of the laws of thermodynamics, this assumption is a "principle of impotence," to use E. T. Whittaker's locution.[4] Einstein's seemingly distinct three reasons are the following: (1) "on the basis of experience," there are no "stationary" light waves, (2) Neither

---

[3] A. Einstein: "Autobiographical Notes," *op. cit.*, p. 53.

[4] Cf. E. T. Whittaker: *From Euclid to Eddington* (Cambridge: Cambridge University Press; 1949), §25, pp. 58–60.

are there any such phenomena on the basis of Maxwell's equations, and (3) At the very beginning, there was intuitive clarity that preferred inertial systems do not exist, the laws of physics, including those of light propagation, being the same in all of them. These three reasons invite the following corresponding three comments.

1. The failure of our experience to have disclosed the existence of stationary light waves is not, of course, presumptive of their non-existence, unless that experience included the circumstances requisite to our observation of such waves if they do exist. What could such circumstances be? Suppose it were physically possible for a star to recede from the earth at the speed c of light. Assuming that there actually is such a star, postulate further that the speed of the light emitted by the star in the direction of the earth is c relatively to the *star*, the light's speed relatively to the earth being given by the *Galilean-Newtonian* velocity addition and hence being *zero*. Then the earth would maintain a *constant distance* from the light wave. And if there were a way for us to register the presence of that stationary light wave, then we could have evidence of its existence. *Mutatis mutandis*, a light source in the laboratory moving at the velocity c might have produced the same kind of phenomenon.

Perhaps Einstein envisaged these kinds of conditions as situations in which our experience ought to have disclosed the existence of stationary light waves.

If so, one wonders, however, how much weight he actually attached to this observational argument on behalf of assumption (ii). For he was undoubtedly cognizant of the contingency of the conditions governing the observable occurrence of the phenomenon in question. In particular, it should be noted that Einstein's mention of "the basis of experience" in this context *cannot* be assumed to be referring to the 1902–1906 experiments by Kaufmann and others on the deflection of electrons (β-rays) in electric and magnetic fields. For if we suppose him to have been familiar with these experiments, they must have left him in a quandary precisely in regard to the truth of assumption (ii): while yielding a mass variation with velocity incompatible with Newtonian dynamics, the results of these experiments were unable to rule out the formulae of *Abraham's* dynamics, which *allowed* particle velocities *exceeding* the velocity of light *in vacuo*.[5]

John Stachel has made the interesting suggestion that the clue to Einstein's reference to "the basis of experience" may be found in his commentary on the significance of the results of Fizeau's experiments on the velocity of light in moving liquids. Writing concerning the empirical formula codifying Fizeau's results, Einstein says: "From the stated formula, one can make the interesting deduction that a liquid which does not refract light at all ($n = 1$) would not affect the propagation of light with respect to it even if the liquid is moving."[6]

2. Can stationary light waves be regarded as ruled out by Maxwell's equations, if one does not already accept the principle of relativity, which guarantees the validity of the usual form of Maxwell's equations in all inertial systems? In other words, can the *second* of Einstein's avowed reasons for his dismissal of the possibility of stationary light waves be regarded as other than a logical consequence of the third? Clearly, if Maxwell's equations *are* coupled with the principle of relativity, then these equations indeed rule out stationary light waves in every inertial system. But since Maxwell's equations are not covariant under *Galilean* transformations, it is far from clear that stationary light waves are precluded by the form assumed by the equations in an inertial system S moving with the velocity c relatively to the primary (aether) frame K *and* having coordinates which are related by the *Galilean* transformations to those of the K-system. Since Einstein does not mention the "Galilean transform" of Maxwell's equations, it would seem that the only reason why he felt justified in regarding Maxwell's equations as support for his repudiation of stationary light waves was that he had already assumed the principle of relativity on intuitive grounds.

3. In view of the presumably flimsy character of the appeal to

---

[5] For details, cf. M. von Laue: *Die Relativitätstheorie, op. cit.*, Vol. I, pp. 26–27. Experiments which can reasonably be interpreted as strongly supporting Einstein's assumption (ii) were not successfully carried out until a good many years after 1905: Cf. C. Møller: *The Theory of Relativity, op. cit.*, pp. 85–89.

[6] A. Einstein: "Die Relativitätstheorie," in: E. Warburg (ed.) *Physik* (Leipzig: Teubner; 1915), p. 704.

experience and of the redundancy of (2) with (3) among the reasons given by Einstein, we are pretty much left with his intuitive confidence in the principle of relativity as the basis for his assumption of (ii).

It is not uncommon to hear it said that Einstein first arrived at assumption (ii) by deducing it from his formulas for the addition of velocities, these formulas thereby allegedly constituting the grounds on which he first convinced himself that assumption (ii) was true. This claim is *not* supported by relevant historical evidence, but has been put forward as the *historical corollary* of an inveterate compound error in which both a formal logical blunder and a philosophical misconception are ingredient: the doubly false supposition that (a) the relativistic addition law for velocities entails assumption (ii), and that (b) assumption (ii) is *not* presupposed by the principle of the constancy of the speed of light (and hence by the Lorentz transformations) but is first derived in the STR from the law for velocity addition.

I shall now demonstrate, however, that both (a) and (b) are grossly in error.

If an object has components of velocity u_x', u_y' and u_z' in an inertial system K′ moving with the velocity v along the positive x-axis of another inertial system K, then the corresponding components of velocity u_x, u_y, and u_z in K are:

$$u_x = \frac{u_x' + v}{1 + u_x' v/c^2}$$

$$u_y = \frac{u_y' (1 - v^2/c^2)^{\frac{1}{2}}}{1 + u_x' v/c^2}$$

$$u_z = \frac{u_z' (1 - v^2/c^2)^{\frac{1}{2}}}{1 + u_x' v/c^2}$$

Whatever may be the components of velocity in one system at a given time, these transformation equations relate them to the corresponding components in the other system. This is shown by the behavior of the function $u_x = f(u_x', v)$ for values of the *independent* variable u_x' which *exceed* c, values whose exclusion from the range of u_x' clearly requires grounds *other than* the above velocity transformations. Thus, if $u_x' = 2c$ while $v = c/2$, then $u_x = 5c/4$. Again, if $v < c$, a finite velocity *greater* than c of magnitude $u_x = c^2/v$ corresponds to an *infinite* velocity u_x'. And the super-c values of the component velocities that are allowed by the velocity-addition laws are, in fact, likewise allowed by the STR as a whole *provided* these velocities pertain to propagations in which no causal influence, energy, or matter is transmitted at super-c speeds. It is *immaterial* that the above equations for the components u_y and u_z do *rule out* that the *system* K′ have a relative velocity v in excess of c by yielding imaginary values for that case. For this result does *not* preclude that accelerating objects have time-dependent *super*-c component velocities in K′.

What the relativistic addition formulas do show, and all that Einstein claimed for them in this connection in Section 5 of his 1905 paper, is that "from a composition of two velocities which are less than c, there always results a velocity less than c" and that "the velocity of light c cannot be altered by composition with a velocity less than that of light."[7] The fallacy of inferring assumption (ii) from the relativistic velocity-addition laws dies hard, however, since it has been explicitly committed in the writings of such eminent authors as E. T. Whittaker[8] and R. C. Tolman.[9]

It is likewise an error to suppose that assumption (ii) is *not presupposed* by the principle of the constancy of the speed of light (and hence by the Lorentz transformations), the result of this error being that assumption (ii) is then wrongly believed to still require logical justification by the time the velocity addition law is derived from the Lorentz transformations. That these suppositions are mistaken can now be readily shown as follows. Einstein deduced the velocity-addition laws in Section 5 of his 1905 paper via the mediation of the Lorentz transformations from the two basic principles of his Section 2, i.e., from the principle of relativity and the principle of the constancy of the speed of light. Now, the latter principle presupposes, as he notes *pointedly* at the start of Section 2, that the *one*-way transit-time ingredient

---

[7] Cf. A. Einstein: *PR, op. cit.*, p. 51.

[8] E. T. Whittaker: *A History of the Theories of Aether and Electricity* (London: Thomas Nelson & Sons, Ltd.; 1953), Vol. II, p. 38.

[9] R. C. Tolman: *Relativity, Thermodynamics and Cosmology* (Oxford: Oxford University Press; 1934), p. 26.

in the *one*-way velocity of light is based on the *definition of simultaneity given in his Section 1,* which in turn rests on an avowedly conventionalist as opposed to absolutist conception of simultaneity. Since the deduction of the velocity-addition laws thus presupposes the *denial* of absolute simultaneity, it clearly presupposes as *one* of its premises the limiting character of the velocity of light *in vacuo,* i.e., assumption (ii). Hence, even if assumption (ii) were deducible from the velocity-addition law, this law could not have been the basis on which Einstein first convinced himself of the truth of assumption (ii).

It is apparent that logical and philosophical confusion concerning the role of assumption (ii) has begotten a historical falsehood as to how Einstein first arrived at that assumption.

We are now in a position to make a conjecture as to Einstein's grounds for assumption (i). We recall that (i) asserts that within the class of events, material clocks do not define relations of absolute simultaneity under transport. As Einstein states explicitly both in our citation from his intellectual autobiography and in his formulation of the principle of the constancy of the speed of light in Section 2 of the 1905 paper, *the absolutistic conception of simultaneity* was the Gordian knot obstructing the resolution of his boyhood paradox, i.e., the reconciliation of the two basic principles of his Section 2. But, assumption (i) was *required* no less than (ii) for the denial of absolute simultaneity! Hence his confidence in (i) must be presumed to have derived from his belief in the correctness of *both* of the two principles in his Section 2. One wonders in this connection what role, if any, was played in the development of Einstein's reflections on simultaneity by Poincaré's *obscurely* conventionalist treatment of simultaneity in his paper "La Mesure du Temps."[1]

Would it be safe to conclude from Einstein's autobiographical statement that actual experimental results bearing on the velocity of light *in vacuo* and on its status as a limiting velocity in fact played no role at all when he groped his way to an espousal of the principle of relativity? If so, there would be the following serious question: can the theoretical guesses of an Einstein be regarded to have been genuinely more *educated,* as opposed to

---

[1] H. Poincaré: "La Mesure du Temps," *op. cit.,* pp. 1–13.

just more *lucky*, than the abortive fantasies of those quixotic scientific thinkers whose names have sunk into oblivion and whose subjective confidence in having made a bona fide discovery was no less passionate than was Einstein's? This important question, which I raised in an earlier publication,[2] was not answered but merely evaded by M. Polanyi, who—on the strength of the wisdom of a half century of hindsight!—offers the following *petitio principii* in criticism of me: "Dr. Grünbaum . . . seems repeatedly to express the view that Einstein had no sufficient reason to adopt the fundamental assumptions of relativity when he in fact did so. But he [Grünbaum] is more puzzled by this than I am, for he does not allow for the unspecifiable clues which justifiably [sic] guided Einstein's formulation."[3] But this declaration immediately prompts the further question of just how Polanyi's account illuminates the epistemological (methodological) attributes of a bona fide scientific discovery as opposed to those of a plausible and cherished but abortive speculation, while maintaining, as he does, both the *fallibility* and the *unspecifiability* of the clues on which scientific theorists rely. Flying the banner of unspecifiability of clues while heaping scorn on the demand for adequate empirical support of knowledge claims in physics does not absolve Polanyi from the necessity of answering this further question. For example, what besides his present, inherently *ex post facto* possession of the wisdom of hindsight enables Polanyi's unspecifiability thesis to discriminate in regard to methodological justifiability between the following two hypotheses at the time of their initial enunciation: first, Einstein's assumption that the velocity of light is independent of the velocity of the emitting source, which has subsequently turned out to be successful (H. Dingle's allegations to the contrary notwithstanding),[4] and second, Ritz's contrary hypothesis, which, to the profound disappointment and chagrin of his followers, turned

[2] Cf. A. Grünbaum: "The Genesis of the Special Theory of Relativity," in H. Feigl & G. Maxwell (eds.) *Current Issues in the Philosophy of Science, op. cit.*, p. 49.

[3] M. Polanyi: "Notes on Professor Grünbaum's Observations," in: *Current Issues in the Philosophy of Science, op. cit.*, pp. 54–55; hereafter this paper by Polanyi will be cited under the abbreviation "Notes, etc."

[4] Cf. A. Grünbaum: "Professor Dingle on Falsifiability: A Second Rejoinder," *British Journal for the Philosophy of Science,* Vol. XII (1961), pp. 153–56.

out to be unsuccessful after his death in 1909? If Polanyi's answer were to be that each of these contrary hypotheses was indeed methodologically justified in the context of its own set of unspecifiable but fallible clues, then his account altogether loses its relevance to its avowed subject. For it then throws no light whatever on the epistemology of scientific discovery as distinct from the psychology of imaginative but pathetically abortive speculation. The conclusion seems inescapable that if Polanyi is at all to avoid the well-known aprioristic pitfalls of classical rationalism, his conception of the logic of scientific discovery compels him to seek refuge in the unspecifiability of clues as an *asylum ignorantiae*. And his answer to my question as to what, besides good fortune, does distinguish *methodologically* between scientific discoveries of genius and brilliant flights of scientific fantasy merely baptizes the difficulty which I raised and gives it the name "unspecifiability."

But is it true that actual experiments on the velocity of light such as the Michelson-Morley experiment did not play any genetic role in the STR? E. T. Bell, writing *before* the publication of Einstein's "Autobiographical Notes," and referring to the influence of the Michelson-Morley experiment on Einstein, claims that "he has stated explicitly that he knew of neither the experiment nor its outcome when he had already convinced himself that the special theory was valid."[5] And M. Polanyi reports[6] that Einstein had authorized him in 1954 to publish the statement that "the Michelson-Morley experiment had a negligible effect on the discovery of relativity."

Yet if *both* of the reasons for the STR adduced by Einstein in the Introduction to his 1905 paper were among the factors which had prompted his initial espousal of the STR, the consonance of the foregoing claims by Bell and Polanyi with Einstein's own text there is quite problematic. For in that Introduction, Einstein cites the following two considerations as suggesting "that the phenomena of electrodynamics as well as of mechanics possess no properties corresponding to the idea of absolute rest":[7] (1)

---

[5] E. T. Bell: *The Development of Mathematics* (New York: McGraw-Hill Book Company; 1945), p. 210.

[6] M. Polanyi: *Personal Knowledge* (Chicago: University of Chicago Press; 1958), pp. 10–11, (hereafter cited under the abbreviation *PK*.)

[7] A. Einstein: *PR, op. cit.*, p. 37; italics are mine.

the lack of *symmetry* in the classical electrodynamic treatment of a current-carrying wire moving relatively to a magnet at rest, on the one hand, and of a magnet moving relatively to such a wire at rest, on the other, (2) *"the unsuccessful attempts to discover any motion of the earth relatively to the 'light medium'* [aether]." Unless they provide some other consistent explanation for the presence of the latter statement in Einstein's text of 1905, it is surely incumbent upon all those historians of the STR who deny the inspirational role of the Michelson-Morley experiments to tell us specifically what other "unsuccessful attempts to discover any motion of the earth relatively to the 'light medium'" Einstein had in mind here. And it was likewise incumbent upon Holton to tell us how he reconciles this explicit reference to experiments, whose identity was presumably clear to Einstein's public in 1905, with the following claims:[8] Einstein's paper "begins with the statement of formal asymmetries or other incongruities of a predominantly esthetic nature (rather than, for example, a puzzle posed by unexplained experimental facts)," [p. 629] and "The [1905 STR] paper does not invoke explicitly any of the several well-known experimental difficulties" [p. 630]. This obligation to take cognizance of the second consideration in the Introduction to the 1905 paper should also have been shouldered by the mature *reminiscing* Einstein himself when authorizing the statement given by Polanyi.

All the more so, since, as early as 1915, Einstein himself gave the following *historical* account in his contribution "Die Relativitätstheorie" to the volume *Physik*:[9] "It is hardly possible to form an independent judgment of the justification of the theory of relativity, if one does not have some acquaintance with the experiences and thought processes which preceded it. Hence these will need to be discussed first. [p. 703] . . . The successes of the Lorentzian theory were so significant that the physicists would have abandoned the principle of relativity without qualms, had it not been for the availability of an important experimental result, . . . namely Michelson's experiment" [p. 706]. It is quite unfortunate that the serious prima facie discrepancies between Einstein's testimony of 1905 and 1915, on the one hand, and of

---

[8] G. Holton: "On the Origins of the Special Theory of Relativity," *op. cit.*, p. 629 and p. 630.

[9] A. Einstein: "Die Relativitätstheorie," *op. cit.*, pp. 703 and 706.

1954 (as transmitted to Polanyi via Dr. N. Balazs), on the other, did not restrain Polanyi from making the following assertion or at least induce him to modify it: "When Einstein discovered rationality in nature, unaided by any observation that had not been available for at least fifty years before, our positivistic textbooks promptly covered up the scandal by an appropriately embellished account of his discovery," i.e., by falsely portraying relativity "as a theoretical response to the Michelson-Morley experiment."[1]

Conceivably our historical conundrum is resolved by Professor Henry S. Frank's suggestion that when writing his STR paper for the scientific public of 1905, Einstein deemed it appropriate to appeal also to the failure of the aether theory in the Michelson-Morley experiments as a justification of the STR, even though that failure had genetically not been a reason for his own initial confidence in the STR. But this suggestion leaves us just as puzzled concerning the logical, as distinct from psychological grounds which would then originally have motivated Einstein to have confidence in the principle of relativity without the partial support of the Michelson-Morley experiment, while that very lack of support would have sufficed, by his own admission, to assure the abandonment of the principle "without qualms" by his colleagues. And our puzzlement is deepened by Bernard Jaffe's report that in a letter to him, Einstein expressed his debt to the American physicist [i.e., Michelson] in these words:

> It is no doubt that Michelson's experiment was of considerable influence upon my work insofar as it strengthened my conviction concerning the validity of the principle of the special theory of relativity. On the other side I was pretty much convinced of the validity of the principle before I did know this experiment and its result. In any case, Michelson's experiment removed practically any doubt about the validity of the principle in optics, and showed that a profound change of the basic concepts of physics was inevitable.[2]

And Jaffe adds that "In 1931, just before the death of Michelson, Einstein publicly attributed his theory to the experiment of Michelson."

---

[1] M. Polanyi: *PK, op. cit.*, p. 11.

[2] B. Jaffe: *Michelson and the Speed of Light* (New York: Doubleday & Company; 1960), pp. 100–101.

Oddly enough, Polanyi quotes Einstein's passage about the non-detection of the earth's aether-motion just after having declared that the outcome of the Michelson-Morley experiment was "on the basis of pure speculation, rationally intuited by Einstein before he had ever heard about it."[3]

On this, my comments are the following. Suppose it were clear, which it is not by any means, that Einstein's frustratingly *non*-specific mention of attempts to confirm the aether theory experimentally is to be explained by Professor H. S. Frank's suggestion and is thus consonant with Polanyi's discounting of the Michelson-Morley experiment. Then we would still have to emphasize that this historical conjecture cannot be used to vitiate an account of both the genesis and justification of scientific theory which has the following two characteristics: first, while being *empiricist,* it is broad enough to accommodate the *valid core* of the Kantian emphasis on the *active,* creative role of the scientific imagination in the postulational elaboration of hypothetico-deductive theories, and second, by being empiricist, it avoids the notorious aprioristic pitfalls of classical rationalism, pitfalls on whose brink Polanyi hovers[4] in the protective twilight of his thesis of the unspecifiability of fallible clues. Can the history of the STR as conjectured by Polanyi be validly adduced here to prove any more than the untenability of a *crudely* empiricist Machian or Aristotelian-Thomist, abstractionist account of theory-construction as a mere codification of the results of experiments? And is the untenability of the latter kind of account not fully recognized by any empiricist conception of scientific knowledge which incorporates the lesson of Kant? Does such an empiricist conception not allow for the difference between knowing of an experimental result in a narrow sense, on the one hand, and speculatively assigning a wider significance to it, on the other?

That the answer to the last two of these questions is indeed "yes" is indicated by a single chapter title in a book by an empiricist writer whose views are presumably anathema to Polanyi: Chapter Six of Reichenbach's *Rise of Scientific Philosophy* is entitled "The Twofold Nature of Classical Physics: Its Empirical and Its Rational Aspect."

[3] Polanyi, *PK, op. cit.,* p. 10.
[4] M. Polanyi: *PK, op. cit.,* p. 15.

In a further attempt to document the fideist nature of allegiance to physical theories, Polanyi refers to the scientific community's reception of D. C. Miller's report of *positive* effects obtained by him in his repetitions of the Michelson-Morley experiment. Thus, Polanyi cites the history of the STR to maintain further that once the "rational" appeal of the theory had captured the minds of the scientific community, contrary evidence hardly acted as a check on its subsequent acceptance. But Polanyi's indictment scores only against those who mistakenly portray evidence-reports as not already theory-laden. For it is the false absolute dichotomy between theory and evidence which would prompt an uncritically hasty *disconfirmatory* use of what is prima facie contrary evidence. As Duhem has emphasized, *collateral* hypotheses pertaining to the conditions of an experiment and to the laws of operation of its test equipment are an essential ingredient in the logical fabric of disconfirmation. And thus there is a mutuality of accreditation between theory and evidence in virtue of the interpenetration of the criteria of credibility which certify evidence as bona fide, on the one hand, and theory as evidentially warranted on the other.

It was therefore hardly a case, as Polanyi would have it, of simply ignoring negative evidence when adherents of the STR surmised—correctly as it seems to have turned out[5]—that the conditions under which Miller obtained his results were different from what he had supposed them to be. And there was also the inductively reasonable suspicion that the Cleveland site at which Miller had obtained his aberrant results was not a terrestrial singularity in regard to delicate optical phenomena, since it had not proven to be a singularity with respect to other (coarse) physical phenomena.

We must examine a final but essential point in Polanyi's critique of the account given by philosophers of science of the relevance of the discovery of the STR to an analysis of its epistemological foundations. Polanyi denies the legitimacy of the distinction between the psychology and the epistemology (logic) of discovery, a distinction which Reichenbach usefully termed as being between the "context of discovery" and the "context of

[5] Cf. R. S. Shankland, S. W. McCuskey, S. C. Leone, and G. Kuerti: *Reviews of Modern Physics,* Vol. XXVII (1955), p. 167.

justification." Quoting H. Mehlberg's statement that "The gist of the scientific method is . . . verification and proof, not discovery," Polanyi writes: "Actually, philosophers deal extensively with induction as a method of scientific discovery; but when they occasionally realize that this is not how discoveries are made, they dispose of the facts to which their theory fails to apply by relegating them to psychology."[6] And while denying that he intends to discount completely the role of experimental results in the genesis of acts of discovery,[7] Polanyi does make the following claims concerning the criteria by which *bona fide* scientific discoveries can legitimately be judged as such:

> Knowledge of the external world is in general acquired by relying on clues which cannot be fully identified.[8] . . . Scientific discoveries are likewise based on clues that are never fully specifiable. This may be called "intuition" but I rarely use this term, for it is traditionally charged with the fallacious connotation of infallibility, which I do not wish to imply.[9] . . . A good problem is a passionate intimation of a hidden truth.[1] . . . I recognize the continued operation of these anticipatory powers in the verification of discovery and holding of knowledge. The scientist's conviction of having arrived at some true knowledge is akin to the powers by which he recognises a problem. It is an anticipation of yet hidden truths, which in this case are expected to emerge in the uncertain future as the unknown consequences of the truth as known at present. When we are told that fruitfulness is the characteristic of a true discovery, the actual facts are obscured. For it is absurd to suggest that we should recognise truth by the wealth of its still unknown future consequences; on the contrary, our recognition of a true piece of knowledge is an anticipation of such unknown consequences, and if we acknowledge that knowledge is rightly held we subscribe to these anticipations. They are an expression of the belief that true knowledge is an aspect of a hidden reality which as such can yet reveal itself in an indeterminate range of future discoveries.[2]

[6] M. Polanyi: *PK, op. cit.*, p. 14 n. 1.
[7] M. Polanyi: "Notes, etc.," *op. cit.*, p. 55.
[8] *Ibid.*, p. 53.
[9] *Ibid.*, p. 54.
[1] *Ibid.*
[2] *Ibid.*

Given his affirmation of the unspecifiability of clues and his admission of their fallibility, this statement of Polanyi's succeeds no more than his earlier ones in providing a consistent articulation of the epistemological attributes of a bona fide discovery as contrasted with those of an initially plausible, passionately espoused but wholly abortive speculation. But in the absence of precisely such an articulation, what is to be our verdict on his indictment of Mehlberg, Reichenbach, and others who do invoke the distinction between the *psychology* of the propounding of scientific hypotheses, on the one hand, and the *epistemological justification* of these hypotheses on the other? It can be none other than that Polanyi's indictment is altogether gratuitous. To repeat, the anticipations and commitments of which he speaks are initially no less passionate in cases of hypotheses which turn out to be untenable than in cases of theoretical conjectures which are abundantly borne out by subsequent evidence. And hence those of Polanyi's remarks which are perceptive pertain not, as he believes, to scientific discovery; instead, they do apply after all only to the *psychology* of the propounding of scientific speculations. For what is it that makes a hypothesis *true,* and what warrants the claim that a bona fide discovery has actually been made? Surely it is not the fact that the scientist has committed himself to the as yet unknown and unverified anticipations flowing from the hypothesis! When maintaining that fruitfulness is the test of a true discovery, philosophers of science are not asserting, as Polanyi supposes incorrectly, that we recognize truth "by the wealth of its still unknown consequences." Fruitfulness in the form of both available evidence *and* future corroborations is indeed the test of the truth of a hypothesis. But we recognize the latter truth at a given time on the basis of supporting evidence then available, this recognition being understood to require the certification of credibility instead of being constituted by some ineffable, thaumaturgical insight.

We conclude against Polanyi that the philosopher's relegation of the promptings of what turn out to be lucky hunches to the psychology of creative work in science is *not* an example of the fallacy of ignoring negative instances. As well condemn the avowed neglect by the literary critic of Friedrich Schiller's reported use of the smell of rotten apples as an aid to his writing of *Die Glocke*.

(D) THE PRINCIPLE OF THE CONSTANCY OF THE SPEED OF LIGHT AND THE FALSITY OF THE AETHER-THEORETIC LORENTZ-FITZGERALD CONTRACTION HYPOTHESIS.

The principle of the constancy of the speed of light—hereafter called the "light principle" for brevity—states that the speed of light is the same constant c in all inertial systems, independent of the relative velocity of the source and observer, and of direction, position, and time.[3] We must inquire in what sense the null-result of the Michelson-Morley experiment can legitimately be held to support this relativistic light principle.

It will be evident from the results of our discussion that it is quite incorrect to suppose, as is done in some quarters, that the Michelson-Morley experiment furnishes a sufficient experimental basis for the light principle. And the statement of the *several* empirical and philosophical constituents of the light principle will then make apparent the *fallaciousness* of the following inference: since two different Galilean observers give formally the same description of the behavior of a single identical light pulse, emitted upon the coincidence of their origins, therefore the respective expanding spherical forms of this disturbance which they are said to "observe" in their respective systems are constituted by the selfsame point-events or the selfsame configuration of photons. Sometimes the ill-conceived attempt is made to render the false conclusion of this fallacious inference plausible by the irrelevant though true remark that while judgments can contradict one another, observed occurrences cannot. As if the

[3] In opposition to the emission theory of light, the classical aether theory had already affirmed the independence of the velocity of light in the aether from that of its *source*, while affirming, however, its dependence on the observer's motion relative to the aether medium.

J. H. Rush has pointed out [*Scientific American*, Vol. CXCIII (August, 1955), p. 62] that the *time* constancy of the velocity of light may have to be questioned: measurements over the past century have yielded three fairly distinct sets of values for c. For a comprehensive recent survey of measurements of the velocity of light, see E. Bergstrand: "Determination of the Velocity of Light," *Handbuch der Physik*, edited by Flügge (Berlin: J. Springer; 1956), Vol. XXIV, pp. 1–43.

claim that the light pulse spreads spherically in each inertial system were warranted by the deliverances of sensory observation alone and did not already presuppose a theory concerning the intersystemic and intrasystemic status of simultaneity.

In order to determine to what extent the light principle draws support from the outcome of the Michelson-Morley experiment, we must see first what logical grounds there are for not interpreting the null outcome of the Michelson-Morley experiment as evidence for the aether-theoretic Lorentz-Fitzgerald contraction hypothesis, an aether-theoretic auxiliary hypothesis to which we shall refer hereafter as the "L-F hypothesis" for brevity. The correct and detailed handling of this prior inquiry is particularly important, since the literature and the classroom instruction in the philosophy of science abound in the citation of the L-F hypothesis as an illustration of a logically *ad hoc* modification of a theory. And the purportedly *ad hoc* status of the L-F hypothesis is widely adduced as the justification for rejecting it as an interpretation of the null outcome of the Michelson-Morley experiment. I wish to show first that this characterization of the L-F hypothesis is as erroneous as it is hackneyed: the L-F hypothesis must be rejected because it is *false* and *not* because it is logically *ad hoc*. We shall also see that on the strength of the mistaken assessment of the aether-theoretic L-F hypothesis as logically *ad hoc,* the historical conclusion has been drawn that the theorizing of Lorentz and Fitzgerald involved a grave infraction of scientific method. My impending explanation of why it is that the L-F hypothesis cannot be regarded as logically *ad hoc* will therefore also demonstrate that philosophical misunderstandings unfoundedly generated a historical allegation concerning Lorentz and Fitzgerald.

An auxiliary hypothesis, introduced in response to new evidence embarrassing to an established theory, is *ad hoc* in the logical sense within the framework of the theory modified by it, if it does not, in principle, lend itself to any independent test whatever. Since the theoretical terms of the postulates of sophisticated empirical theories are given only a partial or incomplete physical interpretation by the rules of correspondence at any given time, these terms (and the concepts they represent) are "open" in the sense of admitting the adjunction of further rules of correspondence to the given theoretical postulates. Hence the class of all observational consequences of a physical theory is not a clearly circumscribed class, unless we specify and delimit the rules of correspondence which, in conjunction with the theoretical postulates, constitute the given established theory. Accordingly, if the question of whether a certain hypothesis auxiliary to an established theory is logically *ad hoc* is to have a well-defined status, it must be understood that this question is relativized to a specific, delimited set of rules of correspondence. In this way, the open-textured character of theoretical terms does not make for any imprecision in the property of being logically *ad hoc*. Now suppose that a certain collateral hypothesis such as the aether-theoretic L-F hypothesis is actually independently testable, at least in principle, but that its advocates fail to be aware of such testability and espouse it nonetheless. Clearly, the auxiliary hypothesis in question can then not be held to be logically *ad hoc* but can be regarded as *ad hoc* only in the psychological sense. For the methodological culpability of those who espouse such an auxiliary hypothesis despite their own (mistaken!) belief that it is not independently testable cannot detract from the actual, purely logical independent testability of the hypothesis in the context of its theoretical framework. Thus, we must distinguish between the logical and psychological senses of being *ad hoc*, much as we distinguish between the logical property possessed by a mathematical proposition that is a theorem within a given axiom system and the psychological attribute possessed by mathematicians who *realize* that the given proposition is indeed a theorem.

The proof which I am about to give that the original L-F hypothesis is not logically *ad hoc* in the context of the aether theory therefore need not take cognizance of the beliefs Lorentz and Fitzgerald actually entertained as to its independent testability.

A comparison of the reasoning underlying the Kennedy-Thorndike experiment[4] with the design of the Michelson-Morley experiment will serve to establish that the aether-theoretic

---

[4] Cf. R. J. Kennedy and E. M. Thorndike: *Physical Review*, Vol. XLII (1932), p. 400, and W. Panofsky and M. Phillips: *Classical Electricity and Magnetism* (Cambridge, Mass.: Addison-Wesley Publishing Company, Inc.; 1955), p. 236.

L-F hypothesis did indeed have independently falsifiable consequences and that it was, in fact, falsified by the null outcome of the Kennedy-Thorndike experiment. Specifically, our analysis will show that: first, although the coupling of the L-F hypothesis with the classical aether theory entails the null outcome produced by the Michelson-Morley experiment, it does rule out the negative result that was actually yielded by the Kennedy-Thorndike experiment, and second, the aether theory as modified by the L-F hypothesis entails a positive outcome of the Kennedy-Thorndike experiment differing quantitatively from the positive result required by the aether theory without the L-F hypothesis.

The essential difference between the apparatus used in the Michelson-Morley and Kennedy-Thorndike experiments is the following: as measured by rods in the laboratory, the horizontal and vertical arms of the Michelson interferometer used in the Kennedy-Thorndike experiment are not equal but are made as different in length as possible, so as to assure a considerable difference in the travel times of the two partial beams from the source to the point at which they recombine to produce interference fringes. By contrast, the horizontal and vertical arms of the apparatus used in the Michelson-Morley experiment are each of the *same* length l as measured by the rods in the laboratory.

Accordingly, in the latter experiment the classically expected round-trip times T_v and T_h for the vertical and horizontal arms are given respectively by

$$T_v = \frac{2l}{(c^2 - v^2)^{\frac{1}{2}}} \text{ and } T_h = \frac{2l}{(c^2 - v^2)^{\frac{1}{2}}} \frac{1}{(1 - \beta^2)^{\frac{1}{2}}}$$

where c represents the velocity of light, v the velocity of the apparatus relative to the aether, and $\beta \equiv v/c$. Now, without a L-F hypothesis, the initial time difference $T_h - T_v$ between the two partial light beams would be expected to change in the course of the rotation of the apparatus through 90° in the Michelson-Morley experiment. And thus a shift in the interference fringes corresponding to this change was anticipated. But, once the aether theory is amended by the introduction of the L-F contraction, the length l in the expression for T_h must be replaced by the length $l(1 - \beta^2)^{\frac{1}{2}}$. As a consequence of the introduction of this auxiliary hypothesis, T_h becomes *equal* to T_v, and the difference between the round-trip times of the two partial beams becomes zero throughout the Michelson-Morley experiment, in conformity with its null result.

It will be noted that the equality of the *terrestrially measured* lengths of the two arms is a necessary condition for the constant vanishing of the difference between the two round-trip times in the L-F account of the Michelson-Morley experiment. But precisely this necessary condition is not fulfilled in the Kennedy-Thorndike experiment, in which the terrestrially measured lengths of the vertical and horizontal arms have the unequal values L and l, respectively. Thus, *upon assuming the L-F hypothesis,* the difference between the travel times of the two light beams of the Kennedy-Thorndike experiment is given by

$$T_v - T_h = \frac{2L}{(c^2 - v^2)^{\frac{1}{2}}} - \frac{2l}{(c^2 - v^2)^{\frac{1}{2}}} = \frac{2}{(c^2 - v^2)^{\frac{1}{2}}}\,(L - l)$$

Instead of vanishing throughout the experiment, this time difference varies with the diurnally and annually changing velocity v of the apparatus relative to the fixed aether. Moreover, if we do not assume a L-F contraction, the difference between the two travel times of the Kennedy-Thorndike experiment has the *different* value given by

$$\begin{aligned} T_v - T_h &= \frac{2L}{(c^2 - v^2)^{\frac{1}{2}}} - \frac{2l}{(c^2 - v^2)^{\frac{1}{2}}} \cdot \frac{1}{(1 - \beta^2)^{\frac{1}{2}}} \\ &= \frac{2}{(c^2 - v^2)^{\frac{1}{2}}}\left(L - \frac{1}{(1 - \beta^2)^{\frac{1}{2}}}\right) \end{aligned}$$

And this difference is likewise a function of the diurnally and annually varying velocity v of the apparatus.

Under what conditions can the changing time difference called for by the L-F modification of the aether theory be expected to give rise to corresponding observable shifts in the interference fringe pattern? On that theory, these fringe shifts should occur if the clocks in the moving system are presumed to have the same rates as the "true" clocks of the aether system and if the frequency (period) of the light source—as measured by either the moving clocks or the aether-system clocks—does not itself depend upon the changing velocity of the apparatus with respect to the aether. But since Kennedy and Thorndike found that the expected fringe shifts failed to materialize, the crucial question before us is whether the L-F hypothesis is falsified by this null result even though the positive result expected on the basis of the L-F form of the aether theory depended for its deduction also upon the assumption that the frequency of the light source is independent of its velocity through the aether. Hence we must inquire whether the null result of the Kennedy-Thorndike experiment could be explained by denying the latter assumption of independence while preserving the L-F hypothesis, so that this hypothesis would then not be falsifiable by this experiment.

Now, the denial of the assumption of independence could take either of the following two forms with a view to explaining the null result of Kennedy and Thorndike while affirming the L-F hypothesis:

a. The velocity-dependent time difference $T_v - T_h$ given above, which can be expressed as

$$T_v - T_h = \frac{2(L-l)}{c(1-\beta^2)^{\frac{1}{2}}},$$

does not give rise to fringe shifts, because the frequency of the moving source does depend on its velocity through the aether as follows: both its frequency as determined by the clocks of the aether system and its frequency as measured by the clocks of the moving system, the latter of which we shall call the "proper" frequency for brevity, are reduced by a factor of $(1-\beta^2)^{\frac{1}{2}}$ as compared with the frequency that would be measured in the aether system, if the moving source came to rest there. Accordingly, *an observational consequence of this way of denying the independence of the frequency from the velocity is that the proper frequency of the moving source would vary with its velocity v (and hence with β).*

b. The velocity-dependence of the time difference $T_v - T_h$ given under (a) does not issue in a fringe shift, because there is also the following compensatory velocity-dependence: both the frequency of the source, as measured by the aether-system clocks, *and* the rate of the clocks in the moving system, as compared with the clocks in the aether system, are reduced by a factor of $(1-\beta^2)^{\frac{1}{2}}$. Thus, the proper frequency is not independent of the velocity of the source, and the values of the round-trip times as measured in the moving system are given by the quantities

$$T_v' = \frac{2L}{c} \quad \text{and} \quad T_h' = \frac{2l}{c}$$

which are independent of β.

We are now ready to state the grounds for a unique choice of the *third* from among the following three rival interpretations of the null result of the Kennedy-Thorndike experiment:

(1) The L-F hypothesis is confirmed *and* the observable *proper* frequency of the moving source *varies* with its velocity.

(2) The L-F contraction *and* the "time dilation" (or "dilatation," i.e., reduction of the rate of the moving clocks) operate together while the *proper* frequency of the source is *constant*, and hence the compound auxiliary hypothesis comprising *both* the L-F contraction *and* the time-dilation is confirmed.

(3) The L-F contraction hypothesis is *falsified* by the outcome of the Kennedy-Thorndike experiment in the sense of being (highly) disconfirmed. And since the Kennedy-Thorndike experiment constitutes a test which is independent of the Michelson-Morley experiment, the former experiment shows that the espousal of the L-F hypothesis in response to the negative result of the Michelson-Morley experiment was not logically *ad hoc*, although we shall see that it happened to have been psychologically *ad hoc*.

The first interpretation is to be rejected: its claim that the observable proper frequency of the moving source varies with the velocity is testable *apart* from the Kennedy-Thorndike experiment and has been found to be false by separate empirical evidence. The issue is therefore whether the fact that interpretation (2) is consistent with the null outcome of the Kennedy-Thorndike experiment renders (2) a *methodologically* acceptable interpretation and inductively a legitimate *alternative* to interpretation (3). That (2) does not, however, pass muster at all methodologically and hence cannot qualify as a rival to (3) emerges from the fact that the *combined* auxiliary hypothesis affirming *both* the L-F contraction *and* the time dilation is plainly *ad hoc*, since it does not lend itself to any independent test whatever. Hence, despite its compatibility with observational findings, methodological grounds prompt us to reject the doubly amended variant of the aether theory offered by (2). And it is now apparent that the attempt to endorse interpretation (2) in order to rescue the L-F hypothesis from the refutation that is claimed by interpretation (3) founders on the following fact: (2) succeeds in preserving the L-F hypothesis from refutation by the null outcome of the Kennedy-Thorndike experiment only by dint of incorporating that hypothesis in an augmented auxiliary hypothesis which is indeed logically *ad hoc.*

We have shown, therefore, that instead of being logically *ad hoc,* the L-F hypothesis has a falsifiable bearing on the outcome of the Kennedy-Thorndike experiment, an experiment which could have detected any existing velocity of the apparatus relative to the aether even on the assumption of a L-F contraction. And we note that the invalidation by the Kennedy-Thorndike experiment of the charge that the L-F hypothesis is logically *ad hoc* is *internal to the aether theory:* this invalidation no more depends logically on the availability of the STR than does the refutation of the original aether theory by the null outcome of the Michelson-Morley experiment. In addition to the Kennedy-Thorndike experiment, the following experiment suggested by C. Møller and others qualifies as a test of the L-F contraction within the framework of *pre*relativistic conceptions: although a *terrestrial* observer could not detect a L-F contraction in a Michelson-Morley experiment, "an observer at rest in the aether outside the earth would, however, in principle, be able to observe the shortening and he would find the earth and all objects on the earth contracted in the direction of motion of the earth."[5]

In view of the *falsity* of the supposition that the L-F hypothesis is logically *ad hoc,* that supposition cannot justifiably serve as the basis for the *historical* contention that Lorentz and Fitzgerald were unable to envision an independent test for their contraction hypothesis and hence were methodologically culpable for espousing it nonetheless.

But it is clear that laboring under the philosophical miscon-

[5] C. Møller: *The Theory of Relativity, op. cit.,* p. 29. In Section F of the present chapter, we shall discuss the *important differences* between the *pre*relativistic and the relativistic conceptions of the status of any contraction disclosed by the extraterrestrial experiment suggested here by Møller.

ception that the L-F hypothesis is logically *ad hoc* could dissuade a historian of science from making the effort to uncover the kind of *historical* evidence which alone can show whether the L-F hypothesis was *psychologically ad hoc.*

H. Dingle[6] has pointed out the existence of historical evidence establishing the biographical fact that, in the case of Lorentz, the contraction hypothesis was psychologically *ad hoc.* But unfortunately this historically documented fact then misled Dingle into inferring that the L-F hypothesis was logically *ad hoc.* As well conclude that a certain mathematical proposition cannot be a theorem in a given axiom system merely because all mathematicians are unaware that the proposition is, in fact, provable!

It is now clear that the null outcome of the Michelson-Morley experiment cannot be construed as support for the aether-theoretic L-F hypothesis, because there is indeed other evidence which shows that this hypothesis is false.

If then we therefore abandon the aether-theoretic framework of interpretation in favor of the relativistic one, we must return to the opening query of this Section and ask: to what extent can the relativistic light principle be claimed to rest on the null-outcome of the Michelson-Morley experiment?

The null-result of the Michelson-Morley experiment merely showed that if *within* an inertial system, light-rays are jointly emitted from a given point in different directions of the system and then reflected from mirrors at equal distances from that point, as measured by rigid rods, then they will return together to their common point of emission. And it must be remembered that the equality of the round-trip times of light in different directions of the system is measured here not, of course, by material clocks but by light itself (absence of an interference fringe shift).[7] The repetition of this experiment at different times of the

---

[6] H. Dingle: *The British Journal for the Philosophy of Science,* Vol. X (1959), pp. 228–29.

[7] Since the classically expected time difference in the second-order terms is only of the order of 10^{-15} second, allowance must be made for the absence of a corresponding accuracy in the measurement of the equality of the two arms. This is made feasible by the fact that, on the aether theory, the effect of any discrepancy in the lengths of the two arms should *vary,* on account of the earth's motion, as the apparatus is rotated. For details, see P. Bergmann: *Introduction to the Theory of Relativity, op. cit.,* pp. 24–26, and J. Aharoni: *The Special Theory of Relativity* (Oxford: Oxford University Press; 1959), pp. 270–73.

year, i.e., in different inertial systems, showed only that there is no difference within *any* given inertial system in the round-trip times as between different directions within that system. But the outcome of the Michelson-Morley experiment does not show at all either that (a) the round-trip (or one-way) time required by light to traverse a closed (or open) path of length 2l (or l) has the same numerical value in different inertial systems, as measured by material clocks stationed in these systems[8] or that (b) if, contrary to the arrangements in the Michelson-Morley and Kennedy-Thorndike experiments, the source of the light is outside the frame K in which its velocity is measured, then the velocity in K will be independent of the velocity of the source relative to K. In fact, the statement about the round-trip times made under (a) was first substantiated by the Kennedy-Thorndike experiment of 1932, as we shall see, and the assertion under (b) received its confirmation by observations on the light from double stars.[9] Yet the light principle certainly affirms both (a) and (b) and thus clearly claims more than is vouchsafed by the Michelson-Morley experiment. Accordingly, we can see that in addition to the result of the Michelson-Morley experiment, the light principle contains at least all of the following theses:

(1) The assertion given under (a), which undoubtedly goes beyond the null-result of the Michelson-Morley experiment. For brevity, we shall call it "the clock axiom" in order to allude to its reference to clock-times of travel. Although lacking experimental corroboration at the time of Einstein's enunciation of the light principle, this "clock axiom" was suggested by the fundamental assumption of special relativity that there are no preferred inertial systems.

---

[8] It is understood that the lengths 2l in the different frames are each ratios of the path to the same unit rod, which is transported from system to system in order to effect these measurements and which remains equal to unity *by definition* in the course of this transport.

[9] Cf. J. Aharoni: *The Special Theory of Relativity, op. cit.*, pp. 269–70. The relativistic assumption stated under (b) was denied by Ritz's emission theory which maintained the following: the velocity of light in free space is always c with respect to the source, but its value in a frame K depends on the velocity of the source relative to K. For a careful account of the body of evidence against Ritz's hypothesis, see W. Pauli: *Theory of Relativity* (New York: Pergamon Press; 1958), pp. 6–9. Additional evidence against Ritz's assumption was furnished by a recent experiment by A. M. Bonch-Bruevich [*Physics Express,* November, 1960, pp. 11–12].

(2) The Einstein postulate regarding the maximal character of the velocity of light, which is a source of the conventionality and relativity of simultaneity within a given inertial system and as between different systems respectively. This postulate *allows* but does *not* entail that we choose the same value $\epsilon = \frac{1}{2}$ for all directions within any given system and also for each different system.

(3) The claim that the velocity of light in any inertial system is independent of the velocity of its source.

To see specifically how the light principle depends for its validity upon the truth of all of these constituent theses, we now consider the consequences of abandoning any one of them while preserving the remaining two. To do so, we direct our attention to the one-way velocity of an outgoing light pulse that traverses a distance l in system S and also to the one-way velocity of such a pulse traversing a distance l in system S′. Let T_s and $T_{s'}$ be the respective outgoing transit times of these light pulses for the distances l in S and S′, the corresponding round-trip times in these frames being τ_s and $\tau_{s'}$. In this notation, our earlier equation $t_2 = t_1 + \epsilon(t_3 - t_2)$ of Section B becomes

$$T_s = \epsilon\tau_s,$$

and similarly for S′. Hence the relevant one-way velocities of light in these systems are, respectively, given by

$$v_s = (l/T_s) = (l/\epsilon\tau_s)$$

and

$$v_{s'} = (l/\epsilon\tau_{s'}).$$

Our problem is to determine, one by one, the consequences of assuming that of the three above ingredients of the light principle, only two can be invoked at a time in an attempt to assure that if $v_s = c$ in virtue of the observed value of τ_s and of a choice of $\epsilon = \frac{1}{2}$ for S, then $v_{s'}$ will also have the value c.

(1) In the absence of the clock axiom, it can well be that $\tau_s \neq \tau_{s'}$. In that case, the relations $v_s = (l/\frac{1}{2}\tau_s) = c$ and $v_{s'} = (l/\epsilon\tau_{s'})$ tell us that it will not be possible to make $v_{s'} = c$ by a choice of $\epsilon = \frac{1}{2}$ in S′. To be sure, it may still be possible then to choose an appropriately different value of ϵ for S′ so as to make $v_{s'} = c$. But, much as the latter choice of ϵ might yield

the value c for the outgoing one-way velocity of light in S′, it would also inescapably entail a value different from c for the return-velocity. And such a result would not be in keeping with the light principle.

(2) If, however, we guarantee that $\tau_s = \tau_{s'}$ by assuming the clock axiom but disallow the freedom to choose a value of ϵ for each inertial system by withdrawing the second constituent thesis above, then $v_{s'}$ could readily be different from c. For the illegitimacy of choosing ϵ to be $\frac{1}{2}$ in S′ would then derive from physical facts incompatible with the conventionality of simultaneity which would objectively fix the value of $T_{s'}$ as different from $\frac{1}{2}\tau_{s'}$.

(3) The need for the third ingredient is obvious in view of what has already been said.

It is of importance to note that even when all of the above constituent principles of the light principle are assumed, no fact of nature independent of our descriptive conventions would be contradicted, if we chose values of ϵ other than $\frac{1}{2}$ for each inertial system, thereby making the velocity of light different from c in both senses along each direction in all inertial systems. I have shown elsewhere[1] that this conclusion is not invalidated either by determinations of the one-way velocity of light on the basis of measurements of the half-wave-length of standing waves in cavity resonators or by measurements of the ratio of the electromagnetic and electrostatic units of charge. The assertion that the invariant velocity of light is c is therefore not, in its entirety, a purely factual assertion. But we saw that it is nonetheless a consequence of presumed physical facts that the specification of the velocity of light involves a stipulational element which, in combination with other factual principles, allows us to say that the velocity of light is c.

(E) THE EXPERIMENTAL CONFIRMATION OF THE KINEMATICS OF THE STR.

We saw that the Michelson-Morley experiment clearly could not be regarded as empirical proof for the "clock axiom" ingredient in the light principle. It is therefore a welcome fact that the

[1] A. Grünbaum: *American Journal of Physics,* Vol. XXIV (1956), pp. 588–90.

Kennedy-Thorndike experiment described in Section D of this chapter can indeed be invoked as support for that axiom's relativistic denial of the existence of preferred inertial systems.[2]

If the Kennedy-Thorndike experiment had yielded a positive effect instead of the null-result which it did actually yield, then it could have been cogently argued that the Michelson-Morley experiment was evidence for a bona fide Lorentz-Fitzgerald contraction, just as a fringe shift produced by heating one of the arms of the interferometer could be held to be evidence for the elongation of that arm. But, in view of the de facto null outcome of the Kennedy-Thorndike experiment, there is very good reason indeed to attribute the absence of a diurnal or annual variation in the time-difference between the two partial beams to a *constancy, as between different inertial systems, of the time required by each of the partial beams to traverse its own closed path.* And thus we are entitled to say that the Kennedy-Thorndike experiment has provided empirical sanction for the clock axiom.[3]

---

[2] J. P. Cedarholm and C. H. Townes have reported their very recent cognate experiment [cf. "A New Experimental Test of Special Relativity," *Nature,* Vol. CLXXXIV (No. 4696, 1959), pp. 1350–51]. They write:

> The experiment compares the frequencies of two maser oscillators with their beams of ammonia molecules pointed in opposite directions, but both parallel to a supposed direction of motion through the aether. . . . A precision of one part in 10^{12} has been achieved in this frequency comparison, and failure to find a frequency change of the predicted type allows setting the upper limit on an aether drift as low as 1/1000 of the orbital velocity of the Earth. . . . The present experiment sets an upper limit on an aether-drift velocity about one-fiftieth that allowed by previous experiments.

[3] It should be mentioned that the experimenters Kennedy and Thorndike themselves conceived of their experiment not as a test of what I have called the "clock axiom" in the light principle but rather as a test of the relativistic clock retardation. The negative outcome of the Kennedy-Thorndike experiment cannot, however, be invoked as evidence for the *relativistic* clock retardation, which is a reciprocal or symmetrical relation between the clocks of relatively moving inertial systems. For the reader will recall my account in Section D of the reasons for *rejecting* the interpretation that the null result of the Kennedy-Thorndike experiment supports the hypothesis that the *aether-theoretic* time-dilation and the L-F contraction operate together. And it is evident from that analysis in Section D that instead of pertaining to the reciprocal *relativistic* clock retardation, the null-outcome of the Kennedy-Thorndike experiment has a bearing on the *ad hoc* assumption of an *aether-theoretic* time dilation, which is *not* a symmetrical or reciprocal relation between the clocks of the moving system and those of the aether-system.

We recall that in 1905, there was no unambiguous experimental evidence supporting Einstein's postulate that light is the fastest signal. But subsequent experiments showed that the mass and kinetic energy of accelerated particles become indefinitely large as their velocity approaches that of light.[4]

There is therefore impressive empirical evidence for all of the constituent theses of the light principle. It would be an error, however, to suppose that the experimental justification of the light principle suffices also to substantiate the Lorentz transformations. For these equations entail the clock retardation, whereas the light principle alone does not. Hence it is a mistake to suppose that upon a suitable choice of zeros of time, the light principle alone permits the deduction of the Lorentz transformations and that all the novel affirmations of relativistic kinematics are thus vouchsafed by the light principle. A lucid demonstration of the fact that the relativistic clock retardation is logically independent of the light principle has recently been given by H. P. Robertson.[5] He considers a linear transformation

$$T: \quad (t', x', y', z') \rightarrow (t, x, y, z)$$

between a kind of primary inertial system Σ and a "moving" inertial system S′. Upon having fixed thirteen of the sixteen coefficients of this transformation by various conventions, symmetry conditions, and a specification of the velocity v of S′ relative to Σ, Robertson is concerned with the experimental warrant for asserting that the remaining three coefficients have the values required by the Lorentz transformations. And he then shows that the Michelson-Morley and Kennedy-Thorndike experiments, which do succeed in completing the confirmation of the light principle, do not suffice to fix the remaining three coefficients of the transformation such that these have the values required by

---

[4] See W. Gerlach: *Handbuch der Physik* (Berlin: Springer-Verlag; 1926), pp. 61 ff., and C. Møller: *The Theory of Relativity, op. cit.,* Chapter iii, Sec. 32.

[5] H. P. Robertson: *Reviews of Modern Physics,* Vol. XXI (1949), p. 378. An earlier proof of the compatibility of the light principle with the *denial* of the clock retardation was given by Reichenbach (see *Axiomatik der relativistischen Raum-Zeit-Lehre, op. cit.,* pp. 79–83, esp. pp. 81–83), who exhibits a consistent set of coordinate transformations embodying both assertions.

the Lorentz transformations. An additional experiment is needed to do so: the laboratory work of Ives and Stilwell (1938) furnished the lacking data by observations on high speed canal rays. And it was their confirmation of the relativistic ("quadratic") Doppler effect[6] that constituted the first experimental proof of the clock retardation affirmed by the Lorentz transformations. Additional confirmation has been provided by data on the rate of disintegration of mesons.[7]

According to the STR, the clock retardation is exhibited by all natural clocks, be they the material clocks which keep astronomical time or atomic clocks like cesium atoms, whose unit the theory assumed to have a time-invariant ratio to the astronomical unit. But this assumption of a constant ratio of the atomic and astronomical units of time has been questioned by Dirac, Milne, Jordan, and others, who have suggested that the ratio increases continuously by amounts of the order of 1/T per year, where T is "the age of the universe" in years. If T is about 4 x 10^9 years, the change at the present epoch is of the order of $1/10^9$. Changes of this order may soon be measurable over intervals of a few years, since Essen and Parry of the National Physical Laboratory in England have recently measured the natural resonant frequency of the cesium atom with a stated precision of 1 in 10^9 and hope for even higher accuracies.[8]

(F) THE PHILOSOPHICAL ISSUE BETWEEN EINSTEIN AND HIS AETHER-THEORETIC PRECURSORS, AND ITS BEARING ON E. T. WHITTAKER'S HISTORY OF THE STR.

In view of our discussion of Newton's conception of the metrics of space and time in Chapter One, it will suffice at this point to say quite briefly that the aether-theoretic version of the New-

[6] Cf. M. von Laue: *Die Relativitätstheorie, op. cit.*, p. 20.

[7] B. Rossi and D. B. Hall: *Physical Review*, Vol. LIX (1941), p. 223. For a proof that in the context of the conditions to which the STR is applicable, the relativistic clock retardation does not give rise to the "clock paradox," see A. Grünbaum: "The Clock Paradox in the STR," *Philosophy of Science*, Vol. XXI (1954), pp. 249–53 and Vol. XXII (1955), pp. 53 and 233.

[8] Cf. L. Essen and J. V. L. Parry: *Nature*, Vol. CLXXVI (1955), p. 280 and the comments by G. M. Clemence: *ibid.*, p. 1230, and in *Science*, Vol. CXXIII (1956), p. 571. See also H. Lyons: *Scientific American*, Vol. CXCVI (February, 1957), p. 71.

tonian theory included the following thesis concerning the absolute rest system of the aether: receptacle space and time each have their own *intrinsic* metric, which exists quite independently of the presence of material rods and clocks in the universe, devices whose function is *at best* the purely epistemic one of enabling us to ascertain the intrinsic metrical relations of the receptacle space and time contingently containing them. Thus, for example, even when material clocks run uniformly, they are merely in accord with but do not "define" the temporal metric.

Awareness of this thesis of the aether-theory will enable us shortly to give a precise statement of the fundamental *logical* differences between the aether-theoretic L-F contraction and the *numerically* equal relativistic "Einstein contraction" of a moving rod, which is deducible from the relativistically interpreted Lorentz transformations. A clear delineation of the logical differences between the L-F and Einstein contractions is especially important since these contractions are often confounded. Thus, we are told that instead of attempting to account for the null result of the Michelson-Morley experiment via the purportedly *ad hoc* L-F contraction hypothesis, Einstein *explained* the outcome of that experiment soundly by deducing the contraction from the Lorentz transformations. To cite a very recent case in point, R. M. Palter, writing on the kinematic results of the STR, credits Einstein with explaining the outcome of the Michelson-Morley experiment by saying: "the contraction of rigid rods was, of course, confirmed by the Michelson-Morley and numerous subsequent aether-drift experiments."[9] But, as is already evident from our account of the constituent theses of the light principle in Section D, far from explaining the outcome of the Michelson-Morley experiment non-trivially as a consequence of more fundamental principles, Einstein incorporated its null result as a physical *axiom* in his light principle, which is a *premise* in his deduction of the Lorentz transformations.

More fundamentally, the erroneous view that the relativistic deduction of the Einstein contraction from the relativity of simultaneity ingredient in the Lorentz transformations *explains* the null-outcome of the Michelson-Morley experiment derives its

---

[9] R. M. Palter: *Whitehead's Philosophy of Science*, *op. cit.*, p. 13.

plausibility from the numerical equality of the contraction factors of the relativistic Einstein contraction and the aether-theoretic L-F contraction, the value being $\sqrt{1-\beta^2}$ in each case. But these contractions are of radically different logical character, because the L-F contraction pertains to the very system in which the contracted arm is *at rest,* whereas the contraction that Einstein derived from the Lorentz transformations pertains to the length measured in a system relative to which the arm is *in motion.* More explicitly, the Lorentz-Fitzgerald contraction hypothesis asserts a comparison of the actual length of the arm, as measured by the round-trip time of light, to the greater length that the travel-time would have revealed, if the classical aether theory were true. Thus, using light as the standard for effecting the comparison, this hypothesis affirms that in the same system and under the same conditions of measurement, the metrical properties of the arm are different from the ones predicted by classical aether theory. And this difference or contraction is clearly quite independent of any contraction based on comparisons of lengths in different inertial systems. By contrast, the contraction which Einstein deduced from the Lorentz transformations is based on a comparison of the length of a rod, as measured from an inertial system relative to which it is in motion, to the length of that same rod, as measured in its own rest system. Unlike the L-F contraction, but like the relativistic clock retardation, this "Einstein contraction" is a *symmetrical* or reciprocal relation between the measurements made in any two inertial systems and is a consequence of the intersystemic relativity of simultaneity, because it relates lengths determined from *different* inertial perspectives of measurement, instead of contrasting conflicting claims concerning the results obtained under the same conditions of measurement. What Einstein did explain, therefore, is this "metrogenic" contraction, a phenomenon which poses no greater logical difficulties than the differences in the angular sizes of bodies that are observed from different distances.

Lest it be thought that the confusion of the two kinds of contraction just discussed is a thing of the past, I cite the following recent statements from Sir E. T. Whittaker's Tarner Lectures:

> The Lorentz transformation . . . supplies at once an explanation of the Fitzgerald contraction. . . . the failure of all attempts

> to determine the velocity of the earth by comparing the Fitzgerald contraction in rods directed parallel and perpendicular to the terrestrial motion . . . is necessitated by the Postulate of Relativity. But there is no impossibility in principle at any rate, in observing the contraction, provided we can make use of an observation-post which is *outside* the moving system.[1]

To be sure, if Whittaker's proviso of an aether-system observation post outside the moving earth be granted, then an observer at that post who interprets his data *pre*-relativistically would confirm the Lorentz-Fitzgerald contraction by finding that the "true" length of the moving arm, which he believes himself to be observing from his vantage point, is smaller than the "spurious" length measured by the rod of a terrestrial observer. But the relativistic explanation of the numerically equal Einstein contraction actually involved here *rejects,* as we shall see in detail, the very conceptions which alone give meaning to (a) construing the findings of the extraterrestrial observer as *equivalent* to a contraction *within* the moving system in the sense of Lorentz and Fitzgerald, and (b) asking within the framework of the STR in the spirit of E. T. Whittaker why there is a L-F contraction within the moving system. And the relativistic deduction of the Einstein contraction from the Lorentz transformations can therefore have no bearing at all in Whittaker's sense on why the Michelson-Morley experiment failed to fulfill the predictions of the classical aether theory.

We see that the locus of the philosophical difference between Lorentz and Einstein has been misplaced by those who point to Einstein's alleged explanation of the null result of the Michelson-Morley experiment. To understand Einstein's philosophical innovation, we must take cognizance of the fact that the L-F contraction hypothesis was not the only addition to the aether theory made by Lorentz in order to account for the available body of experimental data. In addition, he had been driven to postulate with J. Larmor that just as rods are caused to contract in any inertial system moving relatively to the aether, so also clocks are

[1] E. T. Whittaker: *From Euclid to Eddington* (London: Cambridge University Press; 1949), pp. 63–64; italics supplied. The same error is repeated by Whittaker in his *A History of the Theories of Aether and Electricity, op. cit.,* Vol. II, p. 37, which should be read in conjunction with his "G. F. Fitzgerald," *Scientific American,* Vol. CLXXXIX (November, 1953), p. 98.

caused by that very motion to modify their rates and to read a spurious "local" time (as distinct from the "true" time shown by the clocks in the aether-system). The conceptual framework of Lorentz's interpretation of the transformation equations known by his name was the absolutistic one in which the clocks and rods even in the privileged aether system merely *accorded with* but did not "define" the metric of container space and time. On this basis, Lorentz was led to reason somewhat as follows:

(1) Since the horizontal arm of the Michelson-Morley experiment is shorter than a rod lying alongside it but conforming to the expectations of classical optics and its container theory of space, we must infer that when a unit rod in the aether system is transported to a moving system, it can no longer be a true unit rod but becomes shorter than unity in the moving system, and similarly for clocks.

(2) The deviation from the classically expected behavior exhibited by rods and clocks must have a cause in the sense of being due to a perturbational influence. For in the absence of such a cause, the classically expected behavior would have occurred spontaneously.

Einstein left the Lorentz transformations formally unaltered. But the reasoning underlying his radical reinterpretation of their physical meaning was based on the relational conception of space and time discussed in Chapter One. On the relational conception of space, the length of a body AB is an attribute of the relation between two pairs of points: the termini of AB on the one hand and those of the chosen unit rod on the other. Similarly the length of a time-interval is a ratio of that interval to the unit defined by some periodic physical process. And it is inherent in this definition of length as a ratio that the unit rod be at rest relative to AB when performing its metrical function.[2] But, on this relational

[2] It is perfectly clear that relations or relational properties of physical objects (which are expressed numerically as ratios) are fully as objective physically and exist just as independently of the human mind as simple properties of individual objects. Thus, the relation between a copper bar at rest in a system K and the unit rod in K might have the property that the copper bar has a length of 5 units in K. But the different relation of that bar's projection onto the x-axis of a system S, relative to which it is moving along that axis, to the unit rod of S may then yield a length of only

conception of length, one is not entitled to infer with Lorentz that a unit rod in the aether system will no longer be unity, *as a matter of physical fact,* once it has been transported to a moving system, and that such transport renders spurious the "local" readings of clocks in moving systems. For the relational conception allows us to call that same rod unity in the moving system by *definition,* and similarly for the units of time on clocks at rest in a moving system. If Lorentz had realized that the length of this rod in the moving system can be legitimately *decreed* by definition, and similarly for the periods of material clocks, then it would have been clear to him that the ground is cut from under his distinction between "true" ("real") and "local" (i.e., spurious or apparent) lengths and times and thereby from his idea that the horizontal arm in the Michelson-Morley experiment is actually shorter than the vertical arm. Thus, a coupling of the epistemological insights of the relational theory of length with the experimental findings of optics deprives reference to a preferred

---

4.7 S-units. It is incontestable that the differences among the various relations sustained by the bar do *not* render these relations subjective products of the physicist's mind, any more than they do the fact that the bar in question is a copper bar. In a futile attempt to defend a mentalistic metaphysics on the basis of relativity theory, Herbert Dingle denies this fact. Replying to decisive critiques of his views by P. Epstein (*American Journal of Physics,* Vol. X [1942], pp. 1 and 205, and Vol. XI [1943], p. 228), and M. Born (*Philosophical Quarterly,* Vol. III [1953], p. 139), Dingle offers the following argument (*The Sources of Eddington's Philosophy* [London: Cambridge University Press; 1954], pp. 11–12):

> The view that physics is the description of the character of an independent external world was simply no longer tenable. . . . Every relativist will admit that if two rods, A and B, of equal length when relatively at rest, are in relative motion along their common direction, then A is longer or shorter than B, or equal to it, exactly as you please. It is therefore impossible to evade the conclusion that its length is not a property of either rod; and what is true of length is true of every other so-called physical property. Physics is therefore (*sic!*) not the investigation of the nature of the external world.

Far from having demonstrated that relativity physics is subjective, Professor Dingle has merely succeeded in exhibiting his unawareness of the fact that *relational* properties do not cease to be bona fide objective properties just because they involve relations between individuals rather than belong directly to individuals themselves. Only such unawareness can lead to his primitive thesis that the relations of physical entities to one another cannot constitute "the character of an independent external world."

aether system of all objective physical significance and makes possible the enunciation of the principle of relativity.

We can understand Einstein's philosophical departure from the second step in Lorentz's reasoning by giving an analysis of Lorentz's invocation of a cause for the contraction revealed by the Michelson-Morley experiment.

Every physical theory tells us what particular behavior of physical entities or systems it regards as "natural" in the absence of the kinds of *perturbational* influences which it envisions. Concurrently, it specifies the influences or causes which it regards as responsible for any deviations from the assumedly "natural" behavior. But when such deviations are observed and a theory cannot designate the perturbations to which it proposes to attribute them, its assumptions concerning the character of the "natural" or unperturbed behavior become subject to doubt. For the reliability of our conceptions as to what pattern of occurrences is "natural" is no greater than the *scope* of the evidence on which they rest. And a theory's failure to designate the perturbing causes of the nonfulfillment of its expectations therefore demands the envisionment of the possibility that: first, the "natural" behavior of things is indeed different from what the theory in question has been supposing it to be and that, second, deviations from the assumedly natural behavior transpire *without* perturbational causes of the kind previously envisioned by the theory.

Several examples from past and present scientific controversy attest to the mistaken search for the previously envisioned kinds of perturbational factors which are supposed to cause deviations from the pattern which a particular theory unquestioningly and tenaciously affirms to be the natural one. Thus, Aristotelian critics of Galileo, assuming that Aristotle's mechanics describes the natural behavior, asked Galileo to specify the cause which prevents a body from coming to rest and maintains its speed in the same straight line in the manner of Newton's first law of motion. It was axiomatic for them that uniform motion could not continue indefinitely in the absence of net external forces. In our own time, there are those who ask: if the "new cosmology" of Bondi and Gold is true, must there not be a divine interference (perturbation) in the natural order whereby the spontaneous accretion ("creation") of matter is caused?[3] The propounders of this question do not tell us, however, on what basis they take it for granted that the total absence of any accretion of matter, however small or slow, is *cosmically* the natural state of affairs. Curiously enough, some theologians ask us to regard a state of "nothingness" (whatever that is!) as the natural state of the universe and thereby endeavor to create grounds for arguing that the mere existence and conservation of matter or energy require a divine creator and sustainer. But suppose that it were indeed to turn out that the physical world is a three-dimensional expanding spherical space whose radius of curvature had a minimum value at one time and that the latter was a genuine beginning because it was not preceded by an earlier contraction of which the present expansion is an "elastic" rebound. Then within the framework of that theory of the expanding universe, *every* phase of this expansion process must be held to be integral to the *natural* behavior of the universe. In that case, the inception of the expansion does not constitute evidence for the operation of a divine creator as perturbational cause, any more than any other phase of the expansion does. For there is absolutely no logically viable criterion for distinguishing those newly observed or inferred facts which merely compel the revision of previous generalizations and which the proponent of divine creation regards as part of the "natural order" from those to which he gratuitously attributes the status of being "outside" the "natural order" and of therefore being due to divine intervention in that order. We see that the theological concept of miracles attributes a supernatural origin to certain phenomena by the gratuitous declaration that a certain set of limited empirical generalizations which do not allow for these phenomena define with *certainty* what is natural. As if observed events presented themselves to us with identification tags as to their "naturalness"![4]

---

[3] The logical blunder which generates this question was committed in inverted form by Herbert Dingle. One of his reasons for rejecting the cosmology of Bondi and Gold is that it would allegedly require not merely a single act of miraculous divine interference, as Biblical creation *ex nihilo* does, but a continuous series of such acts. See A. Grünbaum: *Scientific American*, Vol. CLXXXIX (December, 1953), pp. 6–8.

[4] For further details on this issue, see A. Grünbaum: *Scientific Monthly*, Vol. LXXIX (1954), pp. 15–16.

The basis for Einstein's philosophical objection to the second step in Lorentz's reasoning is now at hand: it was an error on Lorentz's part to persist, in the face of mounting contrary evidence, in regarding the classically expected behavior as the natural behavior. It was this persistence which forced him to explain the observed deviations from the classical laws by postulating the operation of a physically non-designatable aether as a perturbational cause. Having used the relational theory of length and time to reject the conclusion of the first step in Lorentz's reasoning, Einstein was able to see that the unexpected results of the Michelson-Morley experiment do not require any perturbational causes of the kind envisioned in the aether theory because they are integral to the "natural" behavior of things. This is not to say that what is taken axiomatically as the "natural" or unperturbed behavior of particles and light within the framework of the STR might not be regarded as in need of explanation by reference to perturbing causes within the context of a wider and/or deeper theory such as the GTR.

The character and significance of these fundamental philosophical differences between Einstein's conception of the Lorentz transformations, on the one hand, and the interpretation of earlier versions of these transformations by his predecessors Fitzgerald, Larmor, Poincaré, and Lorentz, on the other, altogether escaped recognition by E. T. Whittaker in his very recent account of the history of the development of the STR. His account of the history of the STR in his monumental *History of the Theories of Aether and Electricity* provides a telling illustration of how philosophical misconceptions can issue in a grievously false historical estimate of the relative contributions of a conceptual innovator vis-à-vis those of his precursors. In particular, we shall now see that a correct philosophical construal of how Einstein understood the space and time coordinates of the Lorentz transformations could have precluded Whittaker's unsound assessment of the contributions made to the STR by Lorentz, Poincaré, Fitzgerald, Larmor, and Voigt, an assessment which lent credence to his disparaging evaluation of Einstein's role.

Entitling his chapter on the history of the development of the STR, "The Relativity Theory of Poincaré and Lorentz," Whittaker gives the following depreciatory evaluation of Einstein's role in the genesis of the STR: "In the autumn of the same year [1905] . . . Einstein published a paper which set forth the relativity theory of Poincaré and Lorentz with some amplifications, and which attracted much attention. . . . In this paper Einstein gave the modifications which must now be introduced into the formulas for aberration and the Doppler effect."[5] This historical assessment of the range of Einstein's originality and of the magnitude of his contribution to the STR derives at least its plausibility (even if perhaps not its inspiration) from Whittaker's *philosophically incorrect* conception of *Einstein's interpretation* of the Lorentz transformations. For Whittaker committed two grievous logical errors. He failed to be cognizant of Einstein's philosophical *repudiation* of the Lorentz-Larmor-Poincaré distinction between "true" or "real" and "spurious" (apparent, local) times and lengths, a repudiation which renders Einstein an authentic conceptual revolutionary of genius in this respect. And correlatively, as we saw earlier, he made no allowance at all for the crucial *logical* differences between the pre-Einsteinian and Einsteinian conceptions of the status of any extraterrestrially observed contraction of a rod moving with the earth and pointing in the direction of the latter's motion.

Thus, only philosophical awareness of the fact that Einstein's conception of the Lorentz transformations is not to be construed along the modified aether-theoretic lines of Lorentz and Poincaré makes possible the discernment of the mistake in Whittaker's historical treatment of the STR.

---

[5] E. T. Whittaker: *A History of the Theories of Aether and Electricity*, *op. cit.*, Vol. II, p. 40. Whittaker himself points out (p. 36) that even at the time of his death in 1928, Lorentz reportedly still favored the concepts of "true" time and absolute simultaneity. For Lorentz's own brief statement on this point, see his *The Theory of Electrons* (New York: Columbia University Press; 1909), pp. 329–30. In regard to the role played by Lorentz and Poincaré as precursors of Einstein, von Laue gives evidence in a more recent publication (*Naturwissenschaften*, Vol. XLIII [1956], p. 4) that Einstein was unaware of the groundwork they had done for the theory of relativity.

Chapter 13

PHILOSOPHICAL APPRAISAL OF E. A. MILNE'S ALTERNATIVE TO EINSTEIN'S STR

E. A. Milne, whose two logarithmically related t and τ scales of time were mentioned in Chapter One, Section C, has attempted to erect the usual space-time structure of the STR on the basis of a light signal kinematics of particle observers purportedly dispensing with the use of rigid solids and isochronous material clocks.[1] In his *Modern Cosmology and the Christian Idea of God*,[2] Milne begins his discussion of time and space by *incorrectly* charging Einstein with failure to realize that the concept of a rigid body as a body whose rest length is invariant under transport contains a conventional ingredient just as much as does the concept of metrical simultaneity at a distance.[3] Milne then proposes to improve upon a rigid body criterion of spatial congruence by proceeding in the manner of radar ranging and using instead the round-trip times required by light to traverse the corresponding closed paths, these times not being measured by material clocks but, in outline, as follows.[4] Each particle is equipped with

[1] E. A. Milne: *Kinematic Relativity, op. cit.*, and *Modern Cosmology and the Christian Idea of God, op. cit.*

[2] *Ibid.*, Chapter iii.

[3] That Einstein was abundantly aware of this point is evident from his definition of the "practically rigid body" in *Geometrie und Erfahrung, op. cit.*, p. 9.

[4] For more detailed summaries of Milne's light-signal kinematics cf. A. G. Walker: "Axioms for Cosmology" in L. Henkin, P. Suppes, and A. Tarski

a device for ordering the genidentical events belonging to it temporally in a linear Cantorean continuum. Such a device is called a "clock," and the single observer at the particle using such a local clock is called a "particle-observer." If now A and B are two particle observers and light signals are sent from one to the other, then the time of arrival $\overline{t'}$ at B can be expressed as a function $\overline{t'} = f(t)$ of the time t of emission at A, and likewise the time of arrival t′ at A is a function $t' = F(\bar{t})$ of the time $\bar{t}$ of emission at B. Particle-observers equipped with clocks as defined are said to be "equivalent," if the so-called signal functions f and F are the same, and the clocks of equivalent particle-observers are said to be *congruent*. It can be shown that if A and B are not equivalent, then B's clock can be regraduated by a transformation of the form $\overline{t'} = \Psi(t)$ so as to render them equivalent.[5] The congruence of the clocks at A and B does not, of course, assure their synchronism. Milne now uses Einstein's definition of simultaneity:[6] the time t_2 assigned by A to the arrival of a light ray at B which is emitted at time t_1 at A and returns to A at time t_3 after instantaneous reflection at B is defined to be

$$t_2 = \tfrac{1}{2}(t_1 + t_3).$$

And he defines the distance r_2 of B, ***by A's clock, upon the arrival of the light from A at B*** to be given by the relation

$$r_2 = \tfrac{1}{2}c(t_3 - t_1),$$

where c is an *arbitrarily* chosen constant.[7] Since

$$\frac{r_2}{t_2 - t_1} = \frac{r_2}{t_3 - t_2} = c,$$

the constant c represents the velocity of the light-signal in terms of the conventions adopted by A for measuring distance and time at a remote point B. Milne gives the following statement of his epistemological objections to Einstein's use of rigid rods and

(eds.) *The Axiomatic Method* (Amsterdam: North Holland Publishing Company; 1959), pp. 309–10, and L. Page and N. I. Adams: *Electrodynamics* (New York: D. Van Nostrand Company; 1940), pp. 78–85.

[5] E. A. Milne: *Modern Cosmology and the Christian Idea of God, op. cit.*, pp. 39–41.

[6] *Ibid.*, p. 42.

[7] *Ibid.*

of his claim that his light-signal kinematics provides a philosophically satisfactory alternative to it:

> the concept of the transport of a rigid body or rigid length measure is itself an indefinable concept. In terms of one given standard metre, we cannot say what we mean by asking that a given "rigid" length measure shall remain "unaltered in length" when we move it from one place to another; for we have no standard of length at the new place. Again, we should have to specify standards of "rest" everywhere, for it is not clear without consideration that the "length" will be the same, even at the same place, for different velocities. The fact is that to say of a body or measuring-rod that it is "rigid" is no definition whatever; it specifies no "operational" procedure for testing whether a given length-measure after transport or after change of velocity is the same as it was before. . . .[8]
>
> It is part of the debt we owe to Einstein to recognize that only "operational" definitions are of any significance in science . . . Einstein carried out his own procedure completely when he analysed the previously undefined concept of simultaneity, replacing it by tests using the measurements which have actually to be employed to recognize whether two distant events are or are not simultaneous. But he abandoned his own procedure when he retained the indefinable concept of the length of a "rigid" body, i.e., a length unaltered under transport. The two indefinable concepts of the transportable rigid body and of simultaneity are on exactly the same footing; they are fog-centres, inhibiting further vision, until analysed and shown to be equivalent to conventions. . . .[9]
>
> It will be one of our major tasks to elucidate the type of graduation employed for graduating our ordinary clocks; that is to say, to inquire what is meant by, and if possible to isolate what is usually understood by, "uniform time." In other words, we wish to inquire which of the arbitrarily many ways in which the markings of our abstract clock may be graduated can be identified with the "uniform time" of physics. . . .[1]
>
> The question now arises: is it possible to arrange that the mode of graduation of observer B's clock corresponds to the

[8] *Ibid.*, p. 35.
[9] *Ibid.*
[1] *Ibid.*, p. 37.

mode of graduation of A's clock in such a way that a meaning can be attached to saying that B's clock is a copy of A's clock? If so, we shall say that B's clock has been made congruent with A's. . . .[2]

It will have been noticed that we have succeeded in making B's clock a copy of A's without bringing B into permanent coincidence with A. We have made a copy of an arbitrary clock *at a distance.* This is something we cannot do with metre-scales or other length-measures. The problem of copying a clock is in principle simpler than the problem of copying a unit of length. We shall see in due course that with the construction of a copy of a clock at a distance we have solved the problem of comparing lengths. . . .[3]

The important point is that epoch and distance (which we shall call coordinates) are purely conventional constructs, and have meaning only in relation to a particular form of clock graduation. . . . But it is to be pointed out that when the mode of clock graduation reduces to that of ordinary clocks in physical laboratories, our coordinate conventions provide measures of epoch and distance which coincide with those based on the standard metre. . . .[4]

The reason why it is more fundamental to use clocks alone rather than both clocks and scales or than scales alone is that the concept of the clock is more elementary than the concept of the scale. The concept of the clock is connected with the concept of "two times at the same place," whilst the concept of the scale is connected with the concept of "two places at the same time." But the concept of "two places at the same time" involves a convention of simultaneity, namely, simultaneous events at the two places, but the concept of "two times at the same place" involves no convention; it involves only the existence of an *ego*. . . .[5]

Length is just as much a conventional matter as an epoch at a distance. Thus the metre-scale is not such a fundamental instrument as the clock. In the first place its length for *any* observer, as measured by the radar method, depends on the

[2] *Ibid.*, p. 39.
[3] *Ibid.*, p. 41.
[4] *Ibid.*, pp. 42–43.
[5] *Ibid.*, p. 46.

> clock used by the observer; in the second place, different observers assign different lengths to it even if their clocks are congruent, owing to the fact that the test of simultaneity is a conventional one. The clock, on the other hand, once graduated, gives epochs at itself which are independent of convention.
>
> Once we have set up a clock, arbitrarily graduated, distances for the observer using this clock become definite. If a rod, moved from one position of rest relative to this observer to another position of rest relative to the same observer, possesses in the two positions the same length, as measured by this observer using his own clock, as graduated, then the rod is said to have undergone a rigid-body-displacement by this clock. In this way we see that once we have fixed on a clock, a rigid-body-displacement becomes definable. But until we have provided a clock, there is no way of saying what we mean by a rigid body under displacement.[6]

Now, if Milne is to make good his criticism of Einstein by erecting the space-time structure of special relativity on alternative epistemological foundations, he must provide us with inertial systems by means of the resources of his light-signal kinematics as well as with the measures of length and time on which the kinematics of special relativity is predicated. This means that he must be able to characterize inertial systems *within* the confines of his epistemological program as some kind of dense assemblage of equivalent particle observers filling space such that each particle observer is at rest relative to and synchronous with every other. We have already seen that he was wholly in error in charging Einstein with lack of awareness of the conventionality of spatial congruence as "defined" by the rigid rod. But that, much more fundamentally, he is mistaken in believing to have erected the kinematics of special relativity on an epistemologically more satisfactory base than Einstein did will now be made clear by reference to the following result pointed out by L. L. Whyte:[7] Using only light signals and temporal succession without either a solid rigid rod or an isochronous material clock, it is *not* possible to construct ordinary measures of length and time. For "a

[6] *Ibid.*, pp. 47–48.

[7] L. L. Whyte: "Light Signal Kinematics," *British Journal for the Philosophy of Science*, Vol. IV (1953), pp. 160–61.

physicist using only light signals cannot discriminate inertial systems from these subjected to arbitrary 4-D similarity transformations.[8] The system of 'resting' mass-points which can be so identified may be arbitrarily expanding and/or contracting relatively to a rod, and these superfluous transformations can only be eliminated by using a rod or a clock."[9]

The significance of the result stated by Whyte is twofold.

First: If Milne dispenses with material clocks and bases his chronometry only on the congruences yielded by his light-signal clocks, then he cannot obtain inertial systems without a rigid rod in the following sense. The rigid rod is *not* needed for the definition of *spatial congruence* within the system but *is* required to assure that the distance between one particular pair of points connected by it *at one time* t_0 is the same as *at some later time* t_1. In other words, the rod is rigid *at a given place* by remaining congruent to itself (by convention) as time goes on. And in this way the rod assures the time-constancy of the distance between the two given points connected by it. This reliance on the rigid rod thus involves the use of the definition of simultaneity. Hence, if Milne were right in charging that the use of a rigid rod is beset by philosophic difficulties, then he indeed would be incurring these liabilities no less than Einstein does. However, let us suppose that:

Second, Milne does use a material clock to define the time-metric at a space point and thereby to particularize his clock graduations to the kind required for the elimination of the unwanted reference systems described by L. L. Whyte. This procedure is a far cry from his purely topological clock which "involves only the existence of an *ego*"[1] in contradistinction to the rigid scale's involvement of a definition of simultaneity. And, in that case, his measurement of the equality of space intervals by means of the equality of the corresponding round-trip time-intervals involves the following conventions: (a) the tacit use of a definition of simultaneity of non-coinciding events. For although

[8] For a brief account of similarity transformations, and a further articulation of Whyte's point here, cf. H. Reichenbach: *PST, op. cit.*, pp. 172–73.

[9] L. L. Whyte: "Light Signal Kinematics," *op. cit.*, p. 161.

[1] E. A. Milne: *Modern Cosmology and the Christian Idea of God, op. cit.*, p. 46.

a *round*-trip time on a given clock does not, of course, itself require such a simultaneity criterion, the measurement of a spatial distance in an inertial system by means of this time does: the distance yielded by the round-trip time on a clock at A is the distance r_2 between A and B *at the time t_2* on the A-clock *when the light pulse from A arrives at B* on its round-trip ABA, (b) successive equal differences in the readings of a given local clock are stipulated to be measures of equal time intervals and thereby of equal space intervals, and (c) equal differences on separated clocks of identical constitution are decreed to be measures of equal time intervals and thereby of equal space intervals.

To what extent then, if any, does Milne have a case against Einstein? It would appear from our analysis that the only justifiable criticism is not at all epistemological but concerns an innocuous point of axiomatic economy: once you grant Milne a material clock, he does not require the rigid rod at all, whereas Einstein utilizes the spatial congruence definition based on the rigid rod *in addition* to all of the conventions needed by Milne. Thus, Milne's kinematics, as supplemented by the use of a material clock, is constructed on a slightly narrower base of conventions than is Einstein's.[2]

It will be recalled from Chapter Four that if measurements of spatial and temporal extension are to be made by means of solid rods and material clocks, allowance must be made computationally for thermal and other perturbations of these bodies so that they can define rigidity and isochronism. Calling attention to this fact and believing Milne's light-signal kinematics to be essentially successful, L. Page deemed Milne's construction more adequate than Einstein's, writing:

> the original formulation of the relativity theory was based on undefined concepts of space and time intervals which could not be identified unambiguously with actual observations. Recently Milne has shown how to supply the desired criterion [of rigidity and isochronism] by erecting the space-time structure on the foundations of a constant light-signal velocity.[3]

[2] The preceding critique of Milne supplants my earlier brief critique ("The Philosophical Retention of Absolute Space in Einstein's General Theory of Relativity," *The Philosophical Review*, Vol. LXVI [1957], pp. 531–33) in which I *misinterpreted* Milne's arguments as indicative of lack of appreciation on his part of the conventionality of *temporal* congruence.

[3] L. Page and N. I. Adams: *Electrodynamics, op. cit.*, pp. 78–79.

It is apparent in the light of our appraisal of Milne's kinematics that Page's claim is vitiated by Milne's need for a rigid rod or material clock as specified.

It should be noted, however, as Professor A. G. Walker has pointed out to me, that if Milne's construction is interpreted as applying *not* to special relativity kinematics but to his cosmological world model, then our criticisms are no longer pertinent. In terms of his logarithmically related τ and t scales of time, it turns out that upon measuring distances by the specified chronometric convention, the galaxies are at relative rest in τ-scale kinematics and in uniform relative motion in the t-scale. Each of these time scales is unique up to a trivial change of units, and their associated descriptions of the cosmological world are equivalent in Reichenbach's sense. In this *cosmological* context, the problem of eliminating the superfluous transformations mentioned by Whyte therefore does not arise.

Chapter 14

HAS THE GENERAL THEORY OF RELATIVITY REPUDIATED ABSOLUTE SPACE?

The literature of recent decades on the philosophy and history of science has nurtured and given wide currency to a myth concerning the present status of the dispute between the absolutistic and relativistic theories of space. In particular, that literature is rife with assertions that the post-Newtonian era has witnessed "the final elimination of the concept of absolute space from the conceptual scheme of modern physics"[1] by Einstein's general theory of relativity and that the Leibniz-Huyghens polemic against Newton and Clarke has thus been triumphantly vindicated.[2] In this vein, Philipp Frank recently reached the following verdict on Einstein's success in the implementation of Ernst Mach's program for a *relativistic* account of the *inertial* properties of matter: "Einstein started a new analysis of Newtonian mechanics which eventually vindicated Mach's reformulation [of Newtonian mechanics]."[3]

I shall now show that the history of the GTR does not at all bear out the widespread view set forth in the quotations from Max Jammer and Philipp Frank. And it will then become apparent in what precise sense there is ample justification for Einstein's own admission of 1953 as follows: the supplanting of the concept

[1] M. Jammer: *Concepts of Space, op. cit.*, p. 2.

[2] A very useful modern edition by H. G. Alexander of *The Leibniz-Clarke Correspondence* has been published in 1956 by the Manchester University Press and by the Philosophical Library in New York.

[3] P. Frank: *Philosophy of Science* (Englewood Cliffs, New Jersey: Prentice-Hall; 1957), p. 153.

of absolute space is "a process which is probably by no means as yet completed."[4]

Mach had urged against Newton that both translational and rotational inertia are intrinsically dependent on the large-scale distribution and relative motion of matter. Assuming the indefinite extensibility of terrestrial axes to form an unlimited Euclidean rigid system S_e, the rotational motion of the stars seemed to be clearly defined with respect to S_e. Unfortunately, however, the GTR was not entitled to make use of S_e: the linear velocity of rotating mass points increases with the distance from the axis of rotation, and hence the existence of a system S_e of unrestricted size would allow *local* velocities greater than that of light, in contravention of the requirement of the local validity of the STR. But to deny, as the GTR therefore must, that S_e can extend even as far as the planet Neptune is to assert that the Machian concept of the *relative* motion of the earth and the stars is no more meaningful physically than the Newtonian bugaboo of the *absolute* rotation of a solitary earth in a space which is structured independently of any matter that it might contain accidentally and indifferently![5] Accordingly, the earth must be held to rotate *not* relative to the stars but with respect to the local "star-compass" formed at the earth by stellar light rays whose paths are determined by the local *metrical field.*

At Einstein's hands, Mach's thesis underwent not only this modification but also the following generalization: Einstein found that *both* the geometry of material rods and clocks *and* the inertial behavior of particles and light in the context of that geometry are functionally related to the same physical quantities. Probably unaware at the time that Riemann had previously conjectured the dependence of the geometry of physical space on the action of matter via a different line of reasoning,[6] Einstein

---

[4] A. Einstein: Foreword to Max Jammer, *Concepts of Space, op. cit.*, p. 15.

[5] For details, see H. Weyl: "Massenträgheit und Kosmos," *Naturwissenschaften*, Vol. XII (1924), p. 197. See also F. E. A. Pirani: "On the Definition of Inertial Systems in General Relativity," in *Bern Jubilee of Relativity Theory*, suppl. IV of *Helvetica Physica Acta* (Basel: Birkhäuser Verlag; 1956), pp. 198–203.

[6] Cf. B. Riemann: *Über die Hypothesen welche der Geometrie zu Grunde liegen*, edited by H. Weyl (3rd ed., Berlin: Julius Springer; 1923), pp. 3 and 20. The reader will find a brief account of the relevant part of Riemann's reasoning in Chapters One and Fifteen of the present book.

named his own organic fusion of Riemann's and Mach's ideas "Mach's Principle."[7] And he sought to implement that principle by requiring that the metrical field given by the quantities g_{ik} be *exhaustively* determined by properties and relations of ponderable matter and energy specified by the quantities T_{ik}. On this conception a single test particle would have no inertia whatever if all other matter and energy were either annihilated or moved indefinitely far away.

But when the problem of solving the nonlinear partial differential equations which connect the derivatives of the g_{ik} to the T_{ik} was confronted, it became apparent that, far from having been exorcised by the GTR, the ghost of Newton's absolute space is nothing less than a haunting incubus. For to obtain a solution of these equations, it is necessary to supply the boundary conditions "at infinity." And to assume, as is done in Schwarzschild's solution, that there are certain preferred coordinate systems in which the g_{ik} have the Lorentz-Minkowski values at infinity is to violate Mach's Principle in the following twofold sense: first, the boundary conditions at infinity then assume the role of Newton's absolute space, since it is not the influence of matter that determines what coordinate systems at infinity are the Galilean ones of special relativity; and second, instead of being the *source* of the *total* structure of space-time, matter then merely *modifies* the latter's otherwise autonomously flat structure. In 1916 Einstein first attempted to avoid this most unwelcome consequence by reluctantly altering the above field equations through the introduction of the cosmological constant λ, which yielded a solution in which space was closed (finite). But this rather forced step did not provide an escape from the troublesome philosophical difficulties that had cropped up in the boundary conditions at infinity, since these difficulties reappeared when W. de Sitter showed that the now *modified* equations violated Mach's Principle by allowing a universe essentially devoid of matter to have a definitely structured space-time. The attempt to dispose of the difficulty at infinity by laying down the finitude of space as a *boundary condition* governing the solution of the *unmodified* field equations is unavailing for the purpose of rescu-

[7] A. Einstein: "Prinzipielles zur allgemeinen Relativitätstheorie," *op. cit.*, p. 241.

ing Mach's Principle *as it was originally conceived,* since such a speculative assumption involves a nonintrinsic connection between the over-all structure of space and the properties of matter. In 1951 the Machian hope of subordinating space-time ontologically to matter was further dashed when A. H. Taub showed that there are conditions under which the *unmodified* field equations yield *curved* space in the absence of matter.[8]

These results inescapably raise the question of whether the failure of the GTR to implement Mach's Principle is to be regarded as an inadequacy on the part of that theory or as a basis for admitting that the GTR was right in philosophically retaining Newton's absolute space to a significant extent, thinly disguised under new structural trappings. Einstein's own attitude in his last years seems to have been one of unmourning abandonment of Mach's Principle. His reason appears to have been that although matter provides the *epistemological* basis for the metrical field, this fact must *not* be held to confer *ontological* primacy on matter over the field: matter is merely part of the field rather than its source.[9]

This is indeed a very far cry from, nay the very antithesis of, Max Jammer's "final elimination of the concept of absolute space from the conceptual scheme of modern physics."[1] In fact, Jammer himself quotes a recent passage from Einstein in which Einstein says that if the *space-time field* were removed, there would be no space.[2] Yet Jammer gives no indication whatever that this is a drastically different thesis from Einstein's earlier one that if all

---

[8] A. H. Taub: "Empty Space-Times Admitting a Three Parameter Group of Motions," *Annals of Mathematics,* Vol. LIII (1951), p. 472.

[9] For a discussion of the status of Einstein's program of field theory, see J. Callaway: "Mach's Principle and Unified Field Theory," *Physical Review,* Vol. XCVI (1954), p. 778. For an alternative theory of gravitation inspired by the aim of strict conformity to Mach's Principle but incomplete in other respects, see D. W. Sciama: "On the Origin of Inertia," *Monthly Notices of the Royal Astronomical Society,* Vol. CXIII (1953), p. 35, and "Inertia," *Scientific American,* Vol. CXCVI (February, 1957), pp. 99–109. Cf. also F. A. Kaempffer: "On Possible Realizations of Mach's Program," *Canadian Journal of Physics,* Vol. XXXVI (1958), pp. 151–59, and O. Klein: "Mach's Principle and Cosmology in their Relation to General Relativity," in *Recent Developments in General Relativity* (Warsaw: Polish Scientific Publishers; 1962), pp. 293–302.

[1] M. Jammer: *Concepts of Space, op. cit.,* p. 2.

[2] *Ibid.,* p. 172.

matter were annihilated, then metric space would vanish as well.[3]

It is now clear that the GTR *cannot* be said to have resolved the controversy between the absolutistic and relativistic conceptions of space in favor of the latter on the issue of the implementation of Mach's Principle. Instead, the current state of knowledge supports the following summary assessment given in 1961 by the physicists C. Brans and R. H. Dicke:

> The . . . view that the geometrical and inertial properties of space are meaningless for an empty space, that the physical properties of space have their origin in the matter contained therein, and that the only meaningful motion of a particle is motion relative to other matter in the universe has never found its complete expression in a physical theory. This picture is . . . old and can be traced from the writings of Bishop Berkeley[4] to those of Ernst Mach.[5] These ideas have found a limited expression in general relativity, but it must be admitted that, although in general relativity spatial geometries are affected by mass distributions, the geometry is not uniquely specified by the distribution. It has not yet been possible to specify boundary conditions on the field equations of general relativity which would bring the theory into accord with Mach's principle. Such boundary conditions would, among other things, eliminate all solutions without mass present.[6]

---

[3] M. Jammer has since taken account of these criticisms on pp. 12 and 195 of the revised Harper Torchbook edition of his book, published in New York in 1960.

[4] "G. Berkeley: *The Principles of Human Knowledge,* Paragraphs 111–17, 1710–*De Motu* (1726)."

[5] "E. Mach: *Conservation of Energy,* note No. 1, 1872 (reprinted by Open Court Publishing Company, LaSalle, Illinois, 1911), and *The Science of Mechanics,* 1883 (reprinted by Open Court Publishing Company, LaSalle, Illinois, 1902, Chapter II, Sec. VI)."

[6] C. Brans and R. H. Dicke: "Mach's Principle and a Relativistic Theory of Gravitation," *The Physical Review,* Vol. CXXIV (1961), p. 925. See also R. H. Dicke: "Mach's Principle and Invariance Under Transformation of Units," *The Physical Review,* Vol. CXXV (1962), p. 2163, and "The Nature of Gravitation," in L. V. Berkner and H. Odishaw (eds.) *Science in Space* (New York: McGraw-Hill Book Company; 1961), Chapter iii, Sec. 3.1, "Mach's Principle," pp. 93–95. For an account of statements which might be regarded as modified versions of Mach's Principle and which are valid in the GTR, see C. H. Brans: "Mach's Principle and the Locally Measured

The difficulties encountered by the attempt to incorporate Mach's Principle as originally conceived into the GTR have most recently prompted two kinds of responses from leading investigators, which illustrate the lack of a uniform conception of this principle. Brans and Dicke[7] have put forward a *modified* relativistic theory of gravitation which is apparently compatible with Mach's principle, and is closely related to the theory of P. Jordan.[8] But J. A. Wheeler has articulated the important modifications which must be made in the original program of Mach's Principle, if Mach's ideas are to preserve their relevance to the GTR in its current state. Wheeler's substantial reformulation of Mach's Principle is as follows: "the specification of a sufficiently regular closed three-dimensional geometry at two immediately succeeding instants, and of the density and flow of mass-energy, is to determine the geometry of space-time, past, present, and future, and thereby the inertial properties of every infinitesimal test particle."[9] On Wheeler's view then, Mach's Principle can be implemented in the GTR in the following drastically altered form: if we are *given* (1) that the *three*-dimensional geometry of space at some initial instant and at some closely succeeding instant does *not* extend to infinity and does *not* show infinite curvature, and (2) the distribution of mass and mass-flow, then the *four*-dimensional geometry of space-time or the "geometrodynamics" and hence the inertial properties of infinitesimal test particles are thereby determined. For Wheeler then, the modified form of Mach's Principle simply *requires ab initio* that the universe be *spatially* closed or finite. In this way, it constitutes a principle for *selecting* out of the many conceivable solutions of Einstein's field equations those for which the three-geometry at a given

---

Gravitational Constant in General Relativity," *The Physical Review,* Vol. CXXV (1962), p. 396.

[7] C. Brans and R. H. Dicke: "Mach's Principle and a Relativistic Theory of Gravitation," *op. cit.*

[8] P. Jordan: *Schwerkraft und Weltall* (Braunschweig: Friedrich Vieweg und Sohn; 1955).

[9] J. A. Wheeler: "Mach's Principle as a Boundary Condition for Einstein's Field Equations and as a Central Part of the 'Plan' of General Relativity," a Report given at the Conference on Relativistic Theories of Gravitation, Warsaw, Poland, July, 1962.

instant is closed and free from singularity, thereby making possible the determination of the four-geometry and of the inertial behavior of infinitesimal test particles.[1]

[1] J. A. Wheeler ("The Universe in the Light of General Relativity," *The Monist*, Vol. XLVII, No. 1 [1962], pp. 40–76) has given a very brief statement of the meaning of his reformulation of Mach's Principle as applied to a universe which is empty of all "real" mass in the sense of the vision of Clifford and Einstein. For details on the latter universe, see C. W. Misner and J. A. Wheeler: "Geometrodynamics," *Annals of Physics*, Vol. II (1957), pp. 525–614; J. A. Wheeler: "Curved Empty Space-Time As the Building Material of the Physical World: An Assessment," in: E. Nagel, P. Suppes, and A. Tarski (eds.) *Logic, Methodology and Philosophy of Science:* Proceedings of the 1960 International Congress (Stanford: Stanford University Press; 1962), pp. 361–74, and J. G. Fletcher: "Geometrodynamics," in: L. Witten (ed.) *Gravitation* (New York: John Wiley & Sons, Inc.; 1962), Chapter xx, pp. 412–37.

Chapter 15

PHILOSOPHICAL CRITIQUE OF WHITEHEAD'S THEORY OF RELATIVITY

Einstein's theory of relativity was probably the most important influence on Whitehead's philosophy of science. But Whitehead's endeavor to reinterpret and modify Einstein's STR and GTR in terms of the categories of his own natural philosophy issued in two important philosophical divergences from Einstein: first, as will be recalled from Chapter Twelve, Whitehead erects the STR on his espousal of a *sensory absolute simultaneity* for any given inertial system in opposition to Einstein's theses on simultaneity, and second, Whitehead repudiates the GTR, because he rejects on epistemological grounds Riemann's conception of the relation between geometry and physics, which Einstein had attempted to weave into the logical fabric of the GTR via Mach's Principle, as explained in Chapter Fourteen.

In Chapter Twelve, I set forth my reasons for regarding Whitehead's rival to Einstein's conception of simultaneity as utterly untenable.[1] It remains therefore in the present chapter to give a critical appraisal of Whitehead's grounds for renouncing Riemann's view of the relation between geometry and physics, a view which permeates Einstein's GTR. Says Whitehead:

> I deduce that our experience requires and exhibits a basis of uniformity. . . . This conclusion entirely cuts away the causal

[1] For a statement of my objections to R. M. Palter's recent defense (*Whitehead's Philosophy of Science, op. cit.*, pp. 34–41) of Whitehead's doctrine of simultaneity, cf. A. Grünbaum: "Whitehead's Philosophy of Science," *The Philosophical Review*, Vol. LXXI (1962), pp. 222–24.

> heterogeneity [of the geometry] . . . which is the essential of Einstein's later theory [of *general* relativity].
>
> It is inherent in my theory to maintain the old division between physics and geometry. Physics is the science of the contingent relations of nature and geometry expresses its uniform relatedness.[2]

Thus, in opposition to the GTR's spatially and temporally *variable* geometry, Whitehead affirms the *uniformity* of the world's geometry, thereby claiming that the geometry is either Euclidean or one of the non-Euclidean geometries of *constant* negative or positive curvature.

To discern the philosophical basis for Whitehead's position, we must be mindful of the reasoning which issues in Riemann's conception of the relation between geometry and physics. Riemann had drawn the revolutionary conclusion that the metric geometry prevailing in physical space with respect to a specified congruence standard is acquired by that space, as Weyl puts it, "*only through the advent of the material content filling it and determining its metric relations.*"[3] In other words, "Riemann rejects the opinion that had prevailed up to his own time, namely, that the metrical structure of space is fixed and inherently independent of the physical phenomena for which it serves as a background, and that the real content takes possession of it as of residential flats."[4] And, as the reader will recall from the statement by Riemann quoted in Chapter One, Section B, Riemann reaches this conclusion *on the basis of his conception of the status of spatial congruence* as follows. In the continuous manifold of space (or of time), the measure of an interval must be provided by a congruence standard which "must come from somewhere else [i.e., from outside the manifold itself]"[5] by convention. *But the coincidence behavior of that standard under transport is left indeterminate by the structure of the spatial (or temporal) manifold itself.* Hence Riemann infers that if "the actual things forming the groundwork of a space" form a continuous manifold, then

[2] A. N. Whitehead: *The Principle of Relativity, op. cit.*, pp. 5 and 6.
[3] H. Weyl: *Space-Time-Matter, op. cit.*, p. 98.
[4] *Ibid.*
[5] B. Riemann: *On The Hypotheses Which Lie At The Foundations of Geometry, op. cit.*, pp. 424–25.

"the basis of metric relations must be sought for outside that actuality, in colligating forces that operate upon it."[6]

It will now become apparent that it was Whitehead's anti-bifurcationist and perceptualistic conception of spatial and temporal congruence—which we had reason to *reject* in Chapter One, Section G—that is the *fons* and *origo* of his renunciation of Riemann's reasoning and hence of Einstein's countenancing of a *variable* geometry in the GTR. For Whitehead's elaboration of the geometry and kinematics ingredient in his rival theory of relativity via his method of extensive abstraction[7] rests on the contention that the unique spatial and temporal congruences provided by our *sensory* contents are such as to require the spatial and temporal *uniformity* of the world's geometry. But if I succeeded in demonstrating the untenability of Whitehead's theses regarding congruence in Chapter One, Section G, then his argument for the uniformity of the world's geometry is clearly gratuitous.

Palter has pointed out[8] that Whitehead adduces his account of sense perception in the mode of pure relatedness as an argument for the uniformity of space, making no mention of congruence in that context. And thus it might seem that I have erred in representing Whitehead as having rested his case for uniformity on the perception of congruence. But there is a compelling mathematical reason for rejecting Palter's view that Whitehead intended his avowal of uniformity to be independent of his account of congruence and that he regarded the latter as relevant only to the determination of which one of the three kinds of uniform geometries obtains in nature. As a mathematician, Whitehead must have been aware that an assertion of uniform metrical relatedness has no meaning without at least tacit reference to a particular criterion of congruence: if space were to have a variable geometry with respect to a certain

---

[6] *Ibid.*, p. 425.

[7] My reasons for believing that Whitehead's method of extensive abstraction is a failure within the framework of his perceptualistic epistemology are given in A. Grünbaum: "Whitehead's Method of Extensive Abstraction," *The British Journal for the Philosophy of Science,* Vol. IV (1953), pp. 215–26.

[8] R. M. Palter: *Whitehead's Philosophy of Science, op. cit.,* pp. 25–27 and 32–33.

congruence, then there would always be infinitely many other congruences that would impart a uniform geometry to it, and conversely. Thus, for example, although the surface of an egg exhibits a variable geometry (Gaussian curvature) with respect to the standard metrization, there are infinitely many other congruences with respect to which that very surface would exhibit the uniform geometry imparted to the surface of a sphere by the standard metrization. Hence, without a specification of the congruence criterion, the mere claim of spatial uniformity places no restrictions on the coincidence behavior of transported rods and does not rule out the variable geometry of Einstein's GTR. It seems unavoidable, therefore, to conclude that Whitehead intended his arguments for uniform relatedness to stand or fall with his theses regarding congruence.

Since I have argued that these latter theses are quite unsound, I must reach the following verdict in regard to Whitehead's rival to Einstein's GTR: if it should turn out that the uniformity of space affirmed by his alternative to Einstein's GTR is strongly supported by future empirical findings, this will *not* be because of the *philosophical* reasons put forward by Whitehead.

Bibliography

Aharoni, J.: *The Special Theory of Relativity.* Oxford: Oxford University Press; 1959.

Akhiezer, N. I.: *The Calculus of Variations.* New York: Blaisdell Publishing Company; 1962.

Alexander, H. G.: *The Leibniz-Clarke Correspondence.* Manchester: Manchester University Press; 1956, and New York: Philosophical Library; 1956.

Alexander, H. L.: *Reactions with Drug Therapy.* Philadelphia: W. B. Saunders Company; 1955.

Aristotle: *On Generation and Corruption.* Book I, Ch. ii, 316^a15–317^a17.

Baker, B. and E. T. Copson: *The Mathematical Theory of Huyghens's Principle.* Oxford: Oxford University Press; 1939.

Baldus, R.: *Nichteuklidische Geometrie.* Edited by F. Löbell. Third revised edition. Berlin: Walter de Gruyter and Company; 1953. Sammlung Göschen, Vol. CMLXX.

Barankin, E. W.: "Heat Flow and Non-Euclidean Geometry." *American Mathematical Monthly,* Vol. XLIX (1942).

Baruch, J. J.: "Horological Accuracy: Its Limits and Implications." *American Scientist,* Vol. XLVI (1958).

Baum, W. A.: "Photoelectric Test of World Models." *Science,* Vol. CXXXIV (1961).

Bell, E. T.: *The Development of Mathematics.* New York: McGraw-Hill Book Company; 1945.

Bergmann, G.: "Review of M. Born's *Natural Philosophy of Cause and Chance.*" *Philosophy of Science,* Vol. XVII (1950).

Bergmann, H.: *Der Kampf um das Kausalgesetz in der jüngsten Physik.* Braunschweig: F. Vieweg and Son; 1929.

Bergmann, P.: *Introduction to the Theory of Relativity.* New York: Prentice-Hall; 1946.

———: "Fifty Years of Relativity." *Science,* Vol. CXXIII (1956).

Bergson, H.: *Creative Evolution.* New York: Random House, Inc.; 1944.

———: *Matière et Mémoire.* Geneva: A. Skira; 1946.

Bergstrand, E.: "Determination of the Velocity of Light." *Handbuch der Physik.* Edited by W. Flügge. Berlin: J. Springer; 1956. Vol. XXIV.

Berkeley, G.: *The Principles of Human Knowledge.* Paragraphs 111–17, 1710-*De Motu* (1726).

Birkhoff, G. D.: *The Origin, Nature and Influence of Relativity.* New York: The Macmillan Company; 1925.

Black, M.: "The 'Direction' of Time." *Analysis,* Vol. XIX (1959).

———: "Review of G. J. Whitrow's *The Natural Philosophy of Time.*" *Scientific American,* Vol. CCVI (1962).

Blank, A. A.: "Analysis of Experiments in Binocular Space Perception." *Journal of the Optical Society of America,* Vol. XLVIII (1958).

———: "Axiomatics of Binocular Vision. The Foundations of Metric Geometry in Relation to Space Perception." *Ibid.*

———: "The Geometry of Vision." *The British Journal of Physiological Optics,* Vol. XIV (1957).

———: "The Luneburg Theory of Binocular Perception." *Psychology, A Study of a Science.* Edited by S. Koch. New York: McGraw-Hill Book Company, Inc.; 1958. Study I, Vol. I.

———: "The Luneburg Theory of Binocular Visual Space." *Journal of the Optical Society of America,* Vol. XLIII (1953).

———: "The non-Euclidean Geometry of Binocular Visual Space." *Bulletin of the American Mathematical Society,* Vol. LX (1954).

Blatt, J. M.: "Time Reversal." *Scientific American* (August 1956).

Blum, H. F.: *Time's Arrow and Evolution.* Second edition. Princeton: Princeton University Press; 1955.

Bohn, D.: *Quantum Theory.* New York: Prentice-Hall; 1951.

Bolza, O.: *Lectures on the Calculus of Variations.* New York: G. E. Stechert; 1946.

Bolzano, B.: *Paradoxes of the Infinite.* Edited by D. A. Steele. New Haven: Yale University Press; 1951.

Bonch-Bruevich, A. M.: "On the Direct Experimental Verification of the Second Postulate of the Special Relativity Theory." *Physics Express* (November 1960), Vol. III.

Bondi, H.: *Cosmology.* Second Edition. Cambridge: Cambridge University Press; 1961.

———: "Relativity and Indeterminacy." *Nature,* Vol. CLXIX (1952).

Bonola, R.: *Non-Euclidean Geometry.* New York: Dover Publications, Inc.; 1955.

Born, M.: "Physical Reality." *Philosophical Quarterly,* Vol. III (1953).

———: *Natural Philosophy of Cause and Chance.* Oxford: Oxford University Press; 1949.

Boyer, C. B.: *The Concepts of the Calculus.* New York: Hafner Publishing Company; 1949.

Braithwaite, R. B.: "Axiomatizing a Scientific System by Axioms in the Form of Identification." *The Axiomatic Method.* Edited by L. Henkin, P. Suppes, and A. Tarski. Amsterdam: North Holland Publishing Company; 1959.

Brans, C. H. and Dicke, R. H.: "Mach's Principle and a Relativistic Theory of Gravitation." *The Physical Review,* Vol. CXXIV (1961).

Brans, C. H.: "Mach's Principle and the Locally Measured Gravitational Constant in General Relativity." *Ibid.*, Vol. CXXV (1962).

Bridgman, P. W.: "The Nature of Physical 'Knowledge.'" *The Nature of Physical Knowledge*. Edited by L. W. Friedrich. Bloomington: Indiana University Press; 1960.

——: *Reflections of a Physicist*. New York: Philosophical Library; 1950.

——: "Reflections on Thermodynamics." *American Scientist*, Vol. XLI (1953).

——: "Some Implications of Recent Points of View in Physics." *Revue Internationale de Philosophie*, Vol. III, No. 10 (1949).

——: *A Sophisticate's Primer of the Special Theory of Relativity*. Middletown, Conn.: Wesleyan University Press; 1962.

Brillouin, L.: *Science and Information Theory*. New York: Academic Press; 1956.

Brouwer, C.: "The Accurate Measurement of Time." *Physics Today*, Vol. IV (1951).

Brown, F. A., Jr.: "Biological Clocks and the Fiddler Crab." *Scientific American*, Vol. CXC (1954).

——: "Living Clocks." *Science*, Vol. CXXX (1959).

——: "Response to Pervasive Geophysical Factors and the Biological Clock Problem." *Cold Spring Harbor Symposia on Quantitative Biology*, Vol. XXV (1960).

——: "The Rhythmic Nature of Animals and Plants." *American Scientist*, Vol. XLVII (1959).

Burington, R. S. and Torrance, C. C.: *Higher Mathematics*. New York: McGraw-Hill Book Company, Inc.; 1939.

Callaway, J.: "Mach's Principle and Unified Field Theory." *Physical Review*, Vol. XCVI (1954).

Cantor, G.: *Gesammelte Abhandlungen*. Edited by E. Zermelo. Berlin: J. Springer; 1932.

Capek, M.: *The Philosophical Impact of Contemporary Physics*. Princeton: D. Van Nostrand Company, Inc.; 1961.

Carathéodory, C.: "Untersuchungen über die Grundlagen der Thermodynamik." *Mathematische Annalen* (1909).

Carnap, R.: *Abriss der Logistik*. Vienna: J. Springer; 1929.

——: "Adolf Grünbaum on The Philosophy of Space and Time." *The Philosophy of Rudolf Carnap*. Edited by P. A. Schilpp. LaSalle: Open Court Publishing Company; 1963.

——: *Der Raum*. Berlin: Reuther und Reichard; 1922.

——: *Introduction to Symbolic Logic and Its Applications*. New York: Dover Publications, Inc.; 1958.

——: *Symbolische Logik*. Vienna: J. Springer; 1954.

——: "Über die Abhängigkeit der Eigenschaften des Raumes von denen der Zeit." *Kantstudien*, Vol. XXX (1925).

Cassirer, E.: *Zur Einsteinschen Relativitätstheorie*. Berlin: Bruno Cassirer Verlag; 1921.

Cedarholm, J. P. and Townes, C. H.: "A New Experimental Test of Special Relativity." *Nature,* Vol. CLXXXIV, No. 4696 (1959).

Chandrasekhar, S. and Wright, J. P.: "The Geodesics in Gödel's Universe." *Proceedings of the National Academy of Sciences,* Vol. XLVIII (1961).

Cherry, E. C.: "The Communication of Information." *American Scientist,* Vol. XL (1952).

Clemence, G. M.: "Astronomical Time." *Review of Modern Physics,* Vol. XXIX (1957).

———: "Dynamics of the Solar System." *Handbook of Physics.* Edited by E. Condon and H. Odishaw. New York: McGraw-Hill Book Company, Inc.; 1958.

———: "Ephemeris Time." *Astronomical Journal,* Vol. LXIV (1959) and *Transactions of the International Astronomical Union,* Vol. X (1958).

———: "Time and Its Measurement." *American Scientist,* Vol. XL (1952).

Clifford, W. K.: *The Common Sense of the Exact Sciences.* New York: Dover Publications, Inc.; 1955.

Courant, R.: *Vorlesungen über Differential und Integralrechnung.* Berlin: J. Springer; 1927.

Courant, R. and Robbins, H.: *What is Mathematics?* New York: Oxford University Press; 1941.

Cramér, H.: *Mathematical Methods of Statistics.* Princeton: Princeton University Press; 1946.

d'Abro, A.: *The Evolution of Scientific Thought from Newton to Einstein.* New York: Dover Publications, Inc.; 1950.

Davies, R. O.: "Irreversible Changes: New Thermodynamics From Old." *Science News,* No. 28 (May 1953).

de Beauregard, O. C.: "Complémentarité et Relativité," *Revue Philosophique,* Vol. CXLV (1955).

———: "L'Irréversibilité Quantique, Phénomène Macroscopique." *Louis de Broglie, Physicien et Penseur.* Paris: Albin Michel; 1953.

———: *Théorie Synthétique de la Relativité Restreinte et des Quanta.* Paris: Gauthier-Villars; 1957.

Dedekind, R.: *Essays on the Theory of Number.* Chicago: Open Court Publishing Company; 1901.

Dicke, R. H.: "Mach's Principle and Invariance under Transformation of Units." *Physical Review,* Vol. CXXV (1962).

———: "The Nature of Gravitation." *Science in Space.* Edited by L. V. Berkner and H. Odishaw. New York: McGraw-Hill Book Company; 1961.

Dingle, H.: "Falsifiability of the Lorentz-Fitzgerald Contraction Hypothesis." *The British Journal for the Philosophy of Science,* Vol. X (1959).

———: *The Sources of Eddington's Philosophy.* London: Cambridge University Press; 1954.

Dingler, H.: "Die Rolle der Konvention in der Physik." *Physikalische Zeitschrift,* Vol XXIII (1922).

Driesch, H.: *Philosophische Gegenwartsfragen.* Leipzig: E. Reinicke; 1933.

du Bois-Reymond, P.: *Die Allgemeine Funktionentheorie.* Tübingen: Lauppische Buchhandlung; 1882. Vol. I.

Duhem, P.: *The Aim and Structure of Physical Theory.* Princeton: Princeton University Press; 1954.

Eddington, A. S.: *The Nature of the Physical World.* New York: The Macmillan Company; 1928.

———: *Space, Time and Gravitation.* Cambridge: Cambridge University Press; 1953.

Edel, A.: *Aristotle's Theory of the Infinite.* New York: Columbia University Press; 1934.

Ehrenfest, P.: "In What Way Does It Become Manifest In The Fundamental Laws of Physics That Space Has Three Dimensions?" *Proceedings of the Amsterdam Academy,* Vol. XX (1917). (German translation: "Welche Rolle spielt die Dreidimensionalität des Raumes in den Grundgesetzen der Physik." *Annalen der Physik,* Vol. LXI [1920].)

Ehrenfest, P. and T.: "Begriffliche Grundlagen der statistischen Auffassung in der Mechanik." *Encyklopädie der mathematischen Wissenschaften,* IV, 2, II.

Einstein, A.: "Autobiographical Notes." *Albert Einstein: Philosopher-Scientist.* Edited by P. A. Schilpp. New York: Tudor Publishing Company; 1949.

———: "Die Relativitätstheorie." *Physik.* Edited by E. Warburg. Leipzig: Teubner; 1915.

———: Foreword to Max Jammer, *Concepts of Space.* Cambridge: Harvard University Press; 1954.

———: "The Foundations of the General Theory of Relativity." *The Principle of Relativity, A Collection of Original Memoirs.* New York: Dover Publications, Inc.; 1952.

———: *Geometrie und Erfahrung.* Berlin: J. Springer; 1921.

———: *The Meaning of Relativity.* Princeton: Princeton University Press; 1955.

———: "On the Electrodynamics of Moving Bodies." *The Principle of Relativity, A Collection of Original Memoirs.* New York: Dover Publications, Inc.; 1952.

———: "On the Influence of Gravitation on the Propagation of Light." *Ibid.*

———: "Prinzipielles zur allgemeinen Relativitätstheorie." *Annalen der Physik,* Vol. LV (1918).

———: "Reply to Criticisms." *Albert Einstein: Philosopher-Scientist.* Edited by P. A. Schilpp. Evanston: The Library of Living Philosophers; 1949.

———: "Über die Entwicklung unserer Anschauungen über die Konstitution und das Wesen der Strahlung." *Physikalische Zeitschrift,* Vol. X (1910).

Eisenhart, L. P.: *Coordinate Geometry.* New York: Dover Publications, Inc.; 1960.

———: *An Introduction to Differential Geometry.* Princeton: Princeton University Press; 1947.

———: *Riemannian Geometry. Ibid.,* 1949.

Epstein, P.: "The Time Concept in Restricted Relativity," and "A Rejoinder." *American Journal of Physics,* Vol. X (1942).

——: "Critical Appreciation of Gibbs' Statistical Mechanics." *A Commentary on the Scientific Writings of J. Willard Gibbs.* Edited by A. Haas. New Haven: Yale University Press; 1936. Vol. II.

Essen, L. and Parry, J. V. L.: "An Atomic Standard of Frequency and Time Interval." *Nature,* Vol. CLXXVI (1955).

Euclid: *The Thirteen Books of Euclid's Elements.* Trans. by T. L. Heath. Cambridge: Cambridge University Press; 1926.

Feyerabend, P. K.: "Comments on Grünbaum's 'Law and Convention in Physical Theory.'" *Current Issues in the Philosophy of Science.* Edited by H. Feigl and G. Maxwell. New York: Holt, Rinehart and Winston; 1961.

Fletcher, J. G.: "Geometrodynamics." *Gravitation.* Edited by L. Witten. New York: John Wiley & Sons, Inc.; 1962.

Frank, P.: *Philosophy of Science.* Englewood Cliffs: Prentice-Hall; 1957.

Fraenkel, A. A.: *Einleitung in die Mengenlehre.* New York: Dover Publications, Inc.; 1946.

Fraenkel, A. A. and Bar-Hillel, Y.: *Foundations of Set Theory.* Amsterdam: North Holland Publishing Company; 1958.

Freudenthal, H.: *Mathematical Reviews,* Vol. XXII (1961).

——: "Zur Geschichte der Grundlagen der Geometric." *Nieuw Archief voor Wiskunde,* Vol. V (1957).

Fürth, R.: "Prinzipien der Statisik." *Handbuch der Physik.* Edited by H. Geiger and K. Scheel. Berlin: J. Springer; 1929. Vol. IV.

Gerhardt, K.: "Nichteuklidische Anschauung und optische Täuschungen." *Naturwissenschaften,* Vol. XXIV (1936).

——: "Nichteuklidische Kinematographie." *Naturwissenschaften,* Vol. XX (1932).

Gerlach, W.: *Handbuch der Physik.* Berlin: Springer-Verlag; 1926.

Gödel, K.: "An Example of a New Type of Cosmological Solutions of Einstein's Equations of Gravitation." *Reviews of Modern Physics,* Vol. XXI (1949).

——: "A Remark About the Relationship of Relativity Theory and Idealistic Philosophy." *Albert Einstein: Philosopher-Scientist.* Edited by P. A. Schilpp. Evanston: The Library of Living Philosophers; 1949.

Gold, T.: "The Arrow of Time." *La Structure et L'Évolution de l'Univers.* Brussels: R. Stoops; 1958.

Goodhard, C. B.: "Biological Time." *Discovery* (December 1957).

Goodman, N.: *Fact, Fiction and Forecast.* Cambridge: Harvard University Press; 1955.

Grünbaum, A.: "A Consistent Conception of the Extended Linear Continuum as an Aggregate of Unextended Elements." *Philosophy of Science,* Vol. XIX (1952).

——: *American Journal of Physics,* Vol. XXIV (1956).

——: "Causality and the Science of Human Behavior." *American Scientist,* Vol. XL (1952). (Reprinted in *Readings in the Philosophy of*

Science. Edited by H. Feigl and M. Brodbeck. New York: Appleton-Century-Crofts, Inc.; 1953.)
——: "The Clock Paradox in the Special Theory of Relativity." *Philosophy of Science,* Vol. XXI (1954), and Vol. XXII (1955).
——: "Complementarity in Quantum Physics and Its Philosophical Generalization." *The Journal of Philosophy,* Vol. LIV (1957).
——: "Das Zeitproblem." *Archiv für Philosophie,* Vol. VII (1957).
——: "Messrs. Black and Taylor on Temporal Paradoxes." *Analysis,* Vol. XII (1952).
——: "Modern Science and the Refutation of the Paradoxes of Zeno." *The Scientific Monthly,* Vol. LXXXI (1955).
——: "Operationism and Relativity." *The Validation of Scientific Theories.* Boston: Beacon Press; 1957.
——: *Philosophy of Science,* Vol. XXII (1955).
——: "The Philosophical Retention of Absolute Space in Einstein's General Theory of Relativity." *The Philosophical Review,* Vol. LXVI (1957).
——: "Professor Dingle on Falsifiability: A Second Rejoinder." *British Journal for the Philosophy of Science,* Vol. XII (1961).
——: "Rejoinder to Feyerabend." *Current Issues in the Philosophy of Science.* Edited by H. Feigl and G. Maxwell. New York: Holt, Rinehart and Winston; 1961.
——: "Relativity and the Atomicity of Becoming." *The Review of Metaphysics,* Vol. IV (1950).
——: "Science and Ideology." *The Scientific Monthly,* Vol. LXXIX (1954).
——: "Science and Man." *Perspectives in Biology and Medicine,* Vol. V (1962).
——: *Scientific American,* Vol. CLXXXIX (December 1953).
——: "Some Highlights of Modern Cosmology and Cosmogony." *The Review of Metaphysics,* Vol. V (1952).
——: "Whitehead's Method of Extensive Abstraction." *The British Journal for the Philosophy of Science,* Vol. IV (1953).
——: "Whitehead's Philosophy of Science." *The Philosophical Review,* Vol. LXXI (1962).
Hadamard, J.: *Lectures on Cauchy's Problem in Linear Partial Differential Equations.* New Haven: Yale University Press; 1923.
Halmos, P. R.: *Measure Theory.* New York: D. Van Nostrand Company, Inc.; 1950.
Hanson, N. R.: "On the Symmetry Between Explanation and Prediction." *The Philosophical Review,* Vol. LXVIII (1959).
Hardy, G. H.: *A Course of Pure Mathematics.* Ninth Edition. New York: The Macmillan Company; 1945.
Hardy, L. H., Rand, G., Rittler, M. C., Blank, A. A., and Boeder, P.: *The Geometry of Binocular Space Perception.* New York: Columbia University College of Physicians and Surgeons; 1953.

Hartshorne, C. and Weiss, P. (eds.): *The Collected Papers of Charles Sanders Peirce*. Cambridge: Harvard University Press; 1935. Vol. VI.

Hasse, H. and Scholz, H.: *Die Grundlagenkrisis der griechischen Mathematik*. Charlottenburg: Pan-Verlag; 1928.

Heath, T. L.: *Mathematics in Aristotle*. Oxford: Oxford University Press; 1949.

sota Studies in the Philosophy of Science, Vol. III. Minneapolis: Uni-

Hempel, C. G.: "Deductive Nomological vs. Statistical Explanation." *Minne-* versity of Minnesota Press; 1962.

Hempel, C. G. and Oppenheim, P.: "Studies in the Logic of Explanation." *Philosophy of Science*, Vol. XV (1948).

Hertz, H.: *The Principles of Mechanics*. New York: Dover Publications, Inc.; 1956.

Hilbert, D. and Bernays, P.: *Grundlagen der Mathematik*. Berlin: J. Springer; 1934.

Hill, E. L. and Grünbaum, A.: "Irreversible Processes in Physical Theory." *Nature*, Vol. CLXXIX (1957).

Hille, E.: *Functional Analysis and Semi-Groups*. New York: Mathematical Society Publications; 1948. Vol. XXXI.

Hjelmslev, J.: "Die natürliche Geometrie." *Abhandlungen aus dem mathematischen Seminar der Hamburger Universität*, Vol. II (1923).

Hoagland, H.: "Chemical Pacemakers and Physiological Rhythms." *Colloid Chemistry*. Edited by J. Alexander. New York. Vol. V (1944).

———: "The Physiological Control of Judgments of Duration: Evidence for a Chemical Clock." *The Journal of General Psychology*, Vol. IX (1933).

Hobson, E. W.: *The Theory of Functions of a Real Variable*. New York: Dover Publications, Inc.; 1957. Vol. I.

Hölder, O.: *Die Mathematische Methode*. Berlin: J. Springer; 1924.

Holton, G.: "On the Origins of the Special Theory of Relativity." *American Journal of Physics*, Vol. XXVIII (1960).

Hood, P.: *How Time is Measured*. London: Oxford University Press; 1955.

Huntington, E. V.: *The Continuum and Other Types of Serial Order*. Second Edition. Cambridge: Harvard University Press; 1942.

———: "Inter-Relations Among the Four Principal Types of Order." *Transactions of the American Mathematical Society*, Vol. XXXVIII (1935).

Huntington, E. V. and K. E. Rosinger: "Postulates for Separation of Point-Pairs (Reversible Order on a Closed Line)." *Proceedings of the American Academy of Arts and Sciences*. Boston. Vol. LXII (1932).

Hutten, E. H.: *The Language of Modern Physics*. London: George Allen and Unwin, Ltd., and New York: The Macmillan Company; 1956.

Jaffe, B.: *Michelson and the Speed of Light*. New York: Doubleday and Company; 1960.

James, W.: *The Principles of Psychology*. New York: Dover Publications, Inc.; 1950.

Jammer, M.: *Concepts of Space*. Cambridge: Harvard University Press; 1954.

Jeffreys, H.: *The Earth*. Cambridge: Cambridge University Press; 1952.

John, F.: "Numerical Solution of the Equation of Heat Conduction for Preceding Times." *Annali di Matematica Pura ed Applicata*, Vol. XL (1955).

Jordan, P.: *Schwerkraft und Weltall*. Braunschweig: F. Vieweg und Sohn; 1955.

Kaempffer, F. A.: "On Possible Realizations of Mach's Program." *Canadian Journal of Physics*, Vol. XXXVI (1958).

Kennedy, R. J. and Thorndike, E. M.: *Physical Review*, Vol. XLII (1932).

Klein, F.: *Elementary Mathematics From An Advanced Standpoint*. New York: Dover Publications, Inc.; 1939. Vol. II.

——: *Vorlesungen über Nicht-Euklidische Geometrie*. Berlin: Springer-Verlag; 1928.

Klein, O.: "Mach's Principle and Cosmology in their Relation to General Relativity." *Recent Developments in General Relativity*. Warsaw: Polish Scientific Publishers; 1962.

Landau, L. and Lifschitz, E.: *Statistical Physics*. Second Edition. Reading: Addison-Wesley; 1958.

Landé, A.: "Axiomatische Begründung der Thermodynamik durch Carathéodory." *Handbuch der Physik*, Vol. IX (1926).

——: "The Logic of Quanta." *British Journal for the Philosophy of Science*, Vol. VI (1956).

——: Optik und Thermodynamik." *Handbuch der Physik*. Berlin: J. Springer; 1928. Vol. XX.

——: "Wellenmechanik und Irreversibilität." *Physikalische Blätter*, Vol. XIII (1957).

Lechalas, G.: *Étude sur l'espace et le temps*. Paris: Alcan Publishing Company; 1896.

——: "L'Axiome de libre Mobilité." *Revue de Métaphysique et de Morale*, Vol. VI (1898).

Leclercq, R.: *Guide Théorique et Pratique de la Recherche Expérimentale*. Paris: Gauthier-Villars; 1958.

Lee, H. D. P.: *Zeno of Elea*. Cambridge: Luzac and Company; 1936.

Lefschetz, S.: *Introduction to Topology*. Princeton: Princeton University Press; 1949.

Leibniz, G. W.: "Initia rerum mathematicorum metaphysica." *Mathematische Schriften*. Edited by C. J. Gerhardt. Berlin: Schmidt's Verlag; 1863.

Lewin, K.: "Die zeitliche Geneseordnung." *Zeitschrift für Physik*, Vol. XIII (1923).

Lewis, C. I.: *Analysis of Knowledge and Valuation*. LaSalle: Open Court Publishing Company; 1946.

——, and Langford, C. H.: *Symbolic Logic*. New York: The Century Company; 1932.

Lorentz, H.: *The Theory of Electrons*. New York: Columbia University Press; 1909.

Loschmidt, J.: "Über das Wärmegleichgewicht eines Systems von Körpern mit Rücksicht auf die Schwere." *Sitzungsberichte der Akademie der Wissenschaften*. Vienna, Vol. LXXIII (1876) and Vol. LXXV (1877).

Ludwig, G.: "Der Messprozess." *Zeitschrift für Physik,* Vol. CXXXV (1953).
——: *Die Grundlagen der Quantenmechanik.* Berlin: J. Springer; 1954.
——: "Die Stellung des Subjekts in der Quantentheorie." *Veritas, Justitia, Libertas.* Festschrift zur 200-Jahr Feier der Columbia University. Berlin: Colloquium Verlag; 1954.
——: "Questions of Irreversibility and Ergodicity." *Ergodic Theories.* Edited by P. Caldivola. New York: Academic Press; 1961.
——: "Zum Ergodensatz und zum Begriff der makroskopischen Observablen. I." *Zeitschrift für Physik,* Vol. CL (1958).
Luneburg: R. K.: *Mathematical Analysis of Binocular Vision.* Princeton: Princeton University Press; 1947.
——: "Metric Methods in Binocular Visual Perception." *Studies and Essays,* Courant Anniversary Volume. New York: Interscience Publishers, Inc.; 1948.
Luria, S.: "Die Infinitesimaltheorie der antiken Atomisten." *Quellen und Studien zur Geschichte der Mathematik, Astronomie, und Physik.* Berlin, 1933. Abteilung B: Studien, II.
Lyons, H.: "Atomic Clocks." *Scientific American,* Vol. CXCVI (1957).
Mach, E.: *Conservation of Energy,* 1872. Reprinted in LaSalle: Open Court Publishing Company; 1911.
——: *The Science of Mechanics.* 1883. Reprinted in LaSalle: Open Court Publishing Company; 1902.
Margenau, H.: "Can Time Flow Backwards?" *Philosophy of Science,* Vol. XXI (1954).
——: "Measurements and Quantum States." *Philosophy of Science,* Vol. XXX (1963).
Margenau, H. and Murphy, G. M.: *The Mathematics of Physics and Chemistry.* New York: D. Van Nostrand Company; 1943.
Maritain, J.: *The Degrees of Knowledge.* London: G. Bles Company; 1937.
Maxwell, J. C.: *Theory of Heat.* Sixth edition. New York: Longman's, Green, and Company; 1880.
Mayr, E.: "Cause and Effect in Biology." *Science,* Vol. CXXXIV (1961).
Mehlberg, H.: "Essai sur le théorie causale du temps." *Studia Philosophica,* Vol. I (1935), and Vol. II (1937).
——: "Physical Laws and Time's Arrow." *Current Issues in the Philosophy of Science.* Edited by H. Feigl and G. Maxwell. New York: Holt, Rinehart, and Winston; 1961.
Menger, K.: *Dimensionstheorie.* Leipzig: B. G. Teubner; 1928.
Michelson, A. A. and Gale, H. G.: "The Effect of the Earth's Rotation on the Velocity of Light." *Astrophysical Journal,* Vol. LXI (1925).
Milham, W. I.: *Time and Timekeepers.* New York: The Macmillan Company; 1929.
Milne, E. A.: *Kinematic Relativity.* Oxford: Oxford University Press; 1948.
——: *Modern Cosmology and the Christian Idea of God.* Oxford: Clarendon Press; 1952.
——: *Sir James Jeans.* Cambridge: Cambridge University Press; 1952.

Misner, C. W. and Wheeler, J. A.: "Geometrodynamics." *Annals of Physics,* Vol. II (1957).

Möbius: *Der Barycentrische Calcul.* Leipzig: Barth; 1827.

Møller, C.: *The Theory of Relativity.* Oxford: Oxford University Press; 1952.

Nagel, E.: "Einstein's Philosophy of Science." *The Kenyon Review,* Vol. XII (1950).

——: "The Formation of Modern Conceptions of Formal Logic in the Development of Geometry." *Osiris,* Vol. VII (1939).

——: *The Structure of Science.* New York: Harcourt, Brace and World; 1961.

Newton, I.: *Principia.* Edited by F. Cajori. Berkeley: University of California Press; 1947.

Northrop, F. S. C.: *The Logic of the Sciences and the Humanities.* New York: The Macmillan Company; 1947.

——: *The Meeting of East and West.* New York: The Macmillan Company; 1946.

——: "Whitehead's Philosophy of Science." *The Philosophy of Alfred North Whitehead.* Edited by P. A. Schilpp. New York: Tudor Publishing Company; 1941.

Page, L.: *Introduction to Theoretical Physics.* New York: D. Van Nostrand Company, Inc.; 1935.

Page, L. and Adams, N. I.: *Electrodynamics.* New York: D. Van Nostrand Company; 1940.

Palter, R. N.: *Whitehead's Philosophy of Science.* Chicago: University of Chicago Press; 1960.

Panofsky, W. and Phillips, M.: *Classical Electricity and Magnetism.* Cambridge, Mass.: Addison-Wesley Publishing Company; 1955.

Penrose, O. and Percival, I. C.: "The Direction of Time." *Proceedings of the Physical Society,* Vol. LXXIX (1962).

Pérard, A.: *Les Mesures Physiques.* Paris: Presses Universitaires de France; 1955.

Pfeffer, W.: "Untersuchungen über die Entstehung der Schlafbewegungen der Blattorgane." *Abhandlungen der sächsischen Akademie der Wissenschaften. Leipzig, Mathematisch-Physicalische Klasse,* Vol. XXX (1907).

Pittendrigh, C. S. and Bruce, V. G.: "Daily Rhythms as Coupled Oscillator Systems and their Relation to Thermoperiodism and Photoperiodism." *Photoperiodism and Related Phenomena in Plants and Animals.* Washington, D. C.: The American Association for the Advancement of Science; 1959.

——: "An Oscillator Model for Biological Clocks." *Rhythmic and Synthetic Processes in Growth.* Edited by D. Rudnick. Princeton: Princeton University Press; 1957.

Poincaré, H.: *Dernières Pensées.* Paris: Flammarion; 1913.

——: "Des Fondements de la Géométrie, à propos d'un Livre de M. Russell." *Revue de Métaphysique et de Morale,* Vol. VII (1899).

——: *The Foundations of Science.* Lancaster: The Science Press; 1946.

———: La Mécanique Nouvelle, cited in R. Dugas: "Henri Poincaré devant les Principes de la Mécanique." *Revue Scientifique,* Vol. LXXXIX (1951).

———: "La Mesure du Temps." *Revue de Métaphysique et de Morale,* Vol. VI (1898).

———: "L'Espace et la Géométrie." *Revue de Métaphysique et de Morale,* Vol. III (1895).

———: *Letzte Gedanken.* Translated by Lichtenecker. Leipzig: Akademische Verlagsgesellschaft; 1913.

———: "The Principles of Mathematical Physics." *The Monist,* Vol. XV (1905). Reprinted in the *Scientific Monthly,* Vol. LXXXII (1956).

———: "Sur les Principes de la Géométrie, Réponse a M. Russell." *Revue de Métaphysique et de Morale,* Vol. VIII (1900).

———: "Sur le problème des trois corps et les equations de la dynamique." *Acta Mathematica,* Vol. XIII (1890).

Polanyi, M.: "Notes on Professor Grünbaum's Observations." *Current Issues in the Philosophy of Science.* Edited by H. Feigl and G. Maxwell. New York: Holt, Rinehart and Winston; 1961.

———: *Personal Knowledge.* Chicago: Chicago University Press; 1958.

Popper, K. R.: *The Logic of Scientific Discovery.* London: Hutchinson and Company, Ltd.; 1959.

———: *Nature,* Vol. CLXXVII (1956), Vol. CLXXVIII (1956), Vol. CLXXIX (1957), Vol. CLXXXI (1958).

Price, D. J.: "The Prehistory of the Clock." *Discovery,* Vol. XVII (1956).

Putnam, H.: "The Analytic and the Synthetic." *Minneosta Studies in the Philosophy of Science.* Edited by H. Feigl and G. Maxwell. Minneapolis: University of Minnesota Press; 1962. Vol. III.

———: "Three-Valued Logic." *Philosophical Studies,* Vol. VIII (1957).

Quine, W. V. O.: *From a Logical Point of View.* Second edition. Cambridge: Harvard University Press; 1961.

Read, J.: *A Direct Entry to Organic Chemistry.* London: Methuen & Company, Ltd.; 1948.

Reichenbach, H.: *Axiomatik der relativistischen Raum-Zeit-Lehre.* Braunschweig: F. Vieweg & Sons; 1924.

———: "Das Kausalproblem in der Physik." *Naturwissenschaften,* Vol. XIX (1931).

———: "Die Bewegungslehre bei Newton, Leibniz und Huyghens." *Kantstudien,* Vol. XXIX (1924).

———: "Die Kausalstruktur der Welt und der Unterschied von Vergangenheit und Zukunft." *Berichte der Bayerischen Akademie München, Mathematisch-Naturwissenschaftliche Abteilung* (November 1925).

———: *The Direction of Time.* Berkeley: University of California Press; 1956.

———: "Discussion of Dingler's paper." *Physikalische Zeitschrift,* Vol. XXIII (1922).

———: "Kant und die Naturwissenschaft." *Die Naturwissenschaften,* 1933, Vol. XXI.

———: "La signification philosophique du dualisme ondes-corpuscules." *Louis de Broglie, Physicien et Penseur.* Edited by A. George. Paris: Editions Albin Michel; 1953.
———: "Les fondements logiques de la mécanique des quanta." *Annales de l'Institut Henri Poincaré,* Vol. XIII (1953).
———: *Modern Philosophy of Science.* London: Routledge and Kegan Paul; 1959.
———: *Philosophic Foundations of Quantum Mechanics.* Berkeley: University of California Press; 1948.
———: "The Philosophical Significance of the Theory of Relativity." *Albert Einstein: Philosopher-Scientist.* Edited by P. A. Schilpp. Evanston: Library of Living Philosophers; 1949.
———: *Philosophie der Raum-Zeit-Lehre.* Berlin: Walter de Gruyter & Company; 1928.
———: *The Philosophy of Space and Time.* Dover Publications, Inc.; 1958.
———: "Planetenuhr und Einsteinsche Gleichzeitigkeit." *Zeitschrift für Physik,* Vol. XXXIII (1925).
———: *The Rise of Scientific Philosophy.* Berkeley: University of California; 1951.
———: "Über die physikalischen Konsequenzen der relativistischen Axiomatik." *Zeitschrift für Physik,* Vol. XXXIV (1925).
———: "Ziele und Wege der physikalischen Erkenntnis." *Handbuch der Physik,* Vol. IV (1929).
Rescher, N.: "On Prediction and Explanation." *British Journal for the Philosophy of Science,* Vol. VIII (1958).
———: "The Stochastic Revolution and The Nature of Scientific Explanation." *Synthèse,* Vol. XIV (1962).
Richter, C. P.: "Biological Clocks in Medicine and Psychiatry: Shock-Phase Hypothesis." *Proceedings of the National Academy of Sciences,* Vol. XLVI (1960).
Riemann, B.: "On the Hypotheses Which Lie at the Foundations of Geometry." *A Source Book in Mathematics.* Edited by David E. Smith. New York: Dover Publications, Inc.; 1959. Vol. II.
Robertson, H. P.: "The Geometries of the Thermal and Gravitational Fields." *American Mathematical Monthly,* Vol. LVII (1950).
———: "Geometry as a Branch of Physics." *Albert Einstein: Philosopher-Scientist.* Edited by P. A. Schilpp. Evanston: Library of Living Philosophers; 1949.
———: *Mathematical Reviews,* Vol. XVI (1955).
———: "Postulate *versus* Observation in the Special Theory of Relativity." *Reviews of Modern Physics,* Vol. XXI (1949).
Rosenfeld, L.: "On the Foundations of Statistical Thermodynamics." *Acta Physica Polonica,* Vol. XIV (1955).
Rossi, B. and Hall, D. B.: *Physical Review,* Vol. LIX (1941).
Rothstein, J.: "Information, Measurement and Quantum Mechanics." *Science,* Vol. CXIV (1951).

Rougier, L.: *La Philosophie Géométrique de Henri Poincaré.* Paris: F. Alcan; 1920.

Rush, J. H.: "The Speed of Light." *Scientific American,* Vol. CXCIII (1955).

Russell, B.: *The Foundations of Geometry.* New York: Dover Publications, Inc.; 1956.

——: *Our Knowledge of the External World.* London: George Allen and Unwin, Ltd.; 1926.

——: *The Philosophy of Leibniz.* London: George Allen and Unwin, Ltd.; 1937.

——: *The Principles of Mathematics.* Cambridge: Cambridge University Press; 1903.

——: "Sur les Axiomes de la Géométrie." *Revue de Métaphysique et de Morale,* Vol. VII (1899).

Saccheri, G.: *Euclides ab omni naevo vindicatus.* Milan: P. A. Montani; 1733.

Sachs, R. G.: "Can the Direction of Flow of Time Be Determined?" *Science,* Vol. CXL (1963).

Salecker, H. and Wigner, E. P.: "Quantum Limitations of the Measurement of Space-Time Distances." *The Physical Review,* Vol. CIX (1958).

Sandage, A.: "Travel Time for Light from Distant Galaxies Related to the Riemannian Curvature of the Universe." *Science,* Vol. CXXXIV (1961).

Scheffler, I.: "Explanation, Prediction and Abstraction." *British Journal for the Philosophy of Science,* Vol. VII (1957).

Schlick, M.: "Are Natural Laws Conventions?" *Readings in the Philosophy of Science.* New York: Appleton-Century-Crofts, Inc.; 1953.

——: *Grundzüge der Naturphilosophie.* Vienna: Gerold & Company; 1948.

Scholz, H.: "Eine Topologie der Zeit im Kantischen Sinne." *Dialectica,* Vol. IX (1955).

Schrödinger, E.: "Irreversibility." *Proceedings of the Royal Irish Academy,* Vol. LIII (1950).

——: "The Spirit of Science." *Spirit and Nature.* Edited by J. Campbell. New York: Pantheon Books; 1954.

——: *What is Life?* New York: The Macmillan Company; 1945.

Sciama, D. W.: "Inertia." *Scientific American,* Vol. CXCVI (February 1957).

——: "On the Origin of Inertia." *Monthly Notices of the Royal Astronomical Society,* Vol. CXIII (1953).

Scribner, C.: "Mistranslation of a Passage in Einstein's Original Paper on Relativity." *American Journal of Physics,* Vol. XXXI (1963).

Scriven, M.: "Comments on Professor Grünbaum's Remarks at the Wesleyan Meeting." *Philosophy of Science,* Vol. XXIX (1962).

——: "Explanation and Prediction in Evolutionary Theory." *Science,* Vol. CXXX (1959).

——: "The Temporal Asymmetry of Explanations and Predictions." *Philosophy of Science.* Edited by Bernard Baumrin. New York: John Wiley and Sons; 1963. Vol. I.

Sellars, W.: "Time and the World Order." *Minnesota Studies in the Philosophy of Science,* Vol. III. Minneapolis: University of Minnesota Press; 1962.

Shankland, R. S., McCuskey, S. W., Leone, S. C., and Kuerti, G.: "New Analysis of the Interferometer Observations of Dayton C. Miller." *Reviews of Modern Physics,* Vol. XXVII (1955).

Shepperd, J. A. H.: "Transitivities of Betweenness and Separation and the Definitions of Betweenness and Separation Groups." *Journal of the London Mathematical Society,* Vol. XXXI (1956).

Silberstein, L.: *The Theory of Relativity.* New York: The Macmillan Company; 1914.

Smart, J. J. C.: "Mr. Mayo on Temporal Asymmetry." *Australasian Journal of Philosophy,* Vol. XXXIII (1955).

——: "Critical Study of H. Reichenbach's *The Direction of Time.*" *Philosophical Quarterly,* Vol. VIII (1958).

——: "Spatializing Time." *Mind,* Vol. CXIV (1955).

——: "The Temporal Asymmetry of the World." *Analysis,* Vol. XIV (1954).

Sokolnikoff, I. S.: *Mathematical Theory of Elasticity.* New York: McGraw-Hill Book Company; 1946.

Sommerfeld, A.: Notes in *The Principle of Relativity,* A Collection of Original Memoirs. New York: Dover Publications, Inc.; 1952. A reprint.

——: *Partial Differential Equations in Physics.* Translated by E. G. Straus. New York: Academic Press, Inc.; 1949.

Sommerville, D. M. Y.: *The Elements of Non-Euclidean Geometry.* New York: Dover Publications, Inc.; 1958.

Stanyukovic, K. P.: "On the Increase of Entropy in an Infinite Universe." *Doklady, Akademiia Nauk* SSSR, N.S., Vol. LXIX (1949).

Stebbing, L. Susan: *Philosophy and the Physicists.* London: Methuen & Company, Ltd.; 1937.

Stille, U.: *Messen und Rechnen in der Physik.* Braunschweig: Vieweg; 1955.

Struik, D. J.: *Classical Differential Geometry.* Cambridge: Addison-Wesley Publishing Company, Inc.; 1950.

Tamm, I. E.: "General Characteristics of Vavilov-Cherenkov Radiation." *Science,* Vol. CXXXI (1960).

——: "Radiation of Particles With Speeds Greater Than that of Light." *American Scientist,* Vol. XLVII (1959).

Tarski, A.: "What is Elementary Geometry?" *The Axiomatic Method.* Edited by L. Henkin, P. Suppes, and A. Tarski. Amsterdam: North Holland Publishing Company; 1959.

Taub, A. H.: "Empty Space-Times Admitting a Three Parameter Group of Motions." *Annals of Mathematics,* Vol. LIII (1951).

Taylor, G. I.: "Tidal Friction in the Irish Sea." *Philosophical Transactions of the Royal Society Academy,* Vol. CCXX (1920).

Taylor, W. B.: *The Meaning of Time in Science and Daily Life.* Doctoral Dissertation. Los Angeles: University of California at Los Angeles; 1953.

Ter Haar, D.: "Foundations of Statistical Mechanics." *Reviews of Modern Physics,* Vol. XXVII (1955).

Thomas, T. Y.: *The Differential Invariants of Generalized Spaces.* Cambridge: Cambridge University Press; 1934.

Timoshenko, S. and Goodier, J. N.: *Theory of Elasticity.* New York: McGraw-Hill Book Company; 1951.

Tolman, R. C.: *Relativity, Thermodynamics and Cosmology.* Oxford: Oxford University Press; 1934.

——: *The Principles of Statistical Mechanics.* Oxford: Oxford University Press; 1938.

Trocheris, M. G.: "Electrodynamics in a Rotating Frame of Reference." *Philosophical Magazine,* Vol. XL (1949).

Truesdell, C.: "Ergodic Theory in Classical Statistical Mechanics." Edited by P. Caldivole. *Ergodic Theories.* New York: Academic Press; 1961.

Tannery, P.: "Le Concept Scientifique du Continu: Zenon d'Élée et Georg Cantor." *Revue Philosophique,* Vol. XX, No. 2 (1885).

Veblen, O. and Whitehead, J. H. C.: *The Foundations of Differential Geometry.* (Cambridge Tracts in Mathematics and Mathematical Physics, No. 29.) Cambridge: Cambridge University Press; 1932.

von Fritz, K.: "The Discovery of Incommensurability by Hippasus of Metapontum." *Annals of Mathematics,* Vol. XLVI (1945).

von Helmholtz, H.: *Schriften zur Erkenntnistheorie.* Edited by P. Hertz and M. Schlick. Berlin: J. Springer; 1921.

von Laue, Max: *Die Relativitätstheorie.* Braunschweig: F. Vieweg; 1952. Vol. I.

——: "A. Einstein und die Relativitätstheorie." *Naturwissenschaften,* Vol. XLIII (1956).

von Neumann, J.: *Mathematische Grundlagen der Quantenmechanik.* Berlin: J. Springer; 1932. Translated by R. T. Beyer. Princeton: Princeton University Press; 1955.

von Weizsäcker, C. F.: "Der zweite Hauptsatz und der Unterschied von Vergangenheit und Zukunft." *Annalen der Physik,* Vol. XXXVI (1939).

Waismann, F.: *Introduction to Mathematical Thinking.* New York: F. Ungar Publishing Company; 1951.

Walker, A. G.: "Axioms for Cosmology." *The Axiomatic Method.* Edited by L. Henkin, P. Suppes, and A. Tarski. Amsterdam: North Holland Publishing Company; 1959.

Ward, F. A. B.: *Time Measurement.* Fourth edition. London: Royal Stationery Office; 1958. Part I.

Watanabe, M. S.: "Le Concept de Temps en Physique Moderne et la Durée Pure de Bergson." *Revue de Métaphysique et de Morale,* Vol. LVI (1951).

——: *Le Deuxième Théorème de la Thermodynamique et la Mécanique Ondulatoire.* Paris: Hermann & Cie.; 1935.

——: "Réversibilité contre Irréversibilité en Physique Quantique." *Louis de Broglie, Physicien et Penseur.* Paris: Albin Michel; 1953.

——: "Symmetry of Physical Laws. Part III, Prediction and Retrodiction." *Reviews of Modern Physics,* Vol. XXVII (1955).

———: "Über die Anwendung Thermodynamischer Begriffe auf den Normalzustand des Atomkerns." *Zeitschrift für Physik,* Vol. CXIII (1939).

Weber, J.: "Phase, Group, and Signal Velocity." *American Journal of Physics,* Vol. XXII (1954).

Weinstein, B.: *Handbuch der Physikalischen Massbestimmungen.* Berlin: J. Springer; 1886, Vol. I, and 1888, Vol. II.

Weiss, P.: *Reality.* New York: Peter Smith; 1949.

Weyl, H.: "50 Jahre Relativitätstheorie." *Naturwissenschaften,* Vol. XXXVIII (1951).

———: *Philosophy of Mathematics and Natural Science.* Princeton: Princeton University Press; 1949.

———: *Space-Time-Matter.* New York: Dover Publications, Inc.; 1950.

Wheeler, J. A.: "Curved Empty Space-Time as the Building Material of the Physical World: An Assessment." Edited by E. Nagel, P. Suppes, and A. Tarski. *Logic, Methodology and Philosophy of Science:* Proceedings of the 1960 International Congress. Stanford: Stanford University Press; 1962.

———: "Mach's Principle as a Boundary Condition for Einstein's Field Equations and as a Central Part of the 'Plan' of General Relativity." A Report given at the Conference on Relativistic Theories of Gravitation, Warsaw, July 1962.

———: "The Universe in the Light of General Relativity." *The Monist,* Vol. XLVII, No. 1 (1962).

Whitehead, A. N.: *The Concept of Nature.* Cambridge: Cambridge University Press; 1926.

———: *Essays in Science and Philosophy.* New York: The Philosophical Library; 1947.

———: *The Principles of Natural Knowledge.* Cambridge: Cambridge University Press; 1955.

———: *The Principle of Relativity.* Cambridge: Cambridge University Press; 1922.

———: *Process and Reality.* New York: The Macmillan Company; 1929.

Whitrow, G. J.: *The Natural Philosophy of Time.* London: Thomas Nelson & Sons, Ltd.; 1961.

Whittaker, E. T.: *From Euclid to Eddington.* London: Cambridge University Press; 1949.

———: "G. F. Fitzgerald." *Scientific American,* Vol. CLXXXIX (November 1953).

———: *A History of the Theories of Aether and Electricity.* London: Thomas Nelson & Sons, Ltd.; 1953.

Whyte, L. L.: "Light Signal Kinematics." *British Journal for the Philosophy of Science,* Vol. IV (1953).

———: "One-Way Processes in Physics and Biophysics." *British Journal for the Philosophy of Science,* Vol. VI (1955).

Wiener, N.: *Cybernetics.* New York: John Wiley & Sons, Inc.; 1948.

Wigner, E. P.: "Relativistic Invariance and Quantum Phenomena." *Reviews of Modern Physics,* Vol. XXIX (1957).

Yanase, M. M.: "Reversibilität und Irreversibilität in der Physik." *Annals of the Japan Association for Philosophy of Science,* Vol. I (1957).

Zawirski, Z.: *L'Évolution de la Notion du Temps.* Cracow: Librairie Gebethner & Wolff; 1936.

Zermelo, E.: "Über einen Satz der Dynamik und der mechanischen Wärmetheorie." *Wiedmannsche Annalen* (*Annalen der Physik und Chemie*), Vol. LVII (1896).

Zilsel, E.: "Über die Asymmetrie der Kausalität und die Einsinnigkeit der Zeit." *Naturwissenschaften,* Vol. XV (1927).

Index

(Note on the index: I have kept topical entries to a minimum by leaving out information that duplicates the Table of Contents. Some exceptions were made, as for topics treated in scattered places in several chapters.

Similarly, I have omitted the names of authors listed in the eighteen-page bibliography. Their inclusion would have required far too many entries followed by long lists of undifferentiated page numbers in the case of the numerous authors who are cited quite frequently.—A.G.)

A NOTE ON THE TYPE

The text of this book is set in Caledonia, a Linotype face designed by W. A. Dwiggins (1880–1956), the man responsible for so much that is good in contemporary book design and typography. Caledonia belongs to the family of printing types called "modern face" by printers—a term used to mark the change in style of type-letters that occurred about 1800. Caledonia borders on the general design of Scotch Modern, but is more freely drawn than that letter.

Composed, printed, and bound by
The Book Press, Brattleboro, Vermont
Typography and binding design by
[illegible]

A NOTE ON THE TYPE

The text of this book is set in Caledonia, a Linotype face designed by W. A. Dwiggins (1880–1956), the man responsible for so much that is good in contemporary book design and typography. Calcdonia belongs to the family of printing types called "modern face" by printers—a term used to mark the change in style of type-letters that occurred about 1800. Caledonia borders on the general design of Scotch Modern but is more freely drawn than that letter.

Composed, printed, and bound by
The Book Press, Brattleboro, Vermont
Typography and binding design by
VINCENT TORRE